NANO OPTICS

NANO OPTICS

Dr. Piyoosh Tyagi

SHREE PUBLISHERS & DISTRIBUTORS
NEW DELHI-110 002

Edition : 2013

ISBN : 978–81–8329–520–8

Published by :
SHREE PUBLISHERS & DISTRIBUTORS
22/4735 Prakash Deep Building,
Ansari Road, Darya Ganj,
New Delhi–110 002

Printed in India.

Preface

The branch of physics concerned with vision and light is known as optics. This is an important branch which has helped the many discoveries, explorations and inventions in the field of astronomical physics etc. in particular and science and technology in general.

Nano-optics aims at the thoughtful of optical phenomena on the nanometer scale, i.e. near or beyond the diffraction limit of light. It is an emerging new field of study, motivated by the rapid advance of nanoscience and nano-technology and by its need for adequate tools and strategies for fabrication, manipulation and characterization at the nanometer scale. Interestingly, nano-optics even predates the trend of nanotechnology by more than a decade. An optical counterpart to the scanning tunneling microscope was demonstrated in 1984 and optical resolutions had been achieved that were significantly beyond the diffraction limit of light. These early experiments sparked a field initially called near-field optics since it was realized quickly that the inclusion of near fields in the problem of optical imaging and associated spectroscopies holds promise for achieving arbitrary spatial resolutions thus providing access for optical experiments on the nanometer scale.

Developments in this field focus on advances in research of transformation optics. Transformation optics is the foundation for exploring a diverse set of theoretical, numerical, and experimental developments, involving the perspectives of the physics and engineering communities. The multi-disciplinary perspectives for inquiry and designing of materials develop understanding of their behaviours, properties, and potential applications for this field.

Integrated optics is a far-reaching attempt to apply thin-film technology to optical circuits and devices, and, by using methods of integrated circuitry, to achieve a better and more economical optical system. The phenomenon of wave propagation in periodic structures or media recurs in many branches of physics and technology. As examples, consider the phenomena of X-ray and electron diffraction in crystals, the diffraction of light from the periodic strain variation accompanying a sound wave, and the band structures and energy gaps of the phonon spectrum and the energy of electrons in a crystal. Among practical devices making use of this phenomenon we may mention the traveling wave tube, the linear particle accelerator, diffraction gratings, and holograms.

An optical microscope is a combination of lenses used to magnify an image. The object is placed close to the lens in order to obtain as great a magnification as desired. The distance from lens to object is adjusted until the object is in the lens's depth of field and is in focus. The field of view is the area seen through the magnifier. With a simple magnifier, the diameter of the field of view is less than its focal length. So selection of a magnifier with the proper field of view is important.

The use of fluorescent dyes makes it possible to selectively obtain images of individual cell components, for example, proteins. Today, the wavelength dogma has been overcome with the development of the Stimulated Emission Depletion (STED) microscope. Now, the German team that developed STED is reporting layer-by-layer light microscopic nanoscale images of cells and without having to prepare thin sections with a technique called *optical 3D far-field microscopy*.

The theory of image charges offers a useful method for solving the electric and magnetic field distributions of either a charge or a magnet near the flat surface of a Perfect Electric Conductor (PEC). In particular, the field distribution for a charge is obtained by replacing the conductor by a fictitious charge placed at the mirror image of the real charge, but with opposite sign. For a magnet near a PEC surface, the mirror magnet has a magnetization vector that is just the specular reflection of the magnetization vector produced by the real source. We therefore conclude that the force between a charge and a PEC surface is always attractive, whereas the force between a magnet and a PEC surface is always repulsive. This simple analysis finds application to electrostatic adsorption and magnetic levitation.

Hopefully, our humble effort would prove to be beneficial for the students, researchers and concerned readers alike.

The present work is purely a research compilation based on authoritative publications/readings/excerpts/notes/reviews/researches authored by eminent information scientists and IT professionals.

Assistance has been sought from Websites through Internet.

The author, humbly, acknowledges the contributions of all those eminent information scientists and IT experts alongwith their respective publishers and Websites in preparation of this book.

Dr. Piyoosh Tyagi

Contents

Preface (*v*)

1. **Introduction** **1**
   - Electromagnetic Radiation
   - Wafer-level Nano-optics
   - The "Age" of Nano-optics
2. **Transformation Optics** **14**
   - Introduction
   - Transformation Optics and Cloaking
   - Research of Transformation Optics
   - Perform of Coordinate Transformations
   - Tensor
   - Unruh Effect
   - Hawking and Black Body Radiation
   - Phenomenon of Optical Black Hole
3. **Integrated Waveguide Optics** **56**
   - Waveguides: Special Form of Transmission
   - Optical Fibre
   - Coupled-Mode Theory
   - Strip Waveguides
4. **Applications to Optical Devices** **74**
   - Inspection Devices
   - Nanostructure-based Optics
   - Near-field Optical Microscopy
   - Near-field Spectroscopy
   - Fibre Probes
   - Scanning Near-Field Optical Microscopy
   - Optical Probes of Near-field
   - Near-Field Scanning Optical Microscope
   - Super-resolution Optics and Super-lenses
   - Superlens: Perfect Lens
   - Image Formation
   - Cylindrical Superlens Via Coordinate Transformation
   - Optical Nano-antennas
   - Plasmonics Nano-devices

- Quantum Dots Lasers and Devices
- Photonic Crystals

5. Nanoscale Optical Microscopy **134**

- Nonlinear Excitation and Saturation
- Nanoscale Resolution in Light Microscopy
- Department of Electronic and Electrical Engineering

6. Theoretical Methods in Nano-optics **156**

- Nano-optics : The Determination of Electromagnetic Field
- Fast Multipole Method
- Optical Microcavity-mirror
- Volume Integral Analysis of Nanoparticles
- Integrating Nano-optics
- Polarizability and Plasmon Resonances
- Resonant Plasmonic Nanoparticle Systems

7. Light Scattering and Nanoplasmonics **190**

- Light Scattering Vacuum
- Nano Plasmonics
- Nano-plasmonic Optics
- Seamless Manufacturing System
- Surface Plasmon Polaritons
- Scalar Wave Equation in one Space Dimension
- Mie Scattering
- Fraunhofer Diffraction Theory and Mie Scattering Theory
- Computational Methods for Nanoscale Optics
- Lasers Raman
- Spin-flip laser Raman
- Raman-active Optical Amplificators
- Applications of Raman
- Resonant-cavity-enhanced Photo Detector

8. Photonic Structures and Quantum Fields **233**

- Anderson Localization in Condensed Matter Physics
- Anderson Localization of Light
- Random Laser
- Random Number Generation
- Optical Microcavity
- Quantization of the Electromagnetic Field
- Short Range Forces in Opto-mechanical Devices
- Spontaneous Emission Enhancement near Nanoparticles
- Cavity Quantum Electrodynamics

Bibliography 277

Index 279

1

Introduction

Nanooptics is a branch of Optical Engineering which measures the optical properties of materials at nanometre scales. Nanooptics can detect light coming from structures with a very high spatial resolution. Therefore, nanooptics is very beneficial for studying the optical properties of nanostructures. One of the most common optical experiments is measuring spectra. A spectrum is basically the Intensity of light versus the wavelength of light, in other words, the brightness of light versus the colour of that light.

Today we encounter a strong trend towards nanoscience and nanotechnology. This trend is motivated by the fact that as we move to smaller and smaller scales the underlying physical laws change from macroscopic to microscopic. The exploitation of quantum effects for technological applications is the most obvious driving force behind the current miniaturization. The recent rapid advances are due in large part to our newly acquired ability to measure and manipulate individual structures on the nanoscale (scanning probe techniques, optical tweezers, high-resolution electron microscopes,...).

The increasing trend towards nanoscience and nanotechnology makes it necessary to address the key issues of optics on the nanometer scale. Since the diffraction limit does not allow us to focus light to dimensions smaller than roughly half a wavelength, traditionally it was not possible to interact selectively with nanoscale features. In recent years several new approaches have been put forth to 'shrink' the diffraction limit (confocal microscopy) or to even overcome it (near-field microscopy). For example, with our tip-enhancement technique we are able to do Raman spectrocopy and multiphoton fluorescence imaging with a spatial resolution of less than 20nm. To date, this is the highest optical resolution of a spectroscopic measurement.

The reason for the hard effort to advance the field of optics to the nanometer scale is the fact that the energy of light lies in the range of electronic and vibrational transitions in matter. Therefore, the interaction of light with matter renders unique information about the structural and dynamical properties of matter. These unique spectroscopic capabilities are of great importance for the study of biological and solid-state nanostructures. We are applying near-field optical techniques to probe complex semiconductor nanostructures as well as individual protein molecules. We tryp to explore the possibility to optically interact with semiconductor nanostructures on length scales smaller than the extent of their quantum wavefunctions. Probing and manipulating these wavefunctions might open up various exciting applications such as data storage and optical switching based on quantum logic. Nanooptics addresses the broad spectrum of optics on the nanometer scale covering technology and basic science. On the technological side, we find topics like nanolithography and high density optical data storage. On the basic sciences end, we might mention atom-photon interactions in the optical near-field and their potential applications for atom trapping and manipulation experiments.

Compared with free propagating light the optical near-field is enriched by so-called virtual photons. These are the same sort of particles responsible for molecular binding (van der Waals/Casimir forces) and are therefore promising for selective probing of atomic structures. The consideration of virtual photons in the field of quantum optics will enlarge the range of fundamental experiments and applications.

Many research topics in nanoscience are interdisciplinary in nature and must be addressed with a collaborative effort. Therefore, we collaborate with many different research groups and are open to new interactions.

NANOOPTICS AND HIGH RESOLUTION SCANNING

The Structural Analysis Team is responsible for the durability and functionality of high-stress structures in all applications. As such, we frequently conduct non-destructive inspections of critical components, searching for small defects. Using fluorescent penetrant techniques (PT), one of the oldest methods of non-destructive inspection, we can detect surface flaws of up to 0.003 mm in size. This technique is time consuming and hazardous because of the penetrant chemicals, developers, and cleaners required to perform the test. Eddy-current (EC) and ultrasonic techniques (UT) are faster and safer, however flaw detection is limited to 0.05 mm for EC and 0.3 mm for UT.

While these resolutions are adequate, preliminary research into the resolution scale of nanooptic sensing supercedes them all, having a resolution of <40 nm. With the ability to detect flaws of this size, it would enable the Aerospace Engineering Department to close-tolerance engineer critical parts. This would cut down expenditures on materials, improve efficiency of the structure in terms of weight, and increase the safety of the structure.

Nanooptics has other benefits to the Aerospace Department, one of which is the generation of nanofilm polarizers and filters, which can greatly improve the canopies and windows of aircraft by cutting glare. In addition, this same technology can be used to assemble miniaturized cameras by bringing the lens assembly closer to the sensor, and allowing one or more optical substrates to be eliminated, saving on weight. This has applications in the consumer electronics field in addition to the use on reconaissance aircraft and satellites.

THE ETHICS OF NANOTECHNOLOGY AND NANOOPTICS

There are many ethical considerations when it comes to nanotechnology. For instance, in the paper "The Ethical and Societal Impact of Nanotechnology," it is stated that:

...some issues are emerging that appear unique to nanotechnology, namely the new environmental, health and safety (EHS) risks arising nanomaterials. For instance, research studies suggest that some nanoparticles are directly harmful to animals, and because they can be taken up by cells, they might enter our food chain to unknown effects on human health. Other research asks whether carbon nanotubes will be the next asbestos, since both have the same whisker-like shape that makes it so difficult to purge from our lungs if inhaled. And the flip side of creating super-strong materials such as carbon nanotubes is their fate at the end of a product life-cycle: will these materials persist indefinitely in our landfills, as is the case with Styrofoam or nuclear waste?

However, privacy is probably the biggest concern when it comes to the field of nanooptics. With the development of technology that can shrink sensing devices to an undetectible scale, the concern is the abuse of these devices. Because of this, I believe it is in the best interest of XYZ Corporation to therefore limit the pursuit of nanooptics to polarizing and filtering films and nanoscale resolution sensors. We can conduct research into the ethical application of these technologies in consumer devices before making the decision to mass market them. It would benefit us to communicate with Nanoethics.org and familiarize ourselves with their work.

ELECTROMAGNETIC RADIATION

Electromagnetic radiation (radio waves, light, *etc.*) consists of interacting, self-sustaining electric and magnetic fields which propagate through empty space at the speed of 299,792 km per second. Thermocuclear reactions in the cores of stars (including the sun) provide the energy that eventually leaves stars primarily in the form of electromagnetic radiation. These waves cover a wide spectrum of frequencies. Sunshine is a familiar example of electromagnetic radiation that is naturally emitted by the sun. Starlight is the same thing from "suns" that are much farther away. When a direct current (DC) is applied to a wire (conductor) the current flow builds an electromagnetic field around the wire, propagating a wave outward from the wire. When the current is removed the field collapses, again propagating a wave. If the current is applied and removed repeatedly over a period of time, or if the applied current is made to alternate its polarity with a uniform period of time, a series of waves is propagated at a discrete frequency. This phenomenon is the basis of electromagnetic radiation.

Electromagnetic radiation is propagated nominally in a straight line at the speed of light in a vacuum, and does not require a medium for transmission. It is slowed as it passes through a medium such as air, water, glass, *etc.* The amount of energy arriving at a detecting device of fixed area located at a given distance from an isotropic source is proportional to the amount of energy passing the surface of an imaginary sphere with a radius of the given distance. Therefore, the amount of electromagnetic energy passing through a unit area decreases with the square of the distance from the source. This relationship is known as the inverse-square law of (electromagnetic) propagation. It accounts for loss of signal strength over space, called space loss. The inverse-square law is significant to the exploration of the universe, because it means that the observable electromagnetic radiation decreases very rapidly as the distance from the emitter is increased. Whether the emitter is a spacecraft with a low-power transmitter or an extremely powerful star, it will deliver only a small amount of electromagnetic energy to a detector on Earth because of the very great distances, and the small area that Earth subtends on the huge imaginary sphere.

ELECTROMAGNETIC SPECTRUM

Light is electromagnetic radiation at those frequencies that can be sensed by the human eye. But the electromagnetic spectrum has a much broader range of frequencies than the human eye can detect, including, in order of increasing frequency: radio frequency (RF), infrared, light, ultraviolet (UV, meaning "above violet"), X rays, and gamma rays. These designations describe only different frequencies of the same phenomenon: electromagnetic radiation.

The speed of light in a vacuum, 299,792 km per second, is the rate of propagation of all electromagnetic waves. The wavelength of a single oscillation of electro-magnetic radiation is the distance that the wave will propagate during the time required for one oscillation. There is a simple relationship between the frequency and wavelength of electromagnetic energy. Since electromagnetic energy is propagated at the speed of light, the wavelength equals the speed of light divided by the frequency of oscillation.

WAFER-LEVEL NANOOPTICS

"Embossing" is an ancient technical term in the English language, traceable to the period 1350 to 1400, and used by Chaucer in its present sense. The technical meaning, creating a raised design or pattern by pressing a die having the raised negative image of the design against the material to be patterned, has not changed during the ensuing 650 years. What has changed are the applications, materials used, and the scale of the design. Today's embossed patterns may have features measured in nanometers, bringing new capabilities to integrated optical systems.

Nano-embossing, often referred to as nano-imprint lithography (NIL), can rapidly pattern relatively large area substrates with feature sizes below 100 nm. Creating patterns at this scale previously required far slower and more costly techniques such as electron-beam writing. NIL, used in combination with related technologies, can create an entire wafer of integrated optical structures with nanometer features in minutes. An array of replicated lenses formed on the surface of a 4-in. silicon wafer.

Figure: Replicated Lens Array on a Silicon Wafer.

Wafer-level NIL promises reductions in the size, cost and complexity of integrated optics, while improving performance. Integrating embossed components on a wafer offers major size reductions. Producing truly "monolithic" optical systems could reduce centimetre-sized systems to millimetre-sized die. Optical performance is improved by locating components physically closer together, reducing the optical path length and the attendant losses. For example, optical elements such as lenses can be fabricated directly onto optical sources, such as lasers. The diffractive lenses formed directly on a VCSEL laser wafer.

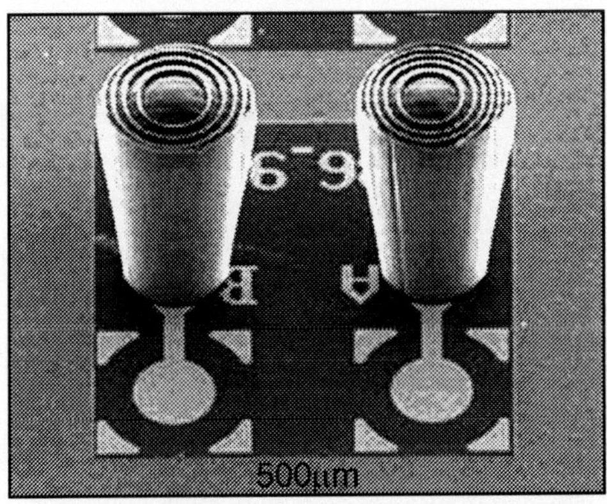

Figure : Portion of a Diffractive Lens Array on a VCSEL Wafer, 200x.

Surprising to the non-expert, the small physical size of the nano-structures also creates optical elements that differ significantly in some characteristics from normal macro-optics. Optical elements with feature sizes smaller than the wavelength of the incident light, sometimes referred to as sub-wavelength elements, interact differently with the light than optical elements with larger feature sizes. More specifically, light interacts in different ways with periodic structures having spacing smaller than its wavelength. The altered interactions create a variety of devices with characteristics that differ from similar but larger structures.

The sub-wavelength optical elements offer some new, desirable characteristics. Sub-wavelength grating structures, for example, may have a wider acceptance angle than conventional gratings. Wider acceptance angles relieve manufacturing tolerances and may permit automated assembly, further reducing costs. Sub-wavelength gratings also perform more consistently over a broader range of wavelengths than conventional gratings.

NIL has demonstrated a growing variety of nano-optical components, including gratings, filters, anti-reflective coatings, refractive and diffractive components, fresnel lenses, waveguides, and lens arrays. With all of these potential products and proven benefits, why has Nanooptics languished so long in the laboratories? Because formidable difficulties must be surmounted to replicate these tiny dimensions in volume manufacturing processes.

NANOEMBOSSING PROCESSES

To understand the manufacturing challenges that are now being overcome, look at the basic processing techniques that have been developed. NIL in its various forms has five processing steps:

- Design & fabricate a master stamp with features patterned by high-resolution techniques.
- Impress the master stamp on a layer of suitably prepared material.
- Maintain the master stamp pressure while the material first flows to duplicate the stamp, and then solidifies to preserve the duplicate pattern.
- Separate the patterned replica from the master stamp, without damaging nanoscale features.
- Post-process the replica further for its desired optical function.

The starting point for NIL is the creation of a master stamp or mold. The ultimate resolution capability of the embossing depends solely on the master stamp. The design and fabrication of the master stamp is critical to minimum feature size, uniformity, fluid flow, and ease of separation.

Designing a stamp to allow quick and complete fluid flow of the softened material is challenging. Once the design is satisfactory, it must be transferred to the master. Fine-line approaches, such as electron beam, are used to create nanoscale features on the master stamp. Electron beams have fine-line capability, but direct e-beam writing has a working area only a few millimetres square, and is relatively slow. Stamps used for hot embossing or for injection moulding are commonly made from nickel or steel. Stamps for UV embossing must be of UV-transparent materials.

While the completed master stamp could be directly used to produce many replicas, wear or deterioration through repeated use would require creating a new master stamp. In practice, to preserve its pristine quality, the completed master stamp generally is cloned, using the replication techniques described here. Cloned sub-masters are used to produce the nanoscale features in volume production, either simultaneously over a whole substrate, or repetitively, at chip size, onto a substrate.

The substrates may be silicon, quartz, glass, or other materials. Replication layers are generally coatings of heat-curable or UV curable polymers, with carefully controlled coating uniformity and thickness. While polymers are common materials for embossing, they may not offer the dimensional precision or the long-term stability under challenging environmental conditions demanded of some system components. This need has been filled by the development of special materials, such as of sol-gel organic/inorganic composite materials. These composites can be UV cured or thermally cured like polymers, and after curing display optical and mechanical properties close to those of glass.

Embossing and curing the pattern on the substrate varies with different processes. Once the coating is cured, the stamp must be separated from the cured material, without damaging the replicated pattern. As feature size becomes smaller, clean separations are more difficult. A special coating may be applied to the stamp before moulding, to facilitate separation.

The embossed wafer may be post-processed by etching, deposition, or by repeating further lithographic processes. The cured patterned material can be the etch resist, for reactive ion etching of the underlying substrate. Alternatively, the patterned material may be used as a resist for metal deposition and lift-off. Post-processing might also include additional lithography, further depositions, and the addition of antireflective or protective coatings.

Hot Embossing

Hot embossing of wafers uses the cloned master die as the wafer stamp. Wafers are first coated with a thermoplastic polymer, that is, a polymer that softens when heated. The wafer is positioned in a controlled-atmosphere stamping chamber. After heating the wafer, the stamp is brought down with pre-determined pressure and temperature to soften the polymer and simultaneously form the image on the whole wafer. After allowing time for polymer flow during heating, the stamp and wafer are cooled to harden the image. To maintain the pattern, pressure must continue to be applied until both stamp and substrate are cooled below the thermoplastic softening point. The hot-embossed wafer may be used for its embossed features, or it may be post-processed as a resist for etching, material deposition, and lift-off patterning. For resist use, curing may be followed by a light etch to remove the remaining traces of materials from the clear areas.

Approach of Step and Stamp

An alternative hot embossing approach is step-and-repeat stamping, using a smaller than wafer size master stamp. Step-and-stamp imitates the action of step-and-repeat optical steppers used in semiconductor fabrication. The smaller master stamp is simpler to design, produce, and control. It is also less costly than a full wafer stamp. The step-and-repeat approach mitigates the dimensional, planarity, flow, and separation anxieties inherent in wafer embossing, by literally reducing the problem. The tradeoffs include the added time required for multiple repetitions of the stamping process, the need to confine heating to the stamp area, and the cost and complexity of a precision step-and-repeat apparatus.

UV Embossing

Cold embossing uses UV light to polymerize and solidify materials after forming by the stamp. The starting materials are thin films of UV-curable materials. Typical materials may be monomers, or inorganic-organic hybrid polymers, which are polymerized by UV to polymers with desirable optical properties. The films may be deposited onto silicon, gallium arsenide, or similar wafers. The UV-transmissive patterned master template is pressed against the film, imprinting the pattern, which is UV cured before pressure is released.

UV embossing allows accurately aligning and overlaying multiple levels of features. UV embossing has a clear advantage over hot embossing in allowing multi-level applications, since the first level is polymerized after exposure, making it irrevocably formed and dimensionally stable, regardless of subsequent layer applications and exposures.

UV embossing is a low pressure, room temperature process, with little shrinkage of the pattern. It also allows partial UV casting, by limiting the illumination to maintain clear areas on the wafer. Cold embossing may be done on one or both sides of the wafer. Double-sided embossing allows micromolding. As with hot embossing, UV embossing may be used with a single wafer size stamp, or with a smaller stamp in a "step and flash" procedure similar to the "step and stamp".

Injection Moulding

Injection moulding of optical components offers throughput and cost advantages for high-volume production. Injection moulding can produce

3-D objects and integrated multicomponent systems. To move beyond conventional injection moulding and capture nanofeatures, electroformed metal inserts are nanopatterned and included as part of the master mold.

Double-sided injection moulding at wafer scale has been used to create optical structures on both sides of a supporting substrate. These structures may be the same on both sides of the substrate, or they may be different; for example, combining a diffractive lens on one side with a refractive lens on the other.

The two sides may be processed separately, or simultaneously. In either case, precise alignment must be made between the features through the intervening substrate.

Injection moulding has been used to produce optical components in hundreds of thousands. However, the high tooling cost can only be justified for these large quantities. Injection moulding draws upon a wider range of materials than direct embossing. Acrylics or polycarbonates may be used for all-plastic systems in low-cost, high-volume injection moulding, producing inexpensive commercial products in the hundreds of thousands.

Comparison of NIL Techniques

Hot embossing, at wafer level or by step-and-stamp, lends itself to single-side patterning on thermoplastic materials. Process temperatures are typically 90°C above the glassivation temperature. Throughput is limited by the time required to heat, hold, and cool the die and substrate. Step and stamp avoids the high cost and tolerances of a wafer-size stamp, at a penalty in multiplying the time delays.

Hot embossing has some well established and perhaps more profitable commercial applications. Music CDs are commonly replicated in volume by embossing.

Roll-to-roll hot embossing has been used to add interesting colour effects to wrapping paper, and to produce material with difficult to counterfeit optical patterns for security documents, such as identification cards.

Cold (UV) embossing may be preferred for critically dimensioned materials.

It can be a room temperature process, eliminating the heating and cooling times and thermal expansions/contractions of hot embossing. It allows single or repeated multiple patterns to be layered on one or both sides of a substrate with accurate alignments. Both sides of the substrate may be embossed with identical or different patterns. For applications with more critical planarity, dimensional tolerances, or environmental requirements, cold embossing of photosensitive organic-inorganic sol-gel materials gives results close to those of glass.

Injection moulding produces 3-D moulding, allowing entire systems to be moulded as a unit. It is not a room temperature process; processing time and temperature depend on the materials used. These may be inexpensive commercial plastics, or highly accurate specialty materials. Patterning can be single- or double-sided, in 3-D, although not generally multilayered as in cold embossing. The high tooling cost makes injection moulding practical for high-volume (*e.g.* 100,000 units and up) products.

MANUFACTURING PROGRESS: TECHNIQUES AND EQUIPMENT

A mixture of techniques and equipment from the semiconductor and micropackaging industries has been adapted to Nanooptics fabrication. Much of the development work reported above used one-of-a-kind laboratory equipment, built for that purpose, or modified equipment, originally developed for other purposes. One report referred to a modified semiconductor mask aligner. Another recent paper reported demonstrating hot embossing using a standard flip chip aligner-bonder. One Nanooptics manufacturer has developed customized robotic embossing systems specifically for volume production. Only recently have major equipment manufacturers announced production equipment intended to meet the challenges of nanometer optics. Dedicated aligner-steppers with sub-20-nm embossing resolution are now available.

Product applications based upon NIL include wavelength monitors, beam splitters, pattern generators, polarization filters, switches, and many others. A nano-optic pattern generator for forming a cross-hair pattern from a single laser beam. Nanooptics allows generating essentially an arbitrary pattern of dots in this manner.

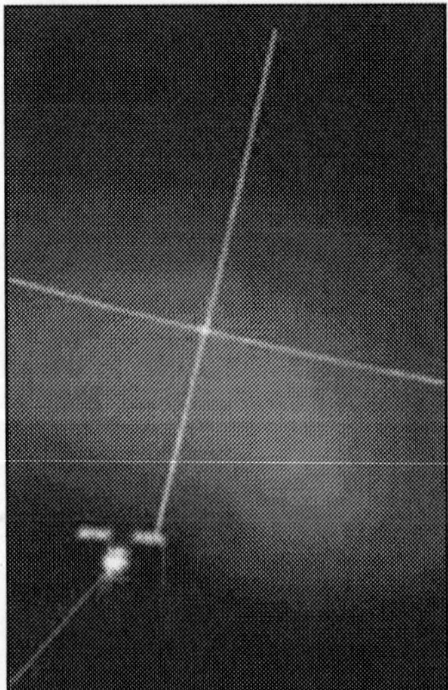

Figure : Replicated Pattern Generator Forming Cross-hair Pattern from a Single Laser Beam.

Also commercially available now are replicated components that are high performance building blocks for fabricating optical assemblies. These offer manufacturing advantages in comprising a common family of similar products, allowing standardization of manufacturing operations and equipment.

Figure : Nano-optical Building Blocks, 1.4mm Square, 0.5mmthick.

Embossing has come a long way since Chaucer. After decades of laboratory development, nano-embossing promises to open new optical frontiers with truly monolithic production systems.

The growing commercialization of nano-optic systems will subject them to the scrutiny and judgment of the marketplace. Success in surmounting the manufacturing challenges will determine whether nano-embossed optical systems remain a technology niche, or become a broad new technology opportunity.

THE "AGE" OF NANOOPTICS

START-UP PHASE (1984–1994)

The systematic exploration of optical near-fields for microscopy began in the early 1980s at the IBM Zurich Research Laboratory and at Cornell University.

The first high-resolution single line scans of test objects, obtained in transmission with a so-called "optical stethoscope" were published in 1984, followed soon by similar results obtained with a near-field scanning optical microscope (NSOM). First real scan images were published from 1985 to 1987, demonstrating a resolution of 20 to 50 nm. The resolution in those early experiments is still among the best achieved with A-probes.

In the next round, based on an A-probe scheme developed by Fischer, NOM imaging *in reflection* was demonstrated. The probe was flat, requiring a curved test sample for close approach. The resolution was comparable to that of transmission NOM. Reflection NOM also was demonstrated with probes that possessed a metallic protrusion instead of an aperture. The scattering intensity of this probe—which in fact was the first apertureless probe—went through a strong resonance upon approach. The resonance was attributed to plasmon excitation in the protrusion probe. It was the first evidence for plasmon effects in NFO, which today is one of the most active branches in this field.

Another interesting form of NOM, called "scanning tunneling optical microscope" (STOM,) or "photon scanning tunneling microscope" (PSTM) uses the pointed end of a *blank* optical fibre as a probe. The sample is transparent and illuminated so that total reflection occurs at the sample surface. The corresponding evanescent wave field outside the sample provides the interaction with the probe tip, in strict analogy to the tunneling electrons in STM. Fischer (1990) devised a similar scheme employing a metal tip instead of glass fibre.

CONSOLIDATION AND GENERALIZATION

NOM began to attract wider interest in the early nineties, marked by a first conference on this topic, the NATO Workshop on Near-Field Optics in Arc-et-Senans, France in 1992.

The field was stimulated in particular by a series of exploratory NOM studies by E. Betzig and coworkers at Bell Labs. Various modes of operation (*e.g.*, fluorescence and polarization) were demonstrated with the help of a technically improved "NSOM," as well as different applications in biology and lithography.

This work triggered further efforts in near-field microscopy. At the same time, other forms of NFO began to raise more interest—in particular, the interaction of light with *single* (more precisely: singled out) molecules including resonant energy transfer and Raman microscopy, trapping of particles by optical forces, localization and waveguiding in so-called photonic crystals, and plasmon excitation at nanometer-sized individual and coupled particles and *optical antennas.*

With increasing numbers of experimental NOM data, the need for better theoretical understanding became obvious. Beginning around 1990, a number of theoretical groups began to work on the problems of field determination in a NOM environment.

Outlook

A major trend in nano-/near-field optics is directed toward the development and manipulation of sophisticated probes, not only for microscopy but also as local sensors of their environment in general. The single molecule (the single quantum dot in particular) will continue to attract attention as ultimate probes. In NOM, main goals are higher resolution on the one hand and more practicable instruments on the other hand. To have an impact on other fields of science and technology, the NOM as an instrument must become more robust and simple than the instruments available today—a nano-engineering task that still waits for its realization. Plasmon phenomena on the nanometer scale will be of considerable interest in view of their potential use in advanced integrated optical devices.

To summarize, it may be expected that NFO will continue to be under investigation, driven by the great potential of optical characterization, inspired by the variety of possible operating modes, and—last but not least—challenged by the great difficulties of the practical application.

REFERENCES

- Aggarwal, S.K. : *A Textbook of Optics*, Wisdom Press, Delhi, 2012.
- Avadhanalu, M.N. : *A Text Book of Optics*, S. Chand Publisher, Delhi, 2006.
- Drexler, K. Eric: *Nanosystems: Molecular Machinery, Manufacturing, and Computation*, New York: Wiley, 1992.
- Evans, C.: *Surface Characterization Techniques*, New Delhi: Long Life Publication, 2003.
- Govind P. Agrawal: *Nonlinear Fiber Optics*, Academic Press, Boston, 2007
- Gupta, Ashish: *Understanding of Nano Science and Technology*, Global Vision, Delhi, 2006.
- Kaushik, N.L. : *Nanotechnology and Micromachines*, Pearl Books, Delhi, 2007.
- Khare, R.P. : *Fiber Optics and Optoelectronics*, Oxford University Press, Delhi, 2004.
- Nakazato, K.: *Extended Abstracts of 15th Symposium on Future Electron Devices*, Tokyo, R&D Association for FED, 1996.
- Pandit, Rajnish: *Basics of Optics*, Anmol, Delhi, 2008.

2

Transformation Optics

INTRODUCTION

Metamaterials give enormous choice of material parameters for electromagnetic applications. So much so that we might ask if there is a new way to design electromagnetic systems exploiting this new flexibility. In an ideal world magnetic and electrical field lines can be placed anywhere that the laws of physics allow and a suitable metamaterial found to accommodate the desired configuration of fields. It was to answer the question of what parameters to choose for the metamaterial that we developed transformation optics.

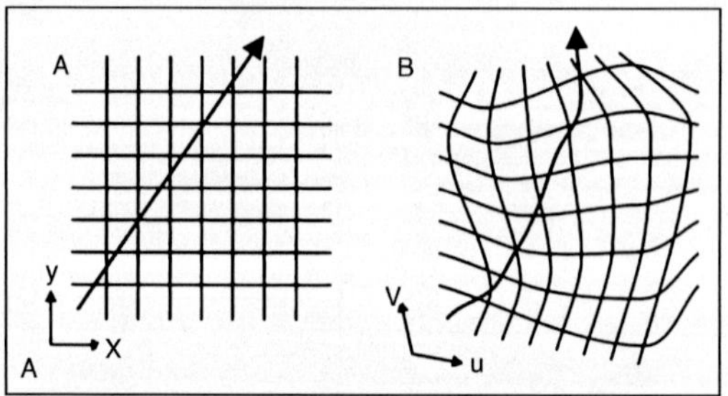

The idea is quite straightforward: start with a field pattern that obeys Maxwell's equations for a system that is topologically similar to the desired configuration but confined either to free space or a simple configuration of permittivity and permeability, then distort the system until the fields are in the desired configuration.

If we imagine that the original system was embedded in an elastic matrix in which Cartesian coordinate lines were drawn, then after distortion the deformed coordinates could be described by a coordinate transformation. Next rewrite Maxwell's equations using the new coordinate system. Some time ago it was shown that Maxwell's equation are of the same form in any coordinate system but the precise values of permittivity and permeability will change.

These new values of permittivity and permeability are the ones we must give to our metamaterial if we want the fields to take up the distorted configuration.

Ward and Pendry used transformations to adapt computer programmes to different coordinate systems: transformation from a Cartesian to a cylindrical system simply modifies the values of e and m used in the codes. Shyroki simplified the formulation using modern techniques an approach we subsequently followed. If the distorted system is described by a coordinate transform

$$x'^{j'}\left(x^j\right)$$

we define,

$$\Lambda_j^{j'} = \frac{\partial x^{j'}}{\partial x^j}$$

Then in the new coordinate system we must use modified values of the permittivity and permeability to ensure that Maxwell's equations are satisfied,

$$\varepsilon'^{i'j'} = \left[\det(\Lambda)\right]^{-1} \Lambda_i^{i'} \Lambda_j^{j'} \varepsilon^{ij}$$

$$\mu'^{i'j'} = \left[\det(\Lambda)\right]^{-1} \Lambda_i^{i'} \Lambda_j^{j'} \mu^{ij}$$

Although the prescription is simple enough it is very powerful in what can be achieved. For example: the perfect lens suffers from the limitation that the image is exactly the same size as the object.

Our new design methodology easily remedies this limitation: imagine a coordinate system that is stretched in the region of the image thus stretching the image itself. Transformation optics then prescribes the metamaterials with which this can be achieved.

A periodic array of radiation sources embedded in a periodic structure comprising a repeated structure of three layers of positively refracting material and one layer of material having an equal and opposite refractive index.

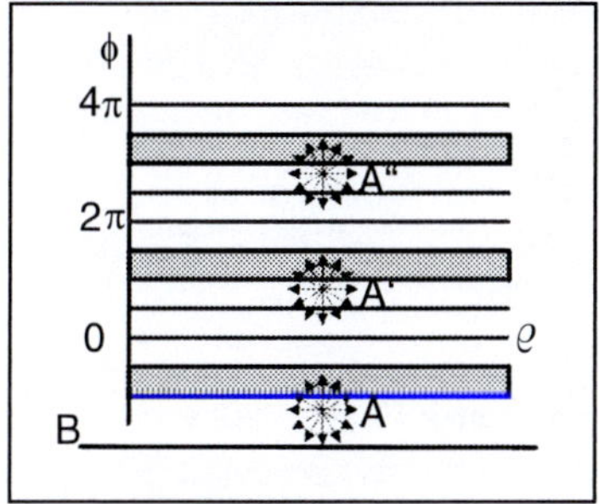

A transformation to cylindrical coordinates,

$$x = r_0 \cos\phi e^{\ell/\ell_0},\ y = r_0 \sin\phi e^{\ell/\ell_0},\ z = Z$$

maps figure B to a new geometry shown in figure C next page.

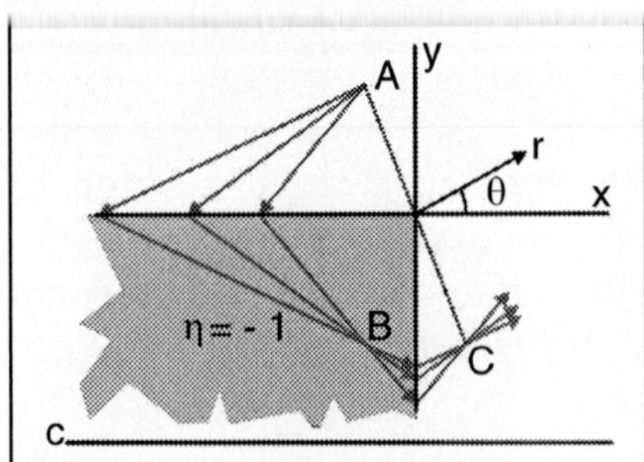

The periodic sources combine into one source and two images are produced: one inside at B and the other outside the material at C. The process of refraction reverses the direction of each ray exactly a corner mirror would do.

In this manner a single device such a lens can generate a whole family of new devices simply by applying new transformations.

As an extreme challenge for transformation optics we set ourselves the problem of constructing a cloak of invisibility. Two problems confronted us: first we must eliminate scattered radiation and hence no radiation must reach the hidden object; second the hidden object must cast no shadow.

TRANSFORMATION OPTICS AND CLOAKING

Negative-index metamaterials are uniform, containing many identical and evenly spaced modules, so their optical properties are uniform as well. Changing the shape and spacing of internal structures within a metamaterial in an arbitrary way opens the door to transforming light in other ways. It's a very powerful concept.

Citing the late Arthur C. Clarke's observation that, "any sufficiently advanced technology is indistinguishable from magic," Ulf Leonhardt at the University of St. Andrews (St. Andrews, Scotland) and David R. Smith of Duke University (Durham, NC) wrote last year that transformation optics were "the first practical demonstrations of 'pure and applied magic.'"

The specific bit of magic that caught their eyes, and those of the popular press, is invisibility cloaking by deflecting light around an object in the middle of an area so the wavefront that reaches an observer seems to have come straight through the cloaked object, as if it wasn't there. Like negative-index materials, cloaking was first seen for microwaves and then demonstrated for light.

Transformation optics determines how the properties of the cloak should vary through the metamaterial to deflect light away from the object. Early demonstrations used metamaterials assembled from conductive modules, but these have important drawbacks. The conductive elements strongly attenuate light, and their properties depend on resonance effects, which vary strongly with wavelength—meaning that they lose a lot of light and work only across a limited spectral range.

Broadband cloaking is possible in other cases. This spring, two groups described broadband "carpet cloaks" that could hide a bump under a rug. Optically, previous cloak designs compress the hidden object into a single point or line from the viewer's standpoint, wrote Valentine, Jensen Li, and Thomas Zentgraf of Zhang's group at Berkeley, but "the carpet cloak compresses an object in only one direction into a conducting sheet." That change in the shape of the wavefront doesn't require singularities or drastic changes in properties through the material, allowing them to avoid resonance and use low-loss dielectric materials to achieve much broader bandwidth than is possible with conductors.

Valentine and colleagues drilled an array of 110 nm holes into a 250-nm-thick silicon slab optical waveguide. The researchers spaced the holes regularly except in a rectangular cloaking region, where the holes were arranged in a two-dimensional pattern designed to cloak a bump on the edge of the slab waveguide.

When they directed a laser beam at a bump "under" the cloak, they saw a single Gaussian peak in the reflected light, as they would from a flat edge. Illuminating an uncloaked bump revealed multiple peaks, as expected for reflections from a bump. They observed the cloaking effect from 1400 to 1800 nm.

Lucas Gabrielli and colleges in Michal Lipson's group at Cornell University (Ithaca, NY) demonstrated a similar dielectric carpet cloak, in which 50 nm silicon posts were embedded in a silica layer. Distribution of the silicon posts determined the optical transformation of the wavefront. The group also reported that its cloak concealed a bump on the edge of the planar waveguide across a 400 nm range in the 1550 nm region. Without the cloak, light reflected from the bump at an angle included a dark shadow-like zone, but with the cloak light reflection from the bump area was as smooth as if the area had been flat.

OPTICAL BLACK HOLES

The power of transformation optics comes from the ability of a nonuniform metamaterial to warp the path of light arbitrarily, like a gravitational field can warp space-time. Zhang is working to extend the analogy so light propagation through metamaterials can be used in laboratory studies of such cosmic phenomena as black holes and strange attractors in the laboratory.

Zhang envisions devices he calls "continuous-index photon traps," which could serve as "perfect" optical cavities that absorb light but don't radiate across a broad range of wavelengths. One possibility is to combine air and the high-index semiconductor indium gallium arsenide phosphide (InGaAsP) to form a "photonic black hole," which would attract light in the same way a gravitational black hole attracts matter. Zhang says transformation optics could design devices "that can mimic the periodic, quasi-periodic, and chaotic motions observed in celestial objects that have been subjected to complex gravitational fields".

One use could be to simulate the chaotic long-term orbits of planets and asteroids. "Unlike astronomers, we will not have to wait 100 years to get experimental results," Zhang says.

Separately and nearly simultaneously, Evgenii E. Narimanova and Alexander V. Kildishev of Purdue proposed using transformation optics to make omnidirectional broadband light absorbers, or "optical black holes." But they weren't thinking of research applications. They envision practical applications including photovolatics, solar-energy harvesting, and optoelectronic detectors.

A similarly intriguing application of metamaterials is an omnidirectional retroreflector, recently demonstrated with microwaves by Leonhardt's group. Like conventional retroreflectors made of multiple mirrors, the devices would reflect all incident light back in the direction from which it came, not just the ray incident on a reflective sphere at a right angle. With conventional optics, such a device would require a dielectric singularity with an infinite refractive index, but Leonhardt's group used transformational optics to convert that singularity into a mere topological defect in a metamaterial.

State of a Fast-moving Art

Although the idea of transformation optics has been around a while, key experiments have been reported within the past few months. As this section was about to go to press, Science published the first report of frequency-tunable "memory metamaterials," in which frequency response can be both tuned and preserved. The tuning, demonstrated in Dimitri Basov's lab at the University of California at San Diego (La Jolla, CA), is based on hysteresis in vanadium oxide structures in the metamaterial, and could lead to many other devices.

With transformation optics, says Shalaev, "you are limited only by your imagination because you really can do anything with light." His excitement at his group's recent demonstration of visible cloaking an area 100 wavelengths wide illustrates the state of the field.

That's only 50 μm, the width of a human hair, and a long way from hiding Harry Potter or even a covert microphone. But it's an important step toward a bright future, and could lead even further. "You could make a cloak for light, you could make a cloak for electron wave functions, and you could make electron black holes," he predicts. Stay tuned.

RESEARCH OF TRANSFORMATION OPTICS

Developments in this field focus on advances in research of transformation optics. Transformation optics is the foundation for exploring a diverse set of theoretical, numerical, and experimental developments, involving the perspectives of the physics and engineering communities. The multi-disciplinary perspectives for inquiry and designing of materials develop understanding of their behaviours, properties, and potential applications for this field.

If a coordinate transformation can be derived or described, a ray of light (in the optical limit) will follow lines of a constant coordinate. There are constraints on the transformations, as listed in the references. In general, however, a particular goal can be accomplished using more than one transformation. The classic cylindrical cloak can be created with many transformations. The simplest, and most often used, is a linear coordinate mapping in the radial coordinate. There is significant ongoing research into determining advantages and disadvantages of particular types of transformations, and what attributes are desirable for realistic transformations. One example of this is the broadband carpet cloak: the transformation used was quasi-conformal. Such a transformation can yield a cloak that uses non-extreme values of permittivity and permeability, unlike the classic cylindrical cloak, which required some parameters to vary towards infinity at the inner radius of the cloak.

General coordinate transformations can be derived which compress or expand space, bend or twist space, or even change the topology (*e.g.* by mimicking a wormhole). Much current interest involves designing invisibility cloaks, event cloaks, field concentrators, or beam-bending waveguides.

MIMICKING CELESTIAL MECHANICS

The interactions of light and matter with spacetime, as predicted by general relativity, can be studied using the new type of artificial optical materials that feature extraordinary abilities to bend light. Light is also known as electromagnetic radiation. This research creates a link between the newly emerging field of artificial optical metamaterials to that of celestial mechanics, thus opening a new possibility to investigate astronomical phenomena in a laboratory setting. The recently introduced, new class, of specially designed optical media can mimic the periodic, quasi-periodic and chaotic motions observed in celestial objects that have been subjected to gravitational fields.

Hence, a new class of metamaterials introduced with the nomenclature "continuous-index photon traps" (CIPTs). CIPTz have applications as optical cavities. As such, CIPTs can control, slow and trap light in a manner similar to celestial phenomena such as black holes, strange attractors, and gravitational lenses.

A composite of air and the dielectric Gallium Indium Arsenide Phosphide (GaInAsP), operated in the infrared spectral range and featured a high refractive index with low absorptions.

This opens an avenue to investigate light phenomena that imitates orbital motion, strange attractors and chaos in a controlled laboratory environment by merging the study of optical metamaterials with classical celestial mechanics.

If a metamaterial could be produced that did not have high intrinsic loss and a narrow frequency range of operation then it could be employed as a type of media to simulate light motion in a curved spacetime vacuum. Such a proposal is brought forward, and metamaterials become prospective media in this type of study. The classical optical-mechanical analogy renders the possibility for the study of light propagation in homogeneous media as an accurate analogy to the motion of massive bodies, and light, in gravitational potentials. A direct mapping of the celestial phenomena is accomplished by observing photon motion in a controlled laboratory environment. The materials could facilitate periodic, quasi-periodic and chaotic light motion inherent to celestial objects subjected to complex gravitational fields.

Twisting the optical metamaterial effects its "space" into new coordinates. The light that travels in real space will be curved in the twisted space, as applied in transformational optics. This effect is analogous to starlight when it moves through a closer gravitational field and experiences curved spacetime or a gravitational lensing effect. This analogue between classic electromagnetism and general relativity, shows the potential of optical metamaterials to study relativity phenomena such as the gravitational lens.

Observations of such celestial phenomena by astronomers can sometimes take a century of waiting. Chaos in dynamic systems is observed in areas as diverse as molecular motion, population dynamics and optics. In particular, a planet around a star can undergo chaotic motion if a perturbation, such as another large planet, is present. However, owing to the large spatial distances between the celestial bodies, and the long periods involved in the study of their dynamics, the direct observation of chaotic planetary motion has been a challenge. The use of the optical-mechanical analogy may enable such studies to be accomplished in a bench-top laboratory setting at any prescribed time.

The study also points toward the design of *novel optical cavities* and *photon traps* for application in microscopic devices and lasers systems.

PRODUCING BLACK HOLES WITH METAMATERIALS

Matter propagating in a curved spacetime is similar to the electromagnetic wave propagation in a curved space and in an in homogeneous metamaterial. Hence a black hole can possibly be simulated using electromagnetic fields and metamaterials. In July 2009 a metamaterial structure forming an effective black hole was theorized, and numerical simulations showed a highly efficient light absorption. The first experimental demonstration of electromagnetic black hole at microwave frequencies occurred in October 2009.

The proposed black hole was composed of non-resonant, and resonant, metamaterial structures, which can absorb electromagnetic waves efficiently coming from all directions due to the local control of electromagnetic fields. It was constructed of a thin cylinder at 21.6 centimeters in diameter comprising 60 concentric rings of metamaterials. This structure created a gradient index of refraction, necessary for bending light in this way. However, it was characterized as being artificially inferior substitute for a real black hole. The characterization was justified by an absorption of only 80% in the microwave range, and that it has no internal source of energy. It is singularly a light absorber. The light absorption capability could be beneficial if it could be adapted to technologies such as solar cells. However, presently the device is limited to the microwave range.

Also in 2009, transformation optics were employed to mimic a black hole of Schwarzschild form. Similar properties of photon sphere were also found numerically for the metamaterial black hole. Several reduced versions of the black hole systems were proposed for easier implementations.

MIT computer simulations by Fung along with lab experiments are designing a metamaterial with a multilayer sawtooth structure that slows and absorbs light over a wide range of wavelength frequencies, and at a wide range of incident angles, at 95% efficiency. This has an extremely wide window for colours of light.

MULTIDIMENSIONAL UNIVERSE

Engineering optical space with metamaterials could be useful to reproduce an accurate laboratory model of the physical multiverse. "*This 'metamaterial landscape' may include regions in which one or two spatial dimensions are compactified.*" Metamaterial models appear to be useful for non-trivial models such as 3D de Sitter space with one compactified dimension, 2D de Sitter space with two compactified dimensions, 4D de Sitter dS4, and anti-de Sitter AdS4 spaces.

Gradient Index Lensing

Transformation optics is employed to increase capabilities of gradient index lenses.

CONVENTIONAL OPTICAL LIMITATIONS

Optical elements (lenses) perform a variety of functions, ranging from image formation, to light projection or light collection. The performance of these systems is frequently limited by their optical elements, which dominate system weight and cost, and force tradeoffs between system parameters such as focal length, field of view (or acceptance angle), resolution, and range.

Conventional lenses are ultimately limited by geometry. Available design parameters are a single index of refraction (n) per lens element, variations in the element surface profile, including continuous surfaces (lens curvature) and/or discontinuous surfaces (diffractive optics). Light rays undergo refraction at the surfaces of each element, but travel in straight lines within the lens. Since the design space of conventional optics is limited to a combination of refractive index and surface structure, correcting for aberrations (for example through the use of achromatic or diffractive optics) leads to large, heavy, complex designs, and/or greater losses, lower image quality, and manufacturing difficulties.

GRIN LENSES

Gradient index lenses (or GRIN lenses) as the name implies, are optical elements whose index of refraction varies within the lens. Control of the internal refraction allows the steering of light in curved trajectories through the lens. GRIN optics thus increase the design space to include the entire volume of the optical elements, providing the potential for dramatically reduced size, weight, element count, and assembly cost, as well as opening up new space to trade between performance parameters. However, past efforts to make large aperture GRIN lenses have had limited success due to restricted refractive index change, poor control over index profiles, and/or severe limitations in lens diameter.

RECENT ADVANCES

Recent steps forward in material science have led to at least one method for developing large (>10 mm) GRIN lenses with 3-dimensional gradient indexes. There is a possibility of adding expanded deformation capabilities to the GRIN lenses. This translates into controlled expansion, contraction, and shear (for variable focus lenses or asymmetric optical variations). These capabilities have been demonstrated. Additionally, recent advances in transformation optics and computational power provide a unique opportunity to design, assemble and fabricate elements in order to advance the utility and availability of GRIN lenses across a wide range of optics-dependent systems, defined by needs. A possible future capability could be to further advance lens design methods and tools, which are coupled to enlarged fabrication processes.

Battlefield Applications

Transformation optics has potential applications for the battlefield. The versatile properties of metamaterials can be tailored to fit almost any practical need, and transformation optics shows that space for light can be bent in almost any arbitrary way. This is perceived as providing new capabilities to soldiers in the battlefield. For battlefield scenarios benefits from metamaterials have both short term and long term impacts. For example, determining whether a cloud in the distance is harmless or an aerosol of enemy chemical or biological warfare is presently very difficult to quickly assess. However, with the new metamaterials being developed, the ability exists to see things smaller than the wavelength of light - something which has yet to be achieved in the far field. Utilizing metamaterials in the creation of a new lens may allow soldiers to be able to see pathogens and viruses that are currently impossible to detect with any visual device.

Harnessing subwavelength capabilities then allow for other advancements which appear to be beyond the battlefied. All kinds of materials could be manufactured with nano-manufacturing, which could go into electronic and optical devices from night vision goggles to distance sensors to other kinds of sensors. Longer term views include the possibility for cloaking materials, which would provide "invisibility" by redirecting light around a cylindrical shape.

PERFORM OF COORDINATE TRANSFORMATIONS

The design and use of holographic optical elements (HOE's) to perform coordinate transformations were first suggested by Bryngdahl.' Transformations with HOE's were exploited for distortion compensation, for angle multiplexing in optical fibres, and for optical data processing. The design of HOE's for these transformations is based on two approximations: the saddle-point approximation and the paraxial approximation. These approximations impose contradicting requirements on the distance between the input and the output planes. The saddle-point approximation calls for a short distance compared to the lateral dimensions, whereas the paraxial approximation requires the distance to be much longer. Such contradicting requirements impose severe limitations on the space-bandwidth product (SBP) of the data that may be transformed without degradations. Furthermore, these design procedures were suitable only for certain types of transformation. For more general transformations, faceted HOE's with even lower SBP capabilities or complex optical systems containing two HOE's in cascade have been suggested.

In this paper we present a new procedure for designing HOE's without the need for paraxial approximation. In this procedure, analytic ray-tracing techniques are first exploited to determine the necessary grating vector, from which the optimal grating function for forming the HOE's is derived by variational methods. With such grating functions, the SBP of the data that can be transformed is significantly improved, and the coordinate transformation need not be limited to a specific type. Furthermore, the Fourier transform lenses that were included in the earlier designs need no longer be added.

TRANSFORMATION-DESIGNED OPTICAL ELEMENTS

The transformation method allows one to design continuous media that exert precise and arbitrary control over a field, particularly the electromagnetic field. Even the challenging problem of reflectionless design in arbitrary geometries is quite straight forward. The resulting material specifications are necessarily complex, being both anisotropic and inhomogenous with specific functional forms. Implementation of such media would be hopeless without recent advances in metamaterials.

These two concepts comprise a great synergy, with metamaterials enabling new media response, and transformation design suggesting new phenomena and devices supported by these media. The recent work on the design and implementation of metamaterial invisibility cloaks points to the potential of these concepts. In this article we apply the transformation method to the design of imaging optics.

A fundamental issue with imaging systems is the constraint on resolution imposed by the diffraction limit. Currently, sub-diffraction images are assembled from scanning probes that can sense near-fields, but the scanning process severely limits image frame rates, and the probes require near direct contact with the object.

The perfect lens began to address these issues by providing an image transfer of both near- and far-field components with a free-space working distance. Shortly there after Pendry transformed the perfect lens to a cylindrical geometry to obtain magnification. More recently implementation strategies for near-field magnification have been explored with layered media (where the proposed devices are sometimes called hyper lenses, so named because they are composed of an indefinite medium with a hyperbolic dispersion relation).

Experimental realizations have also appeared using layered media, and also wire arrays. For a fixed input aperture, near-field magnification actually increases the amount of image information available in the far field. One might even call this a beating of the diffraction limit. If one has both a free-space working distance and near-field magnification one can acquire high resolution, sub-diffraction images using standard resolution detectors, and without making contact with the object.

The transformation method to design slab elements that have a variety of useful functions - including magnification - that operate on both the near- and far- field, and provide free-space working distance. It is worth noting some significant differences between the transformation designed near-field magnifier and previous designs.

One such difference is in the geometry of the focal surfaces. The cylindrical or spherical devices have curved object and image surfaces, whereas the transformation designed element has planar object and image planes that better match typical detector arrays, and display formats. The cylindrical or oblique-cut devices magnify in only one dimension with magnification directly related to the sample geometry, where with the transformation design the magnification can be any reasonable value in each of the two dimensions of the image plane, independent of the element's thickness.

Additionally, we will show that the geometric aberrations of the transformation designed element are theoretically zero, and its substantially flat transfer function has a magnitude indicating excellent coupling to the environment.

For all other recent designs, no analysis is available describing the aberrations or an image transfer function that includes coupling into and out of the environment. Since all recent designs incorporate uncompensated, anisotropic media, which are inherently mismatched to free space, one expects weak coupling to free space fields.

On the down side, the transformation designed device requires magnetic response, making the implementation path to high frequencies, such as visible light, more problematic. However, metamaterials for magnetic response in the visible spectrum are being pursued, and the functionality we describe in this article may increase the incentive for such work. In the nearer term, implementations for sub-diffraction microwave and terahertz imaging may be feasible.

OPTICAL DESIGN GEOMETRY

As is traditional in optical design, we will describe optical elements with reference to an optical axis, which we will take to be the z-axis in our Cartesian coordinate system. Optical elements are considered to extend in the transverse (x,y) directions sufficiently far that the electromagnetic fields at the transverse edges of an element are assumed to be negligible. To verify this condition is a matter of practical importance, but an encumbrance to be shed in a first analysis. This assumption provides a particular freedom when using the transformation method to design perfectly matched, (*i.e.* zero reflecting for all angles of incidence), elements. With this method one must use a coordinate transformation that is everywhere continuous to achieve perfectly-matched behaviour. Since a finite sized material object is by definition completely surround by free space, and free space is described by a "flat" Cartesian coordinate system, a finite-sized, perfectly-matched material object must be described by a coordinate transformation that is continuous with flat Cartesian space, everywhere along its outer boundary. We will relax this condition in the transverse direction of our optical elements, and design elements that are perfectly matched for electromagnetic fields that are near the optic axis and sufficiently far from the transverse edges, using coordinate transformations that cannot be continuous to flat space on their transverse edges. This extra freedom allows for some new functionality in manipulating fields along the optic axis.

TENSOR

Tensors are geometric objects that describe linear relations between vectors, scalars, and other tensors. Elementary examples of such relations include the dot product, the cross product, and linear maps. Vectors and scalars themselves are also tensors. A tensor can be represented as a multidimensional array of numerical values. The *order* (also *degree* or *rank*) of a tensor is the dimensionality of the array needed to represent it, or equivalently, the number of indices needed to label a component of that array. For example, a linear map can be represented by a matrix, a 2-dimensional array, and therefore is a 2nd-order tensor. A vector can be represented as a 1-dimensional array and is a 1st-order tensor. Scalars are single numbers and are thus zeroth-order tensors.

Tensors are used to represent correspondences between sets of geometrical vectors. For example, the stress tensor T takes a direction v as input and produces the stress $T^{(v)}$ on the surface normal to this vector as output and so expresses a relationship between these two vectors. Because they express a relationship between vectors, tensors themselves must be independent of a particular choice of coordinate system. Taking a coordinate basis or frame of reference and applying the tensor to it results in an organized multidimensional array representing the tensor in that basis, or as it looks from that frame of reference. The coordinate independence of a tensor then takes the form of a "covariant" transformation law that relates the array computed in one coordinate system to that computed in another one. This transformation law is considered to be built in to the notion of a tensor in a geometrical or physical setting, and the precise form of the transformation law determines the *type* (or *valence*) of the tensor.

Tensors are important in physics because they provide a concise mathematical framework for formulating and solving physics problems in areas such as elasticity, fluid mechanics, and general relativity. Tensors were first conceived by Tullio Levi-Civita and Gregorio Ricci-Curbastro, who continued the earlier work of Bernhard Riemann and Elwin Bruno Christoffel and others, as part of the *absolute differential calculus*. The concept enabled an alternative formulation of the intrinsic differential geometry of a manifold in the form of the Riemann curvature tensor.

TENSOR DENSITIES

It is also possible for a tensor field to have a "density". A tensor with density *r* transforms as an ordinary tensor under coordinate transformations, except that it is also multiplied by the determinant of the Jacobian to the rth power. Invariantly, in the language of multilinear algebra, one can think of tensor densities as multilinear maps taking their values in a density bundle such as the (1-dimensional) space of *n*-forms (where *n* is the dimension of the space), as opposed to taking their values in just R. Higher "weights" then just correspond to taking additional tensor products with this space in the range.

In the language of vector bundles, the determinant bundle of the tangent bundle is a line bundle that can be used to 'twist' other bundles *r* times. While locally the more general transformation law can indeed be used to recognise these tensors, there is a global question that arises, reflecting that in the transformation law one may write either the Jacobian determinant, or its absolute value. Non-integral powers of the (positive) transition functions of the bundle of densities make sense, so that the weight of a density, in that sense, is not restricted to integer values.

Restricting to changes of coordinates with positive Jacobian determinant is possible on orientable manifolds, because there is a consistent global way to eliminate the minus signs; but otherwise the line bundle of densities and the line bundle of *n*-forms are distinct.

SPINORS

Starting with an orthonormal coordinate system, a tensor transforms in a certain way when a rotation is applied. However, there is additional structure to the group of rotations that is not exhibited by the transformation law for tensors: see orientation entanglement and plate trick. Mathematically, the rotation group is not simply connected. Spinors are mathematical objects that generalize the transformation law for tensors in a way that is sensitive to this fact.

MAXWELL'S EQUATIONS

Maxwell's equations are a set of partial differential equations that, together with the Lorentz force law, form the foundation of classical electrodynamics, classical optics, and electric circuits. These fields in turn underlie modern electrical and communications technologies.

Maxwell's equations have two major variants. The "microscopic" set of Maxwell's equations uses total charge and total current including the difficult-to-calculate atomic level charges and currents in materials. The "macroscopic" set of Maxwell's equations defines two new auxiliary fields that can sidestep having to know these 'atomic' sized charges and currents.

Maxwell's equations are named after the Scottish physicist and mathematician James Clerk Maxwell, since in an early form they are all found in a four-part paper, "On Physical Lines of Force", which he published between 1861 and 1862. The mathematical form of the Lorentz force law also appeared in this paper.

It is often useful to write Maxwell's equations in other forms; these representations are still formally termed "Maxwell's equations". A relativistic formulation in terms of covariant field tensors is used in special relativity, while in quantum mechanics, a version based on the electric and magnetic potentials is preferred. While Maxwell's equations are consistent within special and general relativity, there are some quantum mechanical situations in which Maxwell's equations are significantly inaccurate: including extremely strong fields and extremely short distances. Moreover, various phenomena occur in the world even though Maxwell's equations predicts them to be impossible, such as "nonclassical light" and quantum entanglement of electromagnetic fields. Finally, any phenomenon involving individual photons, such as the photoelectric effect, Planck's law, the Duane–Hunt law, single-photon light detectors, *etc.*, would be difficult or impossible to explain if Maxwell's equations were exactly true, as Maxwell's equations do not involve photons. Maxwell's equations are usually an extremely accurate approximation to the more accurate theory of quantum electrodynamics.

Maxwell's "Macroscopic" Equations

Unlike the "microscopic" equations, "Maxwell's macroscopic equations", also known as Maxwell's equations in matter, factor out the bound charge and current to obtain equations that depend only on the free charges and currents. These equations are more similar to those that Maxwell himself introduced. The cost of this factorization is that additional fields need to be defined: the displacement field D which is defined in terms of the electric field E and the polarization P of the material, and the magnetic-H field, which is defined in terms of the magnetic-B field and the magnetization M of the material.

BOUND CHARGE AND CURRENT

An electric field is applied to a dielectric material its molecules respond by forming microscopic electric dipoles – their atomic nuclei move a tiny distance in the direction of the field, while their electrons move a tiny distance in the opposite direction. This produces a *macroscopic bound charge* in the material even though all of the charges involved are bound to individual molecules.

For example, if every molecule responds the same, similar to that shown in the figure, these tiny movements of charge combine to produce a layer of positive bound charge on one side of the material and a layer of negative charge on the other side. The bound charge is most conveniently described in terms of a polarization, P, in the material. If P is uniform, a macroscopic separation of charge is produced only at the surfaces where P enter and leave the material.

For non-uniform P, a charge is also produced in the bulk. Somewhat similarly, in all materials the constituent atoms exhibit magnetic moments that are intrinsically linked to the angular momentum of the components of the atoms, most notably their electrons. The connection to angular momentum suggests the picture of an assembly of microscopic current loops. Outside the material, an assembly of such microscopic current loops is not different from a macroscopic current circulating around the material's surface, despite the fact that no individual magnetic moment is traveling a large distance. These *bound currents* can be described using the magnetization M.

The very complicated and granular bound charges and bound currents, therefore can be represented on the macroscopic scale in terms of P and M which average these charges and currents on a sufficiently large scale so as not to see the granularity of individual atoms, but also sufficiently small that they vary with location in the material. As such, the *Maxwell's macroscopic equations* ignores many details on a fine scale that can be unimportant to understanding matters on a gross scale by calculating fields that are averaged over some suitable volume.

CONSTITUTIVE RELATIONS

In order to apply 'Maxwell's macroscopic equations', it is necessary to specify the relations between displacement field D and E, and the magnetic H-field H and B. These equations specify the response of bound charge and current to the applied fields and are called constitutive relations. For real-world materials, the constitutive relations are rarely simple, except approximately, see the main article for a fuller description.

The *definitions* (not constitutive relations) of the auxiliary fields are:

$$D(r,t) = \varepsilon_0 E(r,t) + P(r,t)$$

$$H(r,t) = \frac{1}{\mu_0} B(r,t) - M(r,t),$$

where P is the polarization field and M is the magnetization field which are defined in terms of microscopic bound charges and bound current respectively.

The macroscopic forms of Maxwell's equations for different materials. In each case, Faraday's law of induction and Gauss's law for magnetism are always the same.

No Materials (Vacuum)

The constitutive relations are

$$D = \varepsilon_0 E, \quad H = B/\mu_0$$

The currents and charges are free, not total (expected since there are no bound charges nor currents);

Gauss's law:

$$\nabla \cdot D = \rho_f$$

Ampère's circuital law:

$$\nabla \times H = J_f + \frac{\partial D}{\partial t}$$

Linear Materials

The constitutive relations are:

$$D = \varepsilon E, \quad H = B / \mu$$

where ε is the permittivity and μ the permeability of the material.

- For homogeneous materials, ε and μ are constant throughout the material, for inhomogeneous they depend on location within the material (and perhaps time).
- For isotropic materials, ε and μ are independent of the directions of the applied fields to the material, for anisotropic they are tensors (incorporating directional dependence of the medium).
- Materials are generally dispersive, so ε and μ depend on the frequency of any incident EM waves.

The vacuum permittivity and permeability are replaced by those of the material, the charges and currents are free (not total);

Gauss's law:

$$\nabla \cdot (\varepsilon E) = \rho_f$$

Ampère's circuital law:

$$\nabla \times (B / \mu) = J_f + \varepsilon \frac{\partial E}{\partial t}.$$

In the case of non-linear materials, D and P are not proportional to E, similarly B is not proportional to H or M.

Solving Maxwell's equations

Maxwell's equations are partial differential equations that relate the electric and magnetic fields to each other and to the electric charges and currents. Often, the charges and currents are themselves dependent on the electric and magnetic fields via the Lorentz force equation and the constitutive relations. These all form a set of coupled partial differential equations, which are often very difficult to solve. In fact, the solutions of these equations encompass all the diverse phenomena in the entire field of classical electromagnetism. A thorough discussion is far beyond the scope of the article, but some general notes follow:

- Like any differential equation, boundary conditions and initial conditions are necessary for a unique solution. For example, even with no charges and no currents anywhere in spacetime, many solutions to Maxwell's equations are possible, not just the obvious solution E = B = 0. Another solution is E = constant, B = constant, while yet other solutions have electromagnetic waves filling spacetime. In some cases, Maxwell's equations are solved through infinite space, and boundary conditions are given as asymptotic limits at infinity. In other cases, Maxwell's equations are solved in just a finite region of space, with appropriate boundary conditions on that region: For example, the boundary could be a artificial absorbing boundary representing the rest of the universe, or periodic boundary conditions, or (as with a waveguide or cavity resonator) the boundary conditions may describe the walls that isolate a small region from the outside world.

- Jefimenko's equations (or the closely related Liénard–Wiechert potentials) are the explicit solution to Maxwell's equations for the electric and magnetic fields created by any given distribution of charges and currents. It assumes specific initial conditions to obtain the so-called "retarded solution", where the only fields present are the ones created by the charges. Jefimenko's equations are not so helpful in situations when the charges and currents are themselves affected by the fields they create.
- Numerical methods for differential equations can be used to approximately solve Maxwell's equations when an exact solution is impossible. These methods usually require a computer, and include the finite element method and finite-difference time-domain method.
- Maxwell's equations seem overdetermined, in that they involve six unknowns (the three components of E and B) but eight equations (one for each of the two Gauss's laws, three vector components each for Faraday's and Ampere's laws). In fact, Maxwell's equations are somewhat redundant; it can be proven that any system satisfying Faraday's law and Ampere's law *automatically* also satisfies the two Gauss's laws, as long as the system's initial condition does. Therefore numerical algorithms often ignore the two Gauss's laws, although more accurate algorithms take them into account in order to minimize numerical inaccuracy.

UNRUH EFFECT

In the early seventies it was realised that the notion of particles depends on the specific details of the quantum measurement process used to detect them, and that the state of motion of the measuring device can determine whether or not particles are observed. This discovery has created a new viewpoint which was prompted by Fulling, Unruh and Hawking's work demonstrating that the number of particles found in a region depends on the acceleration of the measuring device.

For example, the vacuum, *i.e.* a region that contains no particles at all, would be seen by an accelerated observer as a region with particles. The number of particles and their energy would increase with increased acceleration. This effect is known as the Unruh effect.

Our group in collaboration with Benni Reznik and Ignacio Cirac have studied the feasibility of the experimental realization of these effects in BEC. We have proposed a way to measure the Unruh effect in Bose Einstein Condensates which serves as a simulator for the real vacuum. In this proposal the Unruh temperature would reveal itself by the number of particles of a specific kind.

EXPERIMENTAL OBSERVATION OF THE UNRUH EFFECT

Under experimentally achievable conditions for gravitational systems this effect is too small and its observation is very difficult. It was shown by Bell and Leinaas that if one takes an accelerated observer to be an electron circularly orbiting in a constant external magnetic field, then the experimentally verified Sokolov–Ternov effect coincides with the Unruh effect.

A recent work by Martín-Martínez, Fuentes and Mann showed that accelerated detectors acquire a geometrical phase due to their movement through spacetime and that this can be used for the direct detection of the Unruh effect in regimes physically accessible with current technology.

Explanation

Unruh demonstrated theoretically that the notion of vacuum depends on the path of the observer through spacetime. From the viewpoint of the accelerating observer, the vacuum of the inertial observer will look like a state containing many particles in thermal equilibrium—a warm gas.

Although the Unruh effect would initially be perceived as counter-intuitive, it makes sense if the word *vacuum* is interpreted appropriately.

VACUUM INTERPRETATION

In modern terms, the concept of "vacuum" is not the same as "empty space", as all of space is filled with the quantized fields that make up a universe. Vacuum is simply the lowest *possible* energy state of these fields, a very different definition from "empty".

The energy states of any quantized field are defined by the Hamiltonian, based on local conditions, including the time coordinate. According to special relativity, two observers moving relative to each other must use different time coordinates. If those observers are accelerating, there may be no shared coordinate system. Hence, the observers will see different quantum states and thus different vacua.

In some cases, the vacuum of one observer is not even in the space of quantum states of the other. In technical terms, this comes about because the two vacua lead to unitarily inequivalent representations of the quantum field canonical commutation relations. This is because two mutually accelerating observers may not be able to find a globally defined coordinate transformation relating their coordinate choices.

An accelerating observer will perceive an apparent event horizon forming. The existence of Unruh radiation could be linked to this apparent event horizon, putting it in the same conceptual framework as Hawking radiation. On the other hand, the theory of the Unruh effect explains that the definition of what constitutes a "particle" depends on the state of motion of the observer.

The (free) field needs to be decomposed into positive and negative frequency components before defining the creation and annihilation operators. This can only be done in spacetimes with a timelike Killing vector field. This decomposition happens to be different in Cartesian and Rindler coordinates (although the two are related by a Bogoliubov transformation). This explains why the "particle numbers", which are defined in terms of the creation and annihilation operators, are different in both coordinates.

The Rindler spacetime has a horizon, and locally any non-extremal black hole horizon is Rindler. So the Rindler spacetime gives the local properties of black holes and cosmological horizons. The Unruh effect would then be the near-horizon form of the Hawking radiation.

Calculations

The theory of the Unruh effect involves the Rindler coordinates ρ and τ, which have metric

$$ds^2 = -\rho^2 d\tau^2 + d\rho^2$$

This is just ordinary Minkowski space in relativistic polar coordinates:

$$x = \rho\cosh(\tau)$$

$$t = \rho\sinh(\tau)$$

The orbit in (1+1) space-time is a regular hyperbola. The parametric equations are of the above form, but with ρ equal to the ratio of square of the speed of light to the proper acceleration. It is therefore a constant. There is only one parameter - the proper time τ. In the arguments of the hyperbolic function, this should be multiplied by the ratio of the proper acceleration to the speed of light.

A detector moving along a path of constant ρ is uniformly accelerated, and is coupled to field modes which have a definite steady frequency as a function of ρ. These modes are constantly Doppler shifted relative to ordinary Minkowski time as the detector accelerates, and they change in frequency by enormous factors, even after only a short proper time.

Translation in τ is a symmetry of Minkowski space: It is a boost around the origin. For a detector coupled to modes with a definite frequency in τ, the boost operator is then the Hamiltonian.

In the Euclidean field theory, these boosts analytically continue to rotations, and the rotations close after 2π. So

$$e^{2\pi i H} = 1$$

The path integral for this Hamiltonian is closed with period 2π which guarantees that the H modes are thermally occupied with temperature $(2\pi)^{-1}$. This is not an actual temperature, because H is dimensionless. It is conjugate to the timelike polar angle τ which is also dimensionless. To restore the length dimension, note that a mode of fixed frequency f in τ at position ρ has a frequency which is determined by the square root of the metric at ρ, the redshift factor. The actual inverse temperature at this point is therefore

$$\beta = 2\pi\rho$$

Since the acceleration of a trajectory at constant ρ is equal to 1/a, the actual inverse temperature observed is:

$$\beta = \frac{2\pi}{a}$$

The temperature observed by a uniformly accelerating particle is (in engineering units):

$$kT = \frac{\hbar a}{2\pi c}$$

The Unruh effect could only be seen when the Rindler horizon is visible. If a refrigerated accelerating wall is placed between the particle and the horizon, at fixed Rindler coordinate ρ_0, the thermal boundary condition for the field theory at ρ_0 is the temperature of the wall. By making the positive ρ side of the wall colder, the extension of the wall's state to $\rho > \rho_0$ is also cold. In particular, there is no thermal radiation from the acceleration of the surface of the Earth, nor for a detector accelerating in a circle, because under these circumstances there is no Rindler horizon in the field of view.

The temperature of the vacuum, seen by an isolated observer accelerated at the Earth's gravitational acceleration of g = 9.81 m/s², is only 4×10^{-20} K. For an experimental test of the Unruh effect it is planned to use accelerations up to 10^{26} m/s², which would give a temperature of about 400,000 K.

To put this in perspective, at a vacuum Unruh temperature of 3.978×10^{-20}K, an electron would have a de Broglie Wavelength of $h/\sqrt{(3m_e kT)}$ = 540.85 meters, and a proton at that temperature would have a wavelength of 12.62 meters. If electrons and protons were in intimate contact in a very cold vacuum, they would have rather long wavelengths and interaction distances. At one astronomical unit from the sun, the acceleration is GM s/AU² = 0.005932 m/s². This gives an Unruh temperature of 2.41×10^{-23} kelvin. At that temperature, the electron and proton wavelengths are 21.994 kilometres 513 meters, respectively. Even a uranium atom will have a wavelength of 2.2 meters at such a low temperature.

OTHER IMPLICATIONS

The Unruh effect would also cause the decay rate of accelerated particles to differ from inertial particles. Stable particles like the electron could have nonzero transition rates to higher mass states when accelerated fast enough.

UNRUH RADIATION

Although Unruh's prediction that an accelerating detector would see a thermal bath is not controversial, the interpretation of the transitions in the detector in the non-accelerating frame are. It is widely, although not universally, believed that each transition in the detector is accompanied by the emission of a particle, and that this particle will propagate to infinity and be seen as *Unruh radiation.*

The existence of Unruh radiation is not universally accepted. Some claim that it has already been observed, while others claims that it is not emitted at all. While the skeptics accept that an accelerating object thermalises at the Unruh temperature, they do not believe that this leads to the emission of photons, arguing that the emission and absorption rates of the accelerating particle are balanced.

HAWKING RADIATION

In 1975 Hawking published a shocking result: if one takes quantum theory into account, it seems that black holes are not quite black! Instead, they should glow slightly with "Hawking radiation", consisting of photons, neutrinos, and to a lesser extent all sorts of massive particles. This has never been observed, since the only black holes we have evidence for are those with lots of hot gas falling into them, whose radiation would completely swamp this tiny effect. Indeed, if the mass of a black hole is M solar masses, Hawking predicted it should glow like a blackbody of temperature

$$6 \times 10^{-8}/M \text{ kelvins,}$$

so only for very small black holes would this radiation be significant. Still, the effect is theoretically very interesting, and folks working on understanding how quantum theory and gravity fit together have spent a lot of energy trying to understand it and its consequences. The most drastic consequence is that a black hole, left alone and unfed, should radiate away its mass, slowly at first but then faster and faster as it shrinks, finally dying in a blaze of glory like a hydrogen bomb. However, the total lifetime of a black hole of M solar masses works out to be

$$10^{71} M^3 \text{ seconds}$$

so don't wait around for a big one to give up the ghost.

How does this work? Well, you'll find Hawking radiation explained this way in a lot of "pop-science" treatments:

Virtual particle pairs are constantly being created near the horizon of the black hole, as they are everywhere. Normally, they are created as a particle-antiparticle pair and they quickly annihilate each other. But near the horizon of a black hole, it's possible for one to fall in before the annihilation can happen, in which case the other one escapes as Hawking radiation. The usual computation involves Bogoliubov transformations. The idea is that when you quantize (say) the electromagnetic field you take solutions of the classical equations (Maxwell's equations) and write them as a linear combination of positive-frequency and negative-frequency parts. Roughly speaking, one gives you particles and the other gives you antiparticles. More subtly, this splitting is implicit in the very definition of the vacuum of the quantum version of the theory! In other words, if you do the splitting one way, and I do the splitting another way, our notion of which state is the vacuum may disagree!

This should not be *utterly* shocking, just *pretty darn* shocking. The vacuum, after all, can be thought of as the state of least energy. If we are using really different co-ordinate systems, we'll have really different notions of time, hence really different notions of energy—since energy is defined in quantum theory to be the operator H such that time evolution is given by *exp(-itH)*. So on the one hand, it's quite conceivable that we'll have different notions of positive and negative frequency solutions in classical field theory—a solution that's a linear combination of those with time dependence *exp(–iωt)* is called positive or negative frequency depending on the sign of ω—but of course this depends on a choice of time co-ordinate t. And on the other hand, it's quite conceivable that we'll have different notions of the lowest-energy state.

Now when we are in good old flat Minkowski spacetime, a la special relativity, there are a bunch of "inertial frames" differing by Lorentz transformations. These give different time co-ordinates, but one can check that the difference is never so bad that different co-ordinates give different notions of positive or negative frequency solutions of Maxwell's equations. Nor will different people using these co-ordinate systems ever disagree about what's the lowest-energy state. So all inertial observers agree about what's a particle, what's an antiparticle, and what's the vacuum.

But in curved spacetime there aren't these "best" co-ordinate systems, the inertial ones. So even very reasonable different choices of co-ordinates can give disagreements about particles vs antiparticles, or what's the vacuum. These disagreements don't mean that "everything is relative", because there are nice formulas for how to translate between the descriptions in different co-ordinate systems. These are Bogoliubov transformations.

What is Hawking Radiation?

Hawking radiation (sometimes also called Bekenstein-Hawking radiation) is a theoretical prediction from British physicist Stephen Hawking, which explains thermal properties relating to black hole.

Normally, a black hole is considered to draw all matter and energy in the surrounding region into it, as a result of the intense gravitational fields. However, in 1972 the Israeli physicist Jacob Bekenstein suggested that black holes should have a well-defined entropy, and initiated the development of black hole thermodynamics, including the emission of energy. In 1974, British physicist Stephen Hawking worked out the exact theoretical model for how a black hole could emit black body radiation. In a simplified version of the explanation, Hawking predicted that energy fluctuations from the vacuum causes the generation of particle-antiparticle pairs of virtual particles near the event horizon of the black hole. One of the particles falls into the black hole while the other escapes, before they have an opportunity to annihilate each other. The net result is that, to someone viewing the black hole, it would appear that a particle had been emitted.

Since the particle that is emitted has positive energy, the particle that gets absorbed by the black hole has a negative energy relative to the outside universe. This results in the black hole losing energy, and thus mass (because $E = mc^2$).

Smaller primordial black holes can actually emit more energy than they absorb, which results in them losing net mass. Larger black holes, such as those that are one solar mass, absorb more cosmic radiation than they emit through Hawking radiation.

Hawking radiation was one of the first theoretical predictions which provided insight into how gravity can relate to other forms of energy, which is a necessary part of any theory of quantum gravity.

Though Hawking radiation is generally accepted by the scientific community, there is still some controversy associated with it. There are some concerns that it ultimately results in information being lost, which makes physicists uncomfortable. Alternately, those who don't actually believe that black holes themselves exist are similarly reluctant to accept that they absorb particles.

HAWKING AND BLACK BODY RADIATION

Hawking radiation is black body radiation that is predicted to be emitted by black holes, due to quantum effects near the event horizon. It is named after the physicist Stephen Hawking, who provided a theoretical argument for its existence in 1974, and sometimes also after the physicist Jacob Bekenstein who predicted that black holes should have a finite, non-zero temperature and entropy. Hawking's work followed his visit to Moscow in 1973 where Soviet scientists Yakov Zeldovich and Alexei Starobinsky showed him that according to the quantum mechanical uncertainty principle, rotating black holes should create and emit particles. Hawking radiation reduces the mass and the energy of the black hole and is therefore also known as black hole evaporation. Because of this, black holes that lose more mass than they gain through other means are expected to shrink and ultimately vanish. Micro black holes (MBHs) are predicted to be larger net emitters of radiation than larger black holes and should shrink and dissipate faster.

In September 2010, a signal which is closely related to black hole Hawking radiation was claimed to have been observed in a laboratory experiment involving optical light pulses, however the results remain unverified and debatable. Other projects have been launched to look for this radiation within the framework of analog gravity. In June 2008, NASA launched the GLAST satellite, which will search for the terminal gamma-ray flashes expected from evaporating primordial black holes. In the event that speculative large extra dimension theories are correct, CERN's Large Hadron Collider may be able to create micro black holes and observe their evaporation.

OVERVIEW OF BLACK HOLES

Black holes are sites of immense gravitational attraction. Classically, the gravitation is so powerful that nothing, not even electromagnetic radiation (over-simplified in many descriptions as "light"), can escape from the black hole. It is yet unknown how gravity can be incorporated into quantum mechanics, nevertheless, far from the black hole the gravitational effects can be weak enough for calculations to be reliably performed in the framework of quantum field theory in curved spacetime. Hawking showed that quantum effects allow black holes to emit exact black body radiation, which is the average thermal radiation emitted by an idealized thermal source known as a black body. The electromagnetic radiation is as if it were emitted by a black body with a temperature that is inversely proportional to the black hole's mass.

Physical insight into the process may be gained by imagining that particle-antiparticle radiation is emitted from just beyond the event horizon. This radiation does not come directly from the black hole itself, but rather is a result of virtual particles being "boosted" by the black hole's gravitation into becoming real particles.

A slightly more precise, but still much simplified, view of the process is that vacuum fluctuations cause a particle-antiparticle pair to appear close to the event horizon of a black hole. One of the pair falls into the black hole whilst the other escapes. In order to preserve total energy, the particle that fell into the black hole must have had a negative energy (with respect to an observer far away from the black hole). By this process, the black hole loses mass, and, to an outside observer, it would appear that the black hole has just emitted a particle. In another model, the process is a quantum tunneling effect, whereby particle-antiparticle pairs will form from the vacuum, and one will tunnel outside the event horizon.

An important difference between the black hole radiation as computed by Hawking and thermal radiation emitted from a black body is that the latter is statistical in nature, and only its average satisfies what is known as Planck's law of black body radiation, while the former fits the data better. Thus thermal radiation contains information about the body that emitted it, while Hawking radiation seems to contain no such information, and depends only on the mass, angular momentum, and charge of the black hole (the no-hair theorem). This leads to the black hole information paradox.

However, according to the conjectured gauge-gravity duality (also known as the AdS/CFT correspondence), black holes in certain cases (and perhaps in general) are equivalent to solutions of quantum field theory at a non-zero temperature. This means that no information loss is expected in black holes (since no such loss exists in the quantum field theory), and the radiation emitted by a black hole is probably the usual thermal radiation. If this is correct, then Hawking's original calculation should be corrected, though it is not known how.

A black hole of one solar mass has a temperature of only 60 nanokelvins; in fact, such a black hole would absorb far more cosmic microwave background radiation than it emits. A black hole of 4.5×10^{22} kg (about the mass of the Moon) would be in equilibrium at 2.7 kelvin, absorbing as much radiation as it emits. Yet smaller primordial black holes would emit more than they absorb, and thereby lose mass.

TRANS-PLANCKIAN PROBLEM

The trans-Planckian problem is the observation that Hawking's original calculation requires talking about quantum particles in which the wavelength becomes shorter than the Planck length near the black hole's horizon. It is due to the peculiar behaviour near a gravitational horizon where time stops as measured from far away. A particle emitted from a black hole with a finite frequency, if traced back to the horizon, must have had an infinite frequency there and a trans-Planckian wavelength.

The Unruh effect and the Hawking effect both talk about field modes in the superficially stationary space-time that change frequency relative to other coordinates which are regular across the horizon. This is necessarily so, since to stay outside a horizon requires acceleration which constantly Doppler shifts the modes.

An outgoing Hawking radiated photon, if the mode is traced back in time, has a frequency which diverges from that which it has at great distance, as it gets closer to the horizon, which requires the wavelength of the photon to "scrunch up" infinitely at the horizon of the black hole. In a maximally extended external Schwarzschild solution, that photon's frequency only stays regular if the mode is extended back into the past region where no observer can go. That region doesn't seem to be observable and is physically suspect, so Hawking used a black hole solution without a past region which forms at a finite time in the past. In that case, the source of all the outgoing photons can be identified–it is a microscopic point right at the moment that the black hole first formed.

The quantum fluctuations at that tiny point, in Hawking's original calculation, contain all the outgoing radiation. The modes that eventually contain the outgoing radiation at long times are redshifted by such a huge amount by their long sojourn next to the event horizon, that they start off as modes with a wavelength much shorter than the Planck length. Since the laws of physics at such short distances are unknown, some find Hawking's original calculation unconvincing.

The trans-Planckian problem is nowadays mostly considered a mathematical artifact of horizon calculations. The same effect occurs for regular matter falling onto a white hole solution. Matter which falls on the white hole accumulates on it, but has no future region into which it can go. Tracing the future of this matter, it is compressed onto the final singular endpoint of the white hole evolution, into a trans-Planckian region. The reason for these types of divergences is that modes which end at the horizon from the point of view of outside coordinates are singular in frequency there. The only way to determine what happens classically is to extend in some other coordinates that cross the horizon.

There exist alternative physical pictures which give the Hawking radiation in which the trans-Planckian problem is addressed. The key point is that similar trans-Planckian problems occur when the modes occupied with Unruh radiation are traced back in time. In the Unruh effect, the magnitude of the temperature can be calculated from ordinary Minkowski field theory, and is not controversial.

Emission Process

Hawking radiation is required by the Unruh effect and the equivalence principle applied to black hole horizons. Close to the event horizon of a black hole, a local observer must accelerate to keep from falling in. An accelerating observer sees a thermal bath of particles that pop out of the local acceleration horizon, turn around, and free-fall back in. The condition of local thermal equilibrium implies that the consistent extension of this local thermal bath has a finite temperature at infinity, which implies that some of these particles emitted by the horizon are not reabsorbed and become outgoing Hawking radiation.

A Schwarzschild black hole has a metric;

$$ds^2 = -\left(1 - \frac{2M}{r}\right)dt^2 + \frac{1}{1 - \dfrac{2M}{r}}dr^2 + r^2 d\Omega^2$$

The black hole is the background spacetime for a quantum field theory.

The field theory is defined by a local path integral, so if the boundary conditions at the horizon are determined, the state of the field outside will be specified. To find the appropriate boundary conditions, consider a stationary observer just outside the horizon at position $r = 2M + u^2 / 2M$. The local metric to lowest order is:

$$ds^2 = -\frac{u^2}{4M^2}dt^2 + 4du^2 + dX_\perp^2 = -\rho^2 d\tau^2 + d\rho^2 + dX_\perp^2$$

which is Rindler in terms of $\tau = t/4M$ and $\rho = 2u$. The metric describes a frame that is accelerating to keep from falling into the black hole. The local acceleration diverges as $u \to 0$.

The horizon is not a special boundary, and objects can fall in. So the local observer should feel accelerated in ordinary Minkowski space by the principle of equivalence. The near-horizon observer must see the field excited at a local inverse temperature

$$\beta(u) = 2\pi\rho = (4\pi)u = 4\pi\sqrt{2M(r-2M)}\,,$$

the Unruh effect.

The gravitational redshift is by the square root of the time component of the metric. So for the field theory state to consistently extend, there must be a thermal background everywhere with the local temperature redshift-matched to the near horizon temperature:

$$\beta(r') = 4\pi\sqrt{2M(r-2M)}\sqrt{\frac{1-\frac{2M}{r'}}{1-\frac{2M}{r}}}$$

The inverse temperature redshifted to r′ at infinity is

$$\beta(\infty) = (4\pi)\sqrt{2Mr}$$

and is the near-horizon position, near 2*M*, so this is really:

$$\beta = 8\pi M$$

So a field theory defined on a black hole background is in a thermal state whose temperature at infinity is:

$$T_H = \frac{1}{8\pi M}$$

which can be expressed more cleanly in terms of the surface gravity of the black hole, the parameter that determines the acceleration of a near-horizon observer.

$$T_H = \frac{\kappa}{2\pi}$$

in natural units with $G, c, \hbar$ and equal to 1, and where is the surface gravity of the horizon. So a black hole can only be in equilibrium with a gas of radiation at a finite temperature.

Since radiation incident on the black hole is absorbed, the black hole must emit an equal amount to maintain detailed balance. The black hole acts as a perfect blackbody radiating at this temperature.

In engineering units, the radiation from a Schwarzschild black hole is black-body radiation with temperature:

$$T = \frac{\hbar c^3}{8\pi G M k_b} \quad (\approx \frac{1.227 \times 10^{23}\ \text{kg}}{M}\ \text{K})$$

where is the reduced Planck constant, c is the speed of light, k_b is the Boltzmann constant, G is the gravitational constant, and M is the mass of the black hole.

From the black hole temperature, it is straightforward to calculate the black hole entropy. The change in entropy when a quantity of heat dQ is added is:

$$dS = \frac{dQ}{T} = 8\pi M dQ$$

the heat energy that enters serves increase the total mass:

$$dS = 8\pi M dM = d(4\pi M^2).$$

The radius of a black hole is twice its mass in natural units, so the entropy of a black hole is proportional to its surface area:

$$S = \pi R^2 = \frac{A}{4}.$$

Assuming that a small black hole has zero entropy, the integration constant is zero. Forming a black hole is the most efficient way to compress mass into a region, and this entropy is also a bound on the information content of any sphere in space time. The form of the result strongly suggests that the physical description of a gravitating theory can be somehow encoded onto a bounding surface.

BLACK HOLE EVAPORATION

When particles escape, the black hole loses a small amount of its energy and therefore of its mass (mass and energy are related by Einstein's equation $E = mc^2$).

The power emitted by a black hole in the form of Hawking radiation can easily be estimated for the simplest case of a nonrotating, non-charged Schwarzschild black hole of mass M. Combining the formulas for the Schwarzschild radius of the black hole, the Stefan–Boltzmann law of black-body radiation, the above formula for the temperature of the radiation, and the formula for the surface area of a sphere (the black hole's event horizon), equation derivation:

Stefan–Boltzmann constant:

$$\sigma = \frac{\pi^2 k_B^4}{60\hbar^3 c^2}$$

Schwarzschild radius:

$$r_s = \frac{2GM}{c^2}$$

Black hole surface gravity at the horizon:

$$g = \frac{GM}{r_s^2} = \frac{c^4}{4GM}$$

Hawking radiation has a black-body (Planck) spectrum with a temperature T given by:

$$E = k_B T = \frac{\hbar g}{2\pi c} = \frac{\hbar}{2\pi c}\left(\frac{c^4}{4GM}\right) = \frac{\hbar c^3}{8\pi GM}$$

Hawking radiation temperature:

$$T_H = \frac{\hbar c^3}{8\pi GMk_B}$$

Schwarzschild sphere surface area of Schwarzschild radius r_s:

$$A_s = 4\pi r_s^2 = 4\pi\left(\frac{2GM}{c^2}\right)^2 = \frac{16\pi G^2 M^2}{c^4}$$

Stefan–Boltzmann power law:

$$P = A_s j^* = A_s \epsilon \sigma T^4$$

A black hole is a perfect black-body:

$$\epsilon = 1$$

Stefan–Boltzmann–Schwarzschild–Hawking black hole radiation power law derivation:

$$P = A_s \epsilon \sigma T_H^4 = \left(\frac{16\pi G^2 M^2}{c^4}\right)\left(\frac{\pi^2 k_B^4}{60\hbar^3 c^2}\right)\left(\frac{\hbar c^3}{8\pi GMk_B}\right)^4 = \frac{\hbar c^6}{15360\pi G^2 M^2}$$

Stefan–Boltzmann-Schwarzschild-Hawking power law:

$$P = \frac{\hbar c^6}{15360\pi G^2 M^2}$$

Where P is the energy outflow, $\hbar$ is the reduced Planck constant, c is the speed of light, and G is the gravitational constant. It is worth mentioning that the above formula has not yet been derived in the framework of semiclassical gravity. The power in the Hawking radiation from a solar mass $(M_\odot)$ black hole turns out to be a minuscule 9×10^{-29} watts. It is indeed an extremely good approximation to call such an object 'black'.

$$P = \frac{\hbar c^6}{15360\pi G^2 M_\odot^2} = 9.004 \times 10^{-29}\ \text{W}$$

Under the assumption of an otherwise empty universe, so that no matter or cosmic microwave background radiation falls into the black hole, it is possible to calculate how long it would take for the black hole to dissipate:

$$K_{ev} = \frac{\hbar c^6}{15360\pi G^2} = 3.562 \times 10^{32}\ \mathrm{W{\cdot}kg^2}$$

Given that the power of the Hawking radiation is the rate of evaporation energy loss of the black hole:

$$P = -\frac{dE}{dt} = \frac{K_{ev}}{M^2}$$

Since the total energy E of the black hole is related to its mass M by Einstein's mass-energy formula:

$$E = Mc^2$$

$$P = -\frac{dE}{dt} = -\left(\frac{d}{dt}\right)Mc^2 = -c^2\frac{dM}{dt}$$

We can then equate this to our above expression for the power:

$$-c^2\frac{dM}{dt} = \frac{K_{ev}}{M^2}$$

This differential equation is separable, and we can write:

$$M^2 dM = -\frac{K_{ev}}{c^2}dt$$

The black hole's mass is now a function $M(t)$ of time t. Integrating over M from M_0 (the initial mass of the black hole) to zero (complete evaporation), and over t from zero to t_{ev}:

$$\int_{M_0}^{0} M^2 dM = -\frac{K_{ev}}{c^2}\int_0^{t_{ev}} dt$$

The evaporation time of a black hole is proportional to the cube of its mass:

$$t_{ev} = \frac{c^2 M_0^3}{3K_{ev}} = \left(\frac{c^2 M_0^3}{3}\right)\left(\frac{15360\pi G^2}{\hbar c^6}\right) = \frac{5120\pi G^2 M_0^3}{\hbar c^4} = 8.410\times 10^{-17}\left[\frac{M_0}{\mathrm{kg}}\right]^3 \mathrm{s}$$

The time that the black hole takes to dissipate is:

$$t_{ev} = \frac{5120\pi G^2 M_0^3}{\hbar c^4}$$

Where M_0 is the mass of the black hole.

The lower classical quantum limit for mass for this equation is equivalent to the Planck mass, m_P.

Planck mass quantum black hole Hawking radiation evaporation time:

$$t_{ev} = \frac{5120\pi G^2 m_P^3}{\hbar c^4} = 5120\pi t_p = 5120\pi\sqrt{\frac{\hbar G}{c^5}} = 8.671\times 10^{-40}\ \mathrm{s}$$

$$t_{ev} = 5120\pi\sqrt{\frac{\hbar G}{c^5}}$$

Where t_P is the Planck time.

For a black hole of one solar mass ($M_\odot$ = 1.98892 × 10^{30} kg), we get an evaporation time of 2.098 × 10^{67} years—much longer than the current age of the universe at 13.73 ± 0.12 x 10^9 years.

$$t_{ev} = \frac{5120\pi G^2 M_\odot^3}{\hbar c^4} = 6.617 \times 10^{74} \text{ s}$$

But for a black hole of 10^{11} kg, the evaporation time is 2.667 billion years. This is why some astronomers are searching for signs of exploding primordial black holes.

However, since the universe contains the cosmic microwave background radiation, in order for the black hole to dissipate, it must have a temperature greater than that of the present-day black-body radiation of the universe of 2.7 K = 2.3 × 10^{-4} eV. This implies that must be less than 0.8% of the mass of the Earth.

Cosmic microwave background radiation universe temperature:

$$T_u = 2.725 \text{ K}$$

Hawking total black hole mass:

$$M_H \leq \frac{\hbar c^3}{8\pi G k_B T_u} \leq 1.226 \times 10^{23} \text{ kg}$$

$$M_H \leq \frac{\hbar c^3}{8\pi G k_B T_u}$$

$$\frac{M_H}{M_\oplus} = 7.539 \times 10^{-3} = 0.754\ \%$$

Where, $M_\oplus$ is the total Earth mass.

In common units,

$$p = 3.56345 \times 10^{32} \left[\frac{\text{kg}}{M}\right]^2 \text{W}$$

$$t_{ev} = 8.40716 \times 10^{-17} \left[\frac{M_0}{\text{kg}}\right]^3 \text{s} \approx 2.66 \times 10^{-24} \left[\frac{M_0}{\text{kg}}\right]^3 \text{yr}$$

$$M_0 = 2.28271 \times 10^5 \left[\frac{t_{ev}}{\text{s}}\right]^{1/3} \text{kg} \approx 7.2 \times 10^7 \left[\frac{t_{ev}}{\text{yr}}\right]^{1/3} \text{kg}$$

So, for instance, a 1-second-lived black hole has a mass of 2.28×10^5 kg, equivalent to an energy of 2.05×10^{22} J that could be released by 5×10^6 megatons of TNT. The initial power is 6.84×10^{21} W.

Black hole evaporation has several significant consequences:

- Black hole evaporation produces a more consistent view of black hole thermodynamics, by showing how black holes interact thermally with the rest of the universe.
- Unlike most objects, a black hole's temperature increases as it radiates away mass. The rate of temperature increase is exponential, with the most likely endpoint being the dissolution of the black hole in a violent burst of gamma rays. A complete description of this dissolution requires a model of quantum gravity, however, as it occurs when the black hole approaches Planck mass and Planck radius.
- The simplest models of black hole evaporation lead to the black hole information paradox. The information content of a black hole appears to be lost when it dissipates, as under these models the Hawking radiation is random. A number of solutions to this problem have been proposed, including suggestions that Hawking radiation is perturbed to contain the missing information, that the Hawking evaporation leaves some form of remnant particle containing the missing information, and that information is allowed to be lost under these conditions.

Large Extra Dimensions

Formulae from the previous are only applicable if laws of gravity are approximately valid all the way down to the Planck scale. In particular, for black holes with masses Planck mass (~10^{-5} g), they result in unphysical lifetimes Planck time (~10^{-43} s). This is normally seen as an indication that Planck mass is the lower limit on the mass of a black hole. In the model with large extra dimensions, values of Planck constants can be radically different, and formulas for Hawking radiation have to be modified as well. In particular, the lifetime of a micro black hole is given by

$$\tau \sim \frac{1}{M_*}\left(\frac{M_{BH}}{M_*}\right)^{(n+3)/(n+1)}$$

where M_* is the low energy scale (which could be as low as a few TeV), and n is the number of large extra dimensions. This formula is now consistent with black holes as light as a few TeV, with lifetimes on the order of "new Planck time" ~10^{-26} s.

Experimental observation of Hawking Radiation

Under experimentally achievable conditions for gravitational systems this effect is too small to be observed directly. In September 2010, however, an experimental set-up created a laboratory "white hole event horizon" that the experimenters claimed was shown to radiate Hawking radiation, although its status as a genuine confirmation remains in doubt. Some scientists predict that Hawking radiation could be studied by analogy using sonic black holes, in which sound perturbations are analogous to light in a gravitational black hole and the flow of an approximately perfect fluid is analogous to gravity.

NEW HORIZONS FOR HAWKING RADIATION

Evidence for radiation analogous to that predicted near black holes by Hawking has been found in the emission of optical filaments propagating through glass.

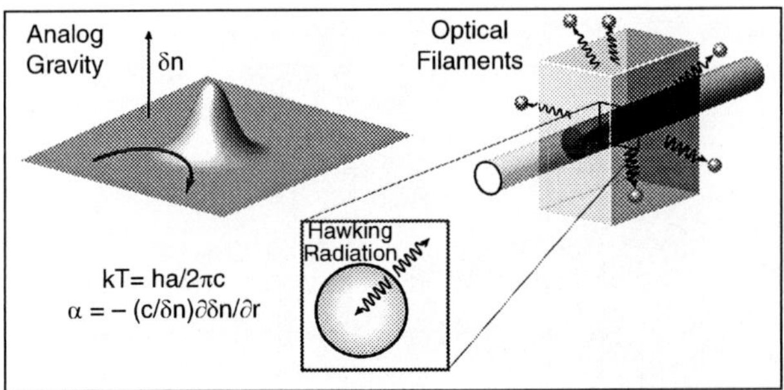

An analog gravitational potential in an optical system. A suitable change in the index of refraction in a moving medium creates an effective event horizon for photons that propagate with it. As indicated by the pink arrow, photons are forbidden from entering beyond the optical event horizon. The analog black body temperature of the optical white hole to how the change in the index varies in time (τ). Belgiorno *et al.* use a high-intensity light filament to perturb the index of refraction in glass and create an optical event horizon and measure the analog Hawking radiation emerging at right angles from the filament.

In 1974, Steven Hawking predicted that black holes were not completely black, but were actually weak emitters of blackbody radiation generated close to the event horizon—the boundary where light is forever trapped by the black hole's gravitational pull. Hawking's insight was to realise how the presence of the horizon could separate virtual photon pairs (constantly being created from the quantum vacuum) such that while one was sucked in, the other could escape, causing the black hole to lose energy. Hawking's idea was significant in suggesting a possible optical signature of a black hole's existence. Yet, even though the prediction created an extensive theoretical literature in cosmology, calculations have since shown that Hawking radiation from black holes is so weak that it would be practically impossible to measure.

It turns out, however, that the physics of how waves interact with a horizon does not depend in a fundamental way on the presence of gravity at all. In principle, an analogous Hawking radiation should occur in other systems. The key requirement is simply that the interaction between waves and the medium in which they propagate causes there to be a boundary between zones where the wave and the medium have different velocities. In a paper in *Physical Review Letters*, Francesco Belgiorno at the Università degli Studi di Milano, in collaboration with researchers at several other institutes, also in Italy, describe a series of experiments where high-intensity filaments of light in glass perturb the optical propagation environment in an analogous manner to the way a gravitational field affects light near a black hole horizon. This perturbation creates the optical equivalent of an event horizon that allows Belgiorno *et al.* to make convincing measurements of analog Hawking radiation at optical frequencies.

These results are highly significant in suggesting a system in which Hawking's prediction can be fully explored in a convenient laboratory environment.

A really useful way to visualize an analog event horizon is to imagine a fish swimming upstream in a river flowing towards a waterfall. The point at which the current flows faster than the fish can swim represents a boundary at which fish cannot escape and they are swept over the waterfall. This "point of no return" for the fish is equivalent to a black hole horizon. A similar analogy exists for white holes, associated with horizons that fish can never enter, namely, if they were attempting to swim upstream towards the bottom of a waterfall.

This simple picture is surprisingly powerful at capturing the physics of horizons, and can be extended rigorously to describe horizon physics in diverse systems including acoustics, cold atoms, and gravity-capillary waves on water. Indeed, horizon effects and a stimulated form of Hawking radiation have recently been explicitly observed in the vicinity of an obstacle placed in an open channel flow. However, the question has remained open as to whether these analog gravity systems also generate the spontaneous thermal radiation Hawking predicted.

Belgiorno *et al.*'s experiments suggest that the answer is yes. They created an optical event horizon by using intense, ultrashort light pulses that change the refractive index of the glass in the vicinity of the moving pulse. The change in the refractive index modifies the effective propagation "geometry" as seen by copropagating light rays such that a trapping horizon forms and, in principle, Hawking radiation should occur. The changing index is an effect called a Kerr nonlinearity, whereby a pulse modifies the refractive index of glass such that it is higher at the centre of the pulse than the wings. Because wave speed depends on refractive index, the propagating pulse induces an effective velocity gradient in the material such that horizons can appear at points on the pulse leading and trailing edges.

It is important to note here that Kerr-induced trapping in itself is not a new effect, but rather one that is well known in nonlinear fibre optics. In fact, the Raman frequency shifting on solitons (where the spectrum of short pulses moves to longer wavelengths) is a deceleration effect that was even shown to lead to an equivalent gravitational potential, but it was only recently that scientists appreciated the possibility of extending the gravitational analogy to test predictions of Hawking radiation.

The particular attraction of experiments in optics is that the intensity and wavelength of the Hawking radiation depends on the induced refractive index gradient, and pulses containing only a few optical cycles or a steep shock front would be expected to generate measureable emission of visible light. Unfortunately, although it is straightforward to demonstrate the existence of an event horizon in an optical fibre, dispersive and dissipative effects present in the fibre appear to prevent the clean formation of the particular pulse profiles needed for the spontaneous generation of Hawking radiation.

The experiments of Belgiorno and co-workers attempt to overcome this problem in a novel way. In fact, they move away from the fibre environment altogether, performing experiments in bulk glass, using laser pulses in the form of needlelike beams known as optical filaments to generate the nonlinear refractive index perturbation.

Optical filament pulses generally have a complex spatiotemporal structure, but it is possible to experimentally synthesize them so that their internal group and phase velocity gradients can be controlled. This is what Belgiorno *et al.* did in their experiments, using properties of what are called Bessel beams, where the filament is preshaped so as to control its group velocity relative to the velocity change induced by the refractive index perturbation. This is a crucial aspect of their experiments because it allows them to fine tune the window of Hawking radiation emission into the near infrared, around 850nm away from any other possible contaminating signals. With a cooled CCD camera, they detect a clear signal above background that they associate with Hawking radiation, and the spectral shift of this signal with incident energy is in good agreement with the predictions of theory. They also performed experiments where the filaments were formed from Gaussian pulses, and although the dynamics are more complicated here, the spectral emission window of the spontaneous radiation is again in agreement with theory.

Overall, this work provides significant evidence for the observation of Hawking radiation in an analog gravity system. The results must nevertheless be interpreted carefully. For example, it is essential to state very plainly that measuring analog Hawking radiation gives no direct insight into quantum gravity because there is no physical gravitational potential involved. Indeed, one might even argue that the description of this emission as "Hawking" radiation is inappropriate, but this seems an unnecessary restriction because the study of analog horizon systems was clearly motivated by Hawking's work over 35 years ago. Of course additional experiments still remain to be carried out. Specifically, the spontaneous photon pairs emitted on either side of the horizon may be detectable with suitable angular resolution, and measurements of their entanglement and correlation will be an essential next step. Moreover, the polarization properties of filaments are very subtle and may need to be taken into account more fully in a complete interpretation of the experiments.

This field of research is at the interface of several areas of physics, and "standardizing" the terminology would be welcome. For example, effects related to Cerenkov radiation are seen both in filament and fibre soliton propagation, and distinguishing similarities and differences is important to avoid confusion. On the other hand, Belgiorno *et al.*'s results now point out the clear need to study the quantum electrodynamics of soliton radiation, as this may have direct bearing on their experiments. This work is also likely to be far reaching in other ways. By showing how tailored spatiotemporal fields provide a high degree of control over the interaction geometry of propagating light pulses, they may represent a new example of experiments in "ultrafast transformation optics," where geometrical modifications of an optical propagation environment can be induced on ultrafast timescales. This may allow a much wider study of other analog physical effects, using a convenient benchtop platform.

'BLACK HOLE' MADE FROM LIGHT

Pulses in an optical fibre can mimic the space-bending physics of collapsed stars.

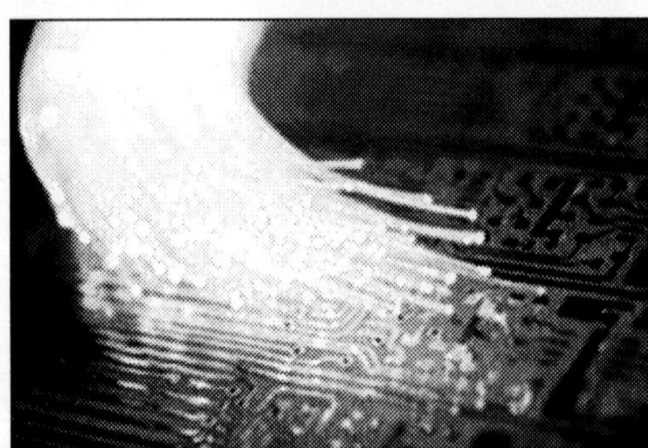

Figure: Black Holes and White Holes are being Mimicked "all the Time" in Optical-fibre Cables.

A black hole has been made in the laboratory. It's not, needless to say, a collapsed star with super-strong gravity, like the black holes thought to exist in deep space. Rather it's a kind of toy black hole made from light — and "completely harmless", its creators say.

It might seem odd to make an analogue of a black hole — so-called because even light cannot escape its gravitational tug — from light itself. But Ulf Leonhardt of the University of St. Andrews, UK, and his colleagues demonstrate that pulses of light travelling down an optical fibre can be made to affect other light waves in much the same way as a black hole does. Laboratory analogues of black holes have been proposed before, including an exotic ultracold gas of metal atoms that might act as a 'sonic black hole', but none has previously been demonstrated. Lab versions of black holes might permit a key aspect of their theory to be put to the test: whether they are not in fact perfectly black, but instead emit some particles and radiation.

Although the model is made in an optical-fibre cable with special properties, the team says that almost any pulse — including those used in fibre-optic telecommunications — should be a black-hole mimic to some degree. "We're making them all the time," says Leonhardt.

A River Runs through It

The group's light pulses provide not only optical black holes but 'white holes' too. Whereas no light can escape from within a region called the event horizon surrounding the centre of a black hole — so light hitting it always falls in and becomes trapped — incoming light is repelled by the event horizon of a white hole.

The St. Andrews team point out that black holes are like rivers flowing over a waterfall, carrying fish with it. The waterfall is the hole's 'singularity', where fish 'disappear'; a little way back upstream lies the event horizon, beyond which no fish can swim fast enough to escape from the brink.

A white hole, meanwhile, is like a fast-flowing river entering the sea: when the speed of the river current exceeds that of fish in the sea, no fish can enter the river mouth. "These are more than analogies," the researchers say. "The propagation of waves in moving fluids is mathematically equivalent to wave propagation in spacetime."

Altered by the Light

The laboratory analogy involves a special, 'non-linear' optical fibre, in which light actually changes the fibre's light-bearing properties. This mimics the way that a black hole's gravity warps the space around it, altering the way light travels through it.

Because of this property of the fibre, a stream of light behind but faster than the 'hole' pulse gets slowed down as it catches up, and then is reflected backwards (like the fish in the sea that cannot enter a stream). So light can never cross into the trailing edge of the pulse, making it like a white hole. Just as with a cosmic white hole, the light also has its wavelength shortened by this process.

The 'black horizon' at the pulse front is more complicated. In a cosmic black hole, incoming light is also frequency-shifted as it approaches, until its wavelength is so small that it requires new and unknown physics to describe how the light behaves. In the optical fibre, the alteration of wavelength is not so extreme. The result is that the light actually bounces away. That sounds as if it's the opposite behaviour to a real black hole, but in fact the maths governing the behaviour is the same.

The Ultimate Test

The researchers speculate about making such pulses act like mirrors — which might then be used to trap light in a way that could lead to 'black-hole lasers'. In the meantime there is a more likely application for the toy: testing an unproven theory about black holes.

In 1974, Stephen Hawking predicted that black holes should actually emit some particles and radiation. This should happen, he said, because empty space is constantly fizzing with 'virtual particles' — pairs of particles and their corresponding antimatter particles, which annihilate one another almost as soon as they are formed. Near the event horizon of a black hole, one member of a pair would occasionally fall across the horizon before annihilation happens, so the other would escape destruction. The surviving particle could then escape from the edge of the black hole.

Hawking Radiation

This 'Hawking radiation' means that black holes should have an intrinsic temperature (albeit very, very low) caused by the emission of these particles. But no one has ever been able to test this, and the prospect for doing so looks bleak: their temperature would be much lower than the pervasive afterglow of the Big Bang.

Leonhardt and his colleagues predict that their optical black holes should emit photons of ultraviolet light that are equivalent to Hawking radiation, which should be detectable. "They are of very low intensity," says Leonhardt, "so seeing them will need great luck. We'd be overjoyed if we found this was happening." It will also be tricky to distinguish these photons from the hole pulse itself, although the photons ought to have a characteristic fingerprint.

PHENOMENON OF OPTICAL BLACK HOLE

An optical black hole is a phenomenon in which slow light is passed through a Bose-Einstein condensate that is itself spinning faster than the local speed of light within to create a vortex capable of trapping the light behind an event horizon just as a gravitational black hole would. Black holes are some of the most exotic objects in the universe. They are incredibly dense, with powerful gravitational fields. One of the key characteristics of a black hole is its event horizon—a boundary that demarcates the region inside the black hole where the gravitational field is so intense that nothing, not even light, can escape.

Physicists and astronomers believe that black holes are formed when huge stars collapse in on themselves at the end of their lives. They exist at the centres of galaxies, where they act as giant engines that drive the motion of stars, according to astronomers. However, studying them is extremely difficult, particularly because in astronomy one can study only the information carried by light. In the case of black holes, the absence of light means astrophysicists have to rely on indirect means, such as inferring the presence of black holes by the way their gravity bends light outside their event horizons—a phenomenon scientists call gravitational lensing.

Unlike other black hole analogs such as a sonic black hole in a Bose-Einstein condensate, a slow light black hole analog is not expected to mimic the quantum effects of a black hole, and thus not emit Hawking radiation. It does, however, mimic the classical properties of a gravitational black hole, making it potentially useful in studying other properties of black holes. More recently, some physicists have developed a fibre optic based system which they believe will emit Hawking radiation.

We developed an approach to broad-band omnidirectional light absorption, based on light propagation in a metamaterial structure forming an effective "black hole." The proposed system does not rely on magnetic response, is nonresonant, and can be fabricated from existing materials. Having access to an artificial black hole in the lab will allow astrophysicists to test predictions made by theorists. Physicists would particularly like to test new theories such as quantum gravity, which seeks to reconcile Einstein's theory of general relativity with quantum mechanics.

MOVING LIGHT

Sacred Geometry is the blueprint of Creation and the genesis of all form. It is an ancient science that explores and explains the energy patterns that create and unify all things and reveals the precise way that the energy of Creation organizes itself. On every scale, every natural pattern of growth or movement conforms inevitably to one or more geometric shapes.

As you enter the world of Sacred Geometry you begin to see as never before the wonderfully patterned beauty of Creation. The molecules of our DNA, the cornea of our eye, snow flakes, pine cones, flower petals, diamond crystals, the branching of trees, a nautilus shell, the star we spin around, the galaxy we spiral within, the air we breathe, and all life forms as we know them emerge out of timeless geometric codes. Viewing and contemplating these codes allow us to gaze directly at the lines on the face of deep wisdom and offers up a glimpse into the inner workings of the Universal Mind and the Universe itself.

The ancients believed that the experience of Sacred Geometry was essential to the education of the soul. They knew that these patterns and codes were symbolic of our own inner realm and the subtle structure of awareness. To them the "sacred" had particular significance involving consciousness and the profound mystery of awareness ..... the ultimate sacred wonder. Sacred

Geometry takes on another whole level of significance when grounded in the experience of *self*-awareness.

The Platonic Solids

As far back as Greek Mystery schools 2500 years ago, we as a species were taught that there are five perfect 3-dimensional forms -The tetrahedron, hexahedron, octahedron, dodecahedron, and icosahedron. Collectively these are known as The Platonic Solids — and are the foundation of everything in the physical world. Modern scholars ridiculed this idea until the 1980's, when Professor Robert Moon at the University of Chicago demonstrated that the entire Periodic Table of Elements — literally everything in the physical world — is based on these same five forms! In fact, throughout modern Physics, Chemistry, and Biology, the sacred geometric patterns of creation are being rediscovered, but often without the greater context of spiritual understanding which protects against their misuse. One of our intentions with light Source, is to provide a bridge to an intuitive spiritual understanding that is in alignment with the appropriate use of this knowledge.

Sacred Geometry

Geometric shapes actually represent the manifest stages of 'becoming'. To see and work with unity and wholeness in geometry can help abolish our false notion of separateness from nature and from each other. Through Sacred Geometry we can discover the inherent proportion, balance and harmony that exists in any situation, all manifest reality and even the circumstances of our day-to-day life.

It was Marcel Proust who said, *" The real voyage of discovery consists not in seeking new lands but seeing with new eyes."*

The Energy in Action

Leonardo Fibonacci was an Italian mathematician who introduced to Europe and popularized the Hindu-Arabic number system. He contributed greatly to number theory, and during his life published many important texts. He is also known for the Fibonacci Series, a numerical series found frequently in the natural world. The Fibonacci sequence is generated by adding the previous two numbers in the list together to form the next and so on and so on. Divide any number in the Fibonacci sequence by the one before it, for example 55/34, or 21/13, and the answer is always close to 1.61803. This is known as the Golden Ratio.

One of the most profound and significant activities encompassed within sacred geometry and light Source is the 'Golden Mean Spiral', derived by using the 'Golden Ratio'.

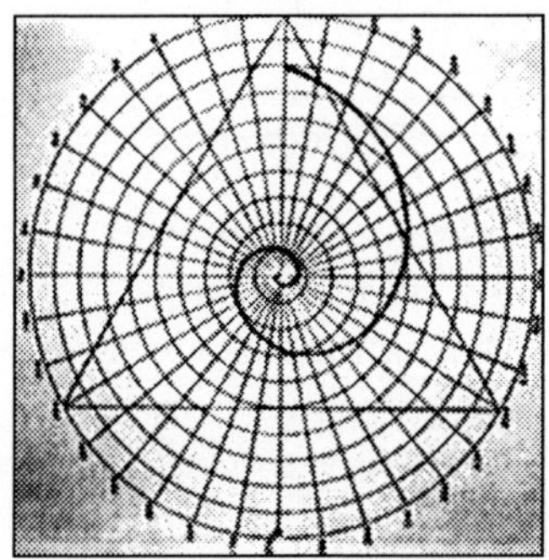

The Golden Mean was used in the design of sacred buildings in ancient architecture to produce spiritual energy that facilitated connectivity with spiritual realms through prayer. Our reality is very structured, and indeed Life is even more structured. This is reflected though Nature in the form of geometry. Geometry is the very basis of our reality, and hence we live in a coherent world governed by unseen laws. These are always manifested in our world. The Golden Mean governs the proportion of our world and it can be found even in the most seemingly proportion-less (active) living forms.

REFERENCES

- Avadhanalu, M. N. : *A Text Book Of Optics*, S. Chand Publisher, Delhi, 2006.
- Brooker, Geoffrey: *Modern Classical Optics*, Oxford University Press, Delhi, 2008.
- Ljubisa Ristic: *Sensor Technology and Devices*, Artech House, Delhi, 1994.
- Load, B.B. : *Lasers and Non-Linear Optics*, New Age International, Delhi, 2011.
- Sharma, P.K. : *NanoMEMS Physics : Applications and Principles*, Vista International Pub, Delhi, 2011.
- Shew, A.: *Discovering the Nanoscale*, Amsterdam: IOS Press, 2004.

3

Integrated Waveguide Optics

Integrated optics is a far-reaching attempt to apply thin-film technology to optical circuits and devices, and, by using methods of integrated circuitry, to achieve a better and more economical optical system.

The phenomenon of wave propagation in periodic structures or media recurs in many branches of physics and technology. As examples, consider the phenomena of X-ray and electron diffraction in crystals, the diffraction of light from the periodic strain variation accompanying a sound wave, and the band structures and energy gaps of the phonon spectrum and the energy of electrons in a crystal. Among practical devices making use of this phenomenon we may mention the traveling wave tube, the linear particle accelerator, diffraction gratings, and holograms.

With the intense pace of recent research in dielectric waveguiding phenomena and devices, a field sometimes referred to as "Integrated Optics," a new generation of optical devices was born, one utilizing periodic thin-film dielectric waveguides.

Some of the devices in this family which have already been demonstrated include thin-film optical filters, distributed feedback lasers (DFB), distributed Bragg reflector lasers (DBR), and input and output grating couplers for dielectric waveguides. This paper attempts to summarize and review the theoretical and experimental progress in periodic thin-film devices with a special effort devoted to bringing out the underlying unifying concepts.

WAVEGUIDES: SPECIAL FORM OF TRANSMISSION

A *waveguide* is a special form of transmission line consisting of a hollow, metal tube. The tube wall provides distributed inductance, while the empty space between the tube walls provide distributed capacitance:

Waveguides are practical only for signals of extremely high frequency, where the wavelength approaches the cross-sectional dimensions of the waveguide. Below such frequencies, waveguides are useless as electrical transmission lines.

When functioning as transmission lines, though, waveguides are considerably simpler than two-conductor cables — especially coaxial cables — in their manufacture and maintenance. With only a single conductor (the waveguide's "shell"), there are no concerns with proper conductor-to-conductor spacing, or of the consistency of the dielectric material, since the only dielectric in a waveguide is air. Moisture is not as severe a problem in waveguides as it is within coaxial cables, either, and so waveguides are often spared the necessity of gas "filling."

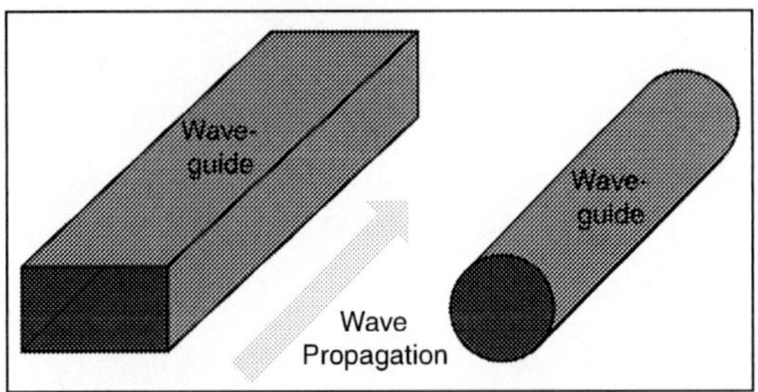

Figure: Wave Guides Conduct Microwave Energy at Lower Loss than Coaxial Cables.

Waveguides may be thought of as conduits for electromagnetic energy, the waveguide itself acting as nothing more than a "director" of the energy rather than as a signal conductor in the normal sense of the word. In a sense, all transmission lines function as conduits of electromagnetic energy when transporting pulses or high-frequency waves, directing the waves as the banks of a river direct a tidal wave. However, because waveguides are single-conductor elements, the propagation of electrical energy down a waveguide is of a very different nature than the propagation of electrical energy down a two-conductor transmission line.

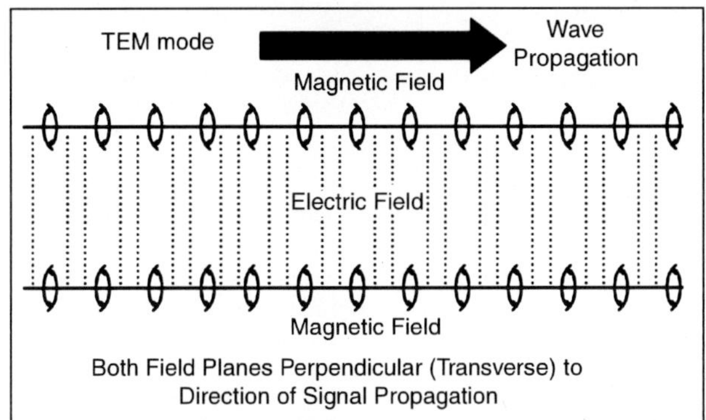

Figure: Twin Lead Transmission Line Propagation.

All electromagnetic waves consist of electric and magnetic fields propagating in the same direction of travel, but perpendicular to each other. Along the length of a normal transmission line, both electric and magnetic fields are perpendicular (transverse) to the direction of wave travel. This is known as the *principal mode*, or *TEM* (Transverse Electric and Magnetic) mode.

This mode of wave propagation can exist only where there are two conductors, and it is the dominant mode of wave propagation where the cross-sectional dimensions of the transmission line are small compared to the wavelength of the signal.

At *microwave* signal frequencies (between 100 MHz and 300 GHz), two-conductor transmission lines of any substantial length operating in standard TEM mode become impractical.

Lines small enough in cross-sectional dimension to maintain TEM mode signal propagation for microwave signals tend to have low voltage ratings, and suffer from large, parasitic power losses due to conductor "skin" and dielectric effects. Fortunately, though, at these short wavelengths there exist other modes of propagation that are not as "lossy," if a conductive tube is used rather than two parallel conductors. It is at these high frequencies that waveguides become practical.

When an electromagnetic wave propagates down a hollow tube, only one of the fields — either electric or magnetic — will actually be transverse to the wave's direction of travel.

The other field will "loop" longitudinally to the direction of travel, but still be perpendicular to the other field. Whichever field remains transverse to the direction of travel determines whether the wave propagates in *TE* mode (Transverse Electric) or *TM* (Transverse Magnetic) mode.

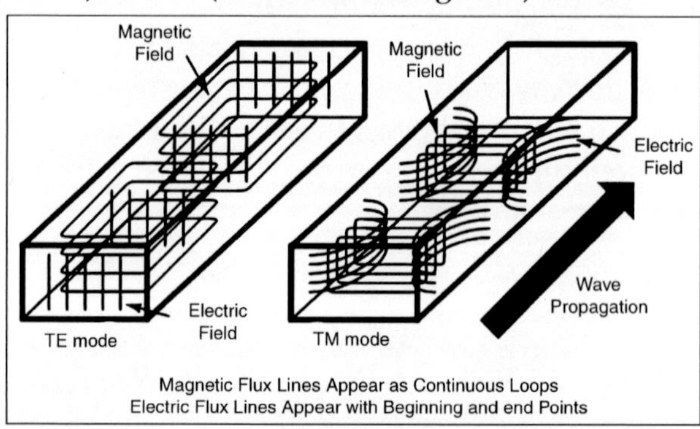

Figure: Waveguide (TE) Transverse Electric and (TM) Transverse Magnetic Modes.

Many variations of each mode exist for a given waveguide, and a full discussion of this is subject well beyond the scope of this book.

Signals are typically introduced to and extracted from waveguides by means of small antenna-like coupling devices inserted into the waveguide. Sometimes these coupling elements take the form of a dipole, which is nothing more than two open-ended stub wires of appropriate length.

Other times, the coupler is a single stub (a half-dipole, similar in principle to a "whip" antenna, $1/4\lambda$ in physical length), or a short loop of wire terminated on the inside surface of the waveguide:

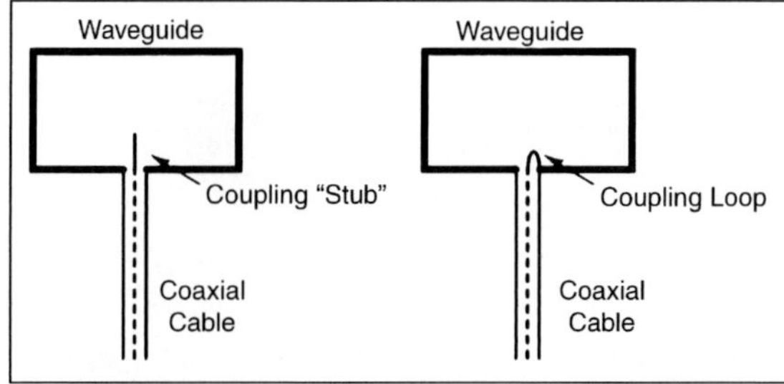

Figure: Stub and Loop Coupling to Waveguide.

In some cases, such as a class of vacuum tube devices called *inductive output tubes* (the so-called *klystron* tube falls into this category), a "cavity" formed of conductive material may intercept electromagnetic energy from a modulated beam of electrons, having no contact with the beam itself:

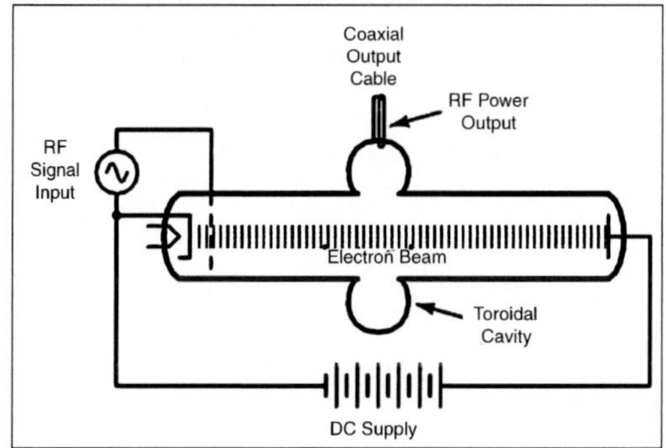

Figure: Klystron Inductive Output Tube.

Just as transmission lines are able to function as resonant elements in a circuit, especially when terminated by a short-circuit or an open-circuit, a dead-ended waveguide may also resonate at particular frequencies. When used as such, the device is called a *cavity resonator*. Inductive output tubes use toroid-shaped cavity resonators to maximize the power transfer efficiency between the electron beam and the output cable.

A cavity's resonant frequency may be altered by changing its physical dimensions. To this end, cavities with movable plates, screws, and other mechanical elements for tuning are manufactured to provide coarse resonant frequency adjustment.

If a resonant cavity is made open on one end, it functions as a unidirectional antenna. The following photograph shows a home-made waveguide formed from a tin can, used as an antenna for a 2.4 GHz signal in an "802.11b" computer communication network. The coupling element is a quarter-wave stub: nothing more than a piece of solid copper wire about 1-1/4 inches in length extending from the centre of a coaxial cable connector penetrating the side of the can:

Figure: Can-tenna Illustrates Stub Coupling to Waveguide.

A few more tin-can antennae may be seen in the background, one of them a "Pringles" potato chip can. Although this can is of cardboard (paper) construction, its metallic inner lining provides the necessary conductivity to function as a waveguide. Some of the cans in the background still have their plastic lids in place. The plastic, being nonconductive, does not interfere with the RF signal, but functions as a physical barrier to prevent rain, snow, dust, and other physical contaminants from entering the waveguide. "Real" waveguide antennae use similar barriers to physically enclose the tube, yet allow electromagnetic energy to pass unimpeded.

OPTICAL WAVEGUIDE

An optical waveguide is a physical structure that guides electromagnetic waves in the optical spectrum. Common types of optical waveguides include optical fibre and rectangular waveguides. Optical waveguides are used as components in integrated optical circuits or as the transmission medium in local and long haul optical communication systems.

Optical waveguides can be classified according to their geometry (planar, strip, or fibre waveguides), mode structure (single-mode, multi-mode), refractive index distribution (step or gradient index) and material (glass, polymer, semiconductor).

DIELECTRIC SLAB WAVEGUIDE

Practical rectangular-geometry optical waveguides are most easily understood as variants of the simple dielectric slab waveguide, also called planar waveguide. The slab waveguide consists of three layers of materials with different dielectric constants, extending infinitely in the directions parallel to their interfaces.

Light may be confined in the middle layer by total internal reflection. This occurs only if the dielectric index of the middle layer is larger than that of the surrounding layers. In practice slab waveguides are not infinite in the direction parallel to the interface, but if the typical size of the interfaces is much larger than the depth of the layer, the slab waveguide model will be an excellent approximation. Guided modes of a slab waveguide can not be excited by light incident from the top or bottom interfaces. Light must be injected with a lens from the side into the middle layer. Alternatively a coupling element may be used to couple light into the waveguide, such as a grating coupler or prism coupler.

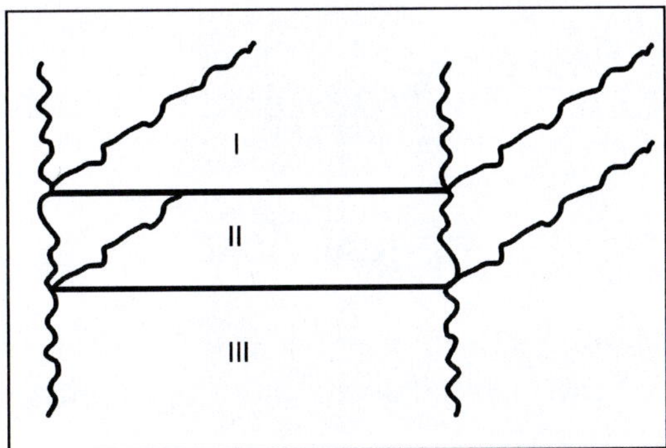

Figure: A Dielectric Slab Waveguide Consists of Three Dielectric Layers with Different Refractive Indices.

One model of guided modes is that of a planewave reflected back and forth between the two interfaces of the middle layer, at an angle of incidence between the propagation direction of the light and the *normal*, or perpendicular direction, to the material interface is greater than the *critical angle*. The critical angle depends on the index of refraction of the materials, which may vary depending on the wavelength of the light. Such propagation will result in a guided mode only at a discrete set of angles where the reflected planewave does not destructively interfere with itself.

This structure confines electromagnetic waves only in one direction, and therefore it has little practical application. Structures that may be approximated as slab waveguides do, however, sometimes occur as incidental structures in other devices.

TWO-DIMENSIONAL WAVEGUIDES

Strip Waveguides

A strip waveguide is basically a strip of the guiding layer confined between cladding layers. The simplest case is a rectangular waveguide, which is formed when the guiding layer of the slab waveguide is restricted in both transverse directions rather than just one. Rectangular waveguides are used in integrated optical circuits, and in laser diodes. They are commonly used as the basis of such optical components as Mach-Zehnder interferometers and wavelength division multiplexers. The cavities of laser diodes are frequently constructed as rectangular optical waveguides. Optical waveguides with rectangular geometry are produced by a variety of means, usually by a planar process.

The field distribution in rectangular waveguide cannot be solved analytically, however approximate solution methods, such as Marcatili's method, are known.

Rib Waveguides

A rib waveguide is a waveguide in which the guiding layer basically consists of the slab with a strip (or several strips) superimposed onto it. Rib waveguides also provide confinement of the wave in two dimensions.

OPTICAL FIBRE

Optical fibre is typically a circular cross-section *dielectric waveguide* consisting of a dielectric material surrounded by another dielectric material with a lower refractive index.

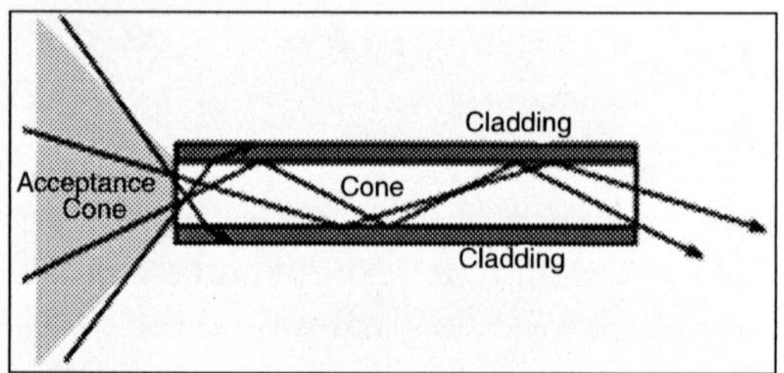

Figure: The Propagation of Light through a Multi-mode Optical Fibre.

Optical fibres are most commonly made from silica glass, however other glass materials are used for certain applications and plastic optical fibre can be used for short-distance applications.

WAVEGUIDE DIRECTIONAL COUPLERS

Waveguide Branch-line Coupler

The branch-line coupler described above can also be implemented in waveguide.

Bethe-hole Directional Coupler

One of the most common, and simplest, waveguide directional couplers is the Bethe-hole directional coupler. This consists of two parallel waveguides, one stacked on top of the other, with a hole between them. Some of the power from one guide is launched through the hole into the other. The Bethe-hole coupler is another example of a backward coupler.

The concept of the Bethe-hole coupler can be extended by providing multiple holes. The holes are spaced λ/4 apart. The design of such couplers has parallels with the multiple section coupled transmission lines. Using multiple holes allows the bandwidth to be extended by designing the sections as a Butterworth, Chebyshev, or some other filter class. The hole size is chosen to give the desired coupling for each section of the filter. Design criteria are to achieve a substantially flat coupling together with high directivity over the desired band.

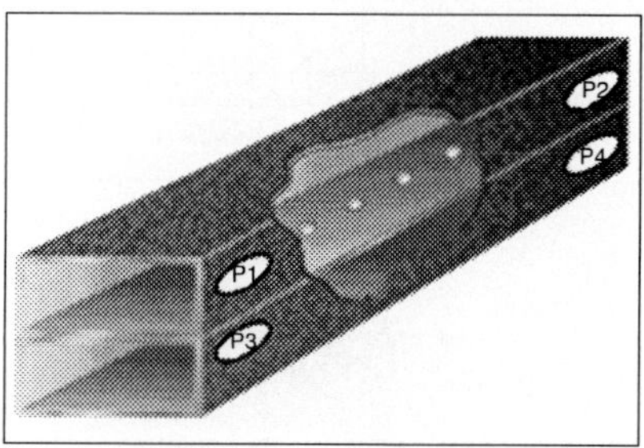

Figure : A Multi-hole Directional Coupler.

Riblet Short-slot Coupler

The Riblet short-slot coupler is two waveguides side-by-side with the side-wall in common instead of the long side as in the Bethe-hole coupler. A slot is cut in the sidewall to allow coupling. This design is frequently used to produce a 3 dB coupler.

Schwinger Reversed-phase Coupler

The Schwinger reversed-phase coupler is another design using parallel waveguides, this time the long side of one is common with the short side-wall of the other. Two off-centre slots are cut between the waveguides spaced λ/4 apart.

The Schwinger is a backward coupler. This design has the advantage of a substantially flat directivity response and the disadvantage of a strongly frequency-dependent coupling compared to the Bethe-hole coupler, which has little variation in coupling factor.

Moreno Crossed-guide Coupler

The Moreno crossed-guide coupler has two waveguides stacked one on top of the other like the Bethe-hole coupler but at right angles to each other instead of parallel. Two off-centre holes, usually cross-shaped are cut on the diagonal between the waveguides a distance $\sqrt{2}\lambda/4$ apart. The Moreno coupler is good for tight coupling applications. It is a compromise between the properties of the Bethe-hole and Schwinger couplers with both coupling and directivity varying with frequency.

WAVEGUIDE POWER DIVIDERS

Magic Tee

Coherent power division was first accomplished by means of simple Tee junctions. At microwave frequencies, waveguide tees have two possible forms– the E-plane and H-plane. These two junctions split power equally, but because of the different field configurations at the junction, the electric fields at the output arms are in phase for the H-plane tee and are 180° out of phase for the E-plane tee. The combination of these two tees to form a hybrid tee known as the Magic Tee. The Magic Tee is a four-port component which can perform the vector sum (Σ) and difference (Δ) of two coherent microwave signals.

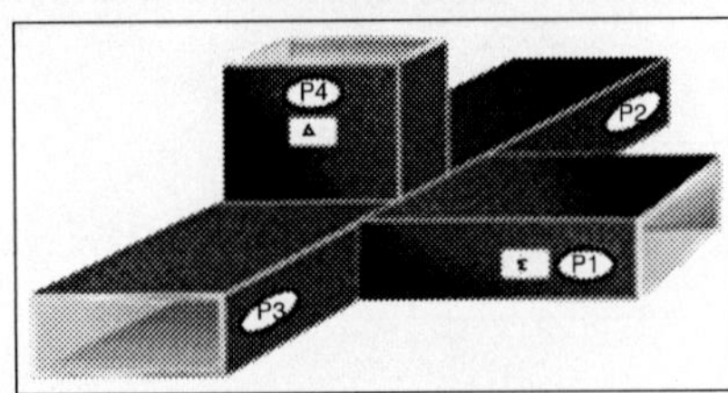

Figure : Magic Tee.

RF WAVEGUIDE COUPLERS INFORMATION

RF waveguide couplers are circuits that sample radio frequency (RF) transmissions by coupling signals asymmetrically. There are three basic types of products: unidirectional, bidirectional, and dual directional. Unidirectional couplers are three-port devices with two lines: a main line with a separate input and output, and a coupled line with a coupled output and internal termination. Power that passes along the main line is combined with the coupled output.

Power that is reflected from the main line output is coupled with the termination. Bidirectional RF waveguide couplers are similar to unidirectional devices, but remove termination from the fourth port. Main line power is coupled to the forward output of the secondary line. Reflected power is coupled with the reflected output. Dual directional couplers are also available. These four-port devices can be connected back-to-back in series or integrated into one device with a single main line and two secondary lines. The use a single main line reduces insertion loss and increases isolation.

Materials Manufactured From

RF waveguide couplers are made of aluminum, brass, bronze, copper, or silver and plated with cadmium, nickel, rhodium, silver, or tin. Flanges are used to connect waveguide sections or terminate waveguide components. RF waveguide couplers use several different types of flanges. Union guide (UG) is the military standard MIL-DTL-3922 for a range of flange types. Cover or plate flanges are flat, square UG-style flanges. Choke flanges are UG-style devices with an O-ring groove and choke cavity. They are often used to minimize the effects of flange discontinuity. Connector pressurized rectangular (CPR) flanges meet Electronic Industry Association (EIA) specifications and are used in many non-military applications. CPR variants include miniature (CMR), grooved (CPRG), and flat (CPRF) flanges. Mill-back flanges are made with the opening of the flange face equal in size to the inside of the waveguide, which is soldered into a larger milled opening at the rear of the flange.

Performance Specifications

Performance specifications for RF waveguide couplers include operating frequency range, EIA waveguide size, length or height, coupling, directivity, and voltage standing wave ratio. Operating frequency range is the range for which RF waveguide couplers meet all guaranteed specifications. EIA waveguide sizes have a prefix of WR and indicate the widest inside dimension of the waveguide in hundredths of an inch.

For example, WR650 is 6.50 in. wide, WR90 is 0.90 in. wide, and WR75 is 0.75 in. wide. Length or height is also measured in inches. Coupling is a measure of the amount by which the signal in the main waveguide is greater than the signal coupled to the secondary waveguide.

Directivity is a measure of the ratio between the power coupled to the secondary guide and signals traveling in the two possible directions along the main guide.

Voltage standing wave ratio (VSWR) is a unit-less ratio ranging from 1 to infinity that expresses the amount of reflected energy at the input or output of the device. A value of one indicates that all energy passes. Any other value indicates that a portion of the energy is reflected.

THE WAVE GUIDE, MODES AND EVANESCENT TAILS

Basics

We are taught that "Light travels in straight lines" as "rays" and this is a valid statement for traditional geometrical and physical optics where light propagates linearly in bulk media. Light is also described as a transverse electromagnetic wave composed of an electric vector (E) and orthogonal magnetic vector. The energy in a light beam may be described by the E field distribution orthogonal to the direction of propagation.

Total internal reflection (TIR) occurs when light travelling through a medium is reflected from the interface with a less dense medium at an angle exceeding the critical angle. The 90 degree prism is a simple example of this effect.

As a convex surface of similar density material is brought into close proximity to the reflecting surface, there occurs some transmission of light leaving a partial internal reflection.

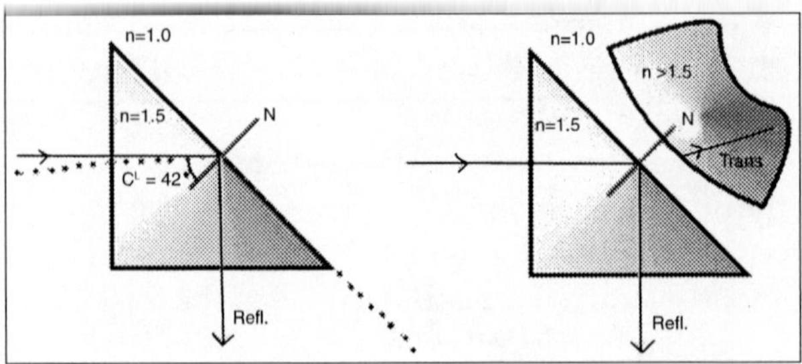

Figure: Total Internal Reflection **Figure**: Partial Internal Reflection

This demonstrates that there is an evanescent energy field outside the cube material allowing interaction of the two surfaces across the small air gap.

If the surfaces contact, all the light is transmitted as a refracted beam. The media have the same refractive index the surface effectively disappears and the ray is unbent showing no refraction.

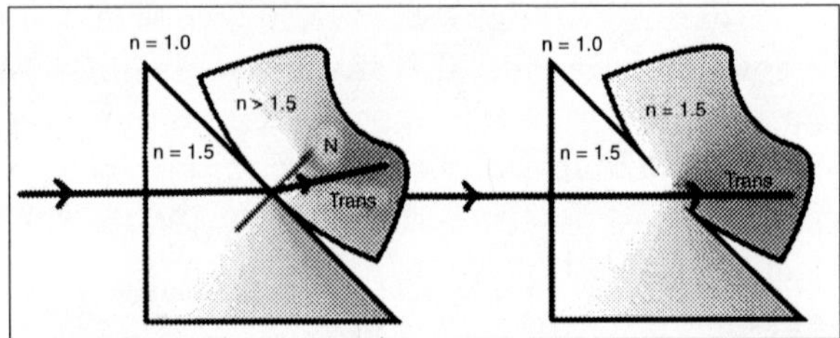

Figure: Optical Contact Unmatched Refractive Indices **Figure**: Optical Contact Matched Refractive Indices

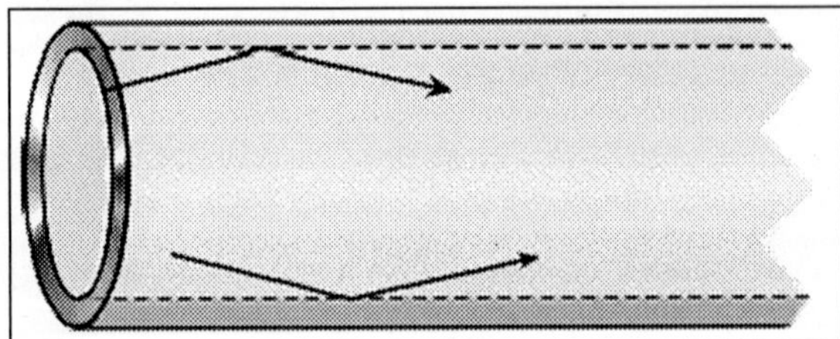

Figure: Rays Propagating in Multimode Fibre

In multimode fibre, which has a relatively large denser core diameter, the modes may be considered to be numerous traditional rays.

A transverse wave propagating in the positive/x/direction has the equation: Y=/a/sin {2ptn - b/x/} where/a/is the amplitude, t is time, n is the frequency and b a propagation constant, defined as 2pn/l, where l is the wavelength and n is the refractive index of the medium.

Guided Waves occur when light propagates along or is constrained by the physical boundaries of a waveguide. This is the case for a singlemode fibre where the denser core has refractive index n1 and the cladding is less dense with refractive index n2. When the core diameter is small enough that the number of possible totally internally reflected rays is reduced to one, thus allowing only a single mode of guided light, the concept of rays changes to modes which bend with or are guided by the core.

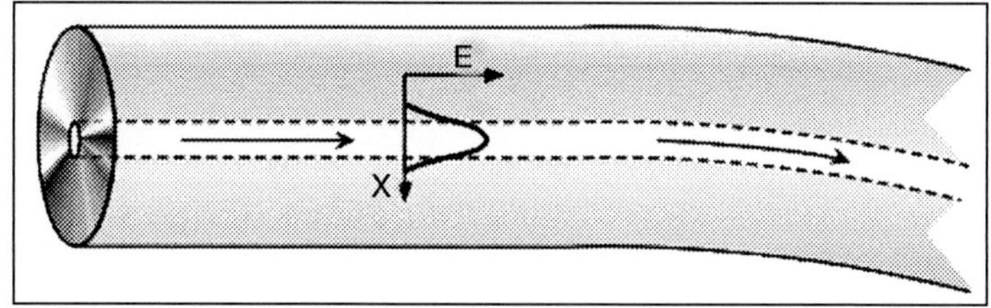

Figure: Mode Guided by Core of Singlemode Fibre

The propagation constant has a value 2pn1/l < b < 2pn2/l, where n1 is the index of the core and n2 is the index of the cladding.

For cores of intermediate diameter that support several modes there are a range of b values. The energy in any one mode is distributed between the core and the cladding where that in the cladding is called the evanescent tail.

Eigenmodes (Supermodes) and Coupling

Consider two parallel singlemode waveguides in close proximity. If the evanescent tails of each waveguide have considerable overlap, it can be shown that there are two possible solutions for mode propagation in the two waveguide structure. These are the Supermodes or Eigenmodes. The two solutions have symmetric and antisymmetric energy distributions and differing propagation constant (b) values.

As the relative phases of the modes change the energy is shared between the two waveguides and at matching and mismatched phase the energy is alternately maximized in each waveguide. *i.e.* the energy beats back and forth between the waveguides, dependent on the waveguide separation and the interaction length. This gives the term "evanescent wave coupling".

Matching and Over Coupling

The polished evanescent wave coupler is based on bringing the cores of two fibres close together by removing part of the cladding and optically contacting the polished faces. By this process, the two cores behave as if they are contained within the same cladding.

Bent in their supporting substrates to allow fabrication, the cores are not parallel, so the coupling is progressive as the cores become closer. The interaction length and core separation are therefore averages or estimates.

A 3dB coupler is one which is made such that the energy is shared equally. In percentage terms this is a 50/50 split. A 100% coupler is where all the energy has transferred from the input fibre to the secondary output fibre. This is the first beat condition and is constrained by the matching of the individual b values for the polished fibres in the interaction zone. This would imply that the polished surfaces have identical refractive indices at the contact if identical fibres are coupled. As polishing a fibre and removing cladding changes the refractive index profile of the fibre 100% coupling is sensitive to polishing depth.

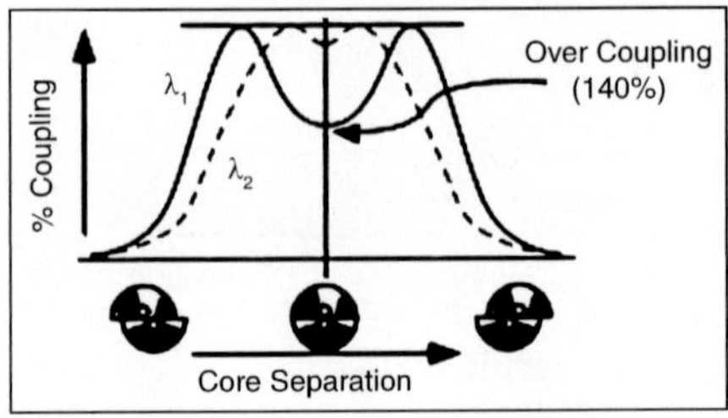

Figure: Coupling and Overcoupling as a Function of Fibre Core Seperation

Overcoupling is defined as greater than 100% when the interaction length or core separation allows the first beat to be exceeded. A 150% coupler is therefore equivalent to a 50% coupler but shows increased wavelength dependence.

COUPLED-MODE THEORY

Coupled semiconductor laser arrays are of particular interest for applications that require high-power, narrow divergence laser light or electrically tunable beam-steering. For these applications, the coupled elements of the array must be highly coherent, having a mutual degree of coherence at or near unity. The second-order correlation functions of the fields emitted by the array elements describe the coherence of the array. In order to understand and design highly-coherent laser arrays, the development and analysis of these correlation functions is especially illuminating.

Semiconductor laser arrays can be modelled as coupled oscillators or coupled waveguides. Both approaches can be formulated as derivatives of coupled-mode theory. However, traditional coupled-mode theory is entirely deterministic and predicts all fields to be fully coherent. By introducing random sources or boundary conditions and looking at the second-order correlations, we have demonstrated that partial coherence is expected under certain conditions. This result has been verified experimentally.

History

The first mention of the CMT was made in papers from Miller and Pierce in the domain of the microwave transmission lines and in electron beams. They put in place the mathematical foundations for the actual formulation.

In 1985, the subject of integrated optics have seen a new interest in this theoretical approach, specially by introducing the notion of coupling between modes.

Formulation

The formulation of the coupled mode theory is based on the development of the solution to an electromagnetic problem into modes. Most of the time it is eigenmodes which are taken in order to form a complete base. The choice of the base and the adoption of certain hypothesis like parabolic approximation differs from formulation to formulation. The classification proposed by of the different formulation is as follows:

- The choice of starting differential equation, some of the coupled mode theories are derived directly from the Maxwell differential equations (here) although others use simplifications in order to obtain an Helmholtz equation.
- The choice of principle to derive the equations of the CMT. Either the reciprocity theorem or the variational principle have been used.
- The choice of orthogonality product used to establish the eigenmode base. Some references use the unconjugated form and others the complex-conjugated form.
- Finally, the choice of the form of the equation, either vectorial or scalar.

RECTANGULAR WAVEGUIDES

Rectangular waveguides are important for two reasons. First of all, the Laplacian operator separates nicely in Cartesian coordinates, so that the boundary value problem that must be solved is both familiar and straightforward. Second, they are extremely common in actual application in physics laboratories for piping *e.g.* microwaves around as experimental probes.

In Cartesian coordinates, the wave equation becomes:

$$\left(\frac{\partial^2}{\partial x^2}+\frac{\partial^2}{\partial y^2}+(\mu\epsilon\omega^2-k^2)\right)\psi=0$$

This wave equation separates and solutions are products of sin, cos or exponential functions in each variable separately. To determine which combination to use it suffices to look at the BC's being satisfied. For TM waves, one solves for $\psi=E_z$ subject to $E_z\,|_S=0$, which is automatically true if:

$$E_z(x,y)=\psi_{mn}(x,y)=E_0\sin\left(\frac{m\pi x}{a}\right)\sin\left(\frac{n\pi y}{b}\right)$$

where a and b are the dimensions of the x and y sides of the boundary rectangle and where in principle m, n = 0, 1, 2....

However, the wave number of any given mode (given the frequency) is determined from:

$$k^2 = \mu\epsilon\omega^2 - \pi^2\left(\frac{m^2}{a^2} + \frac{n^2}{b^2}\right)$$

where $k^2 > 0$ for a "wave" to exist to propagate at all. If either index m or n is zero, there is no wave, so the first mode that can propagate has a dispersion relation of:

$$k_{11}^2 = \mu\epsilon\omega^2 - \pi^2(\frac{1}{a^2} + \frac{1}{b^2})$$

so that:

$$h\omega \geq \frac{\pi}{\sqrt{\mu\epsilon}}\sqrt{\frac{1}{a^2} + \frac{1}{b^2}} = \omega_{c,\text{TM}}$$

Each combination of permitted and is associated with a cutoff of this sort-waves with frequencies greater than or equal to the cutoff can support propogation in all the modes with lower cutoff frequencies.

If we repeat the argument above for TE waves you will be led by nearly identical arguments to the conclusion that the lowest frequency mode cutoff occurs for $a > b$, $m = 1$ and $n = 1$ to produce the solution to the $H_z(x, y) = \psi(x, y)$ wave equation above. The cutoff in this case is:

$$h\omega \geq \frac{\pi}{\sqrt{\mu\epsilon}}\frac{1}{a} = \omega_{c,\text{TE}}(10) < \omega_{c,\text{TM}}$$

There exists, therefore, a range of frequencies in between where *only* one TE mode is supported with dispersion:

$$k^2 = k_{10}^2 = \mu\epsilon\omega^2 - \frac{\pi^2}{a^2}.$$

Note well that this mode and cutoff corresponds to exactly one-half a free-space wavelength across the long dimension of the waveguide. The wave solution for the right-propagating TE mode is:

$$H_z = H_0 \cos\left(\frac{\pi x}{a}\right) e^{ikz - i\omega t}$$

$$H_z = \frac{ik}{\mu\epsilon\omega^2 - k^2}\frac{\partial H_z}{\partial x} = -\frac{ika}{\pi} H_0 \sin\left(\frac{\pi x}{a}\right) e^{ikz - i\omega t}$$

$$H_y = \frac{\mu\omega}{k} H_x = \frac{i\mu\omega a}{\pi} H_0 \sin\left(\frac{\pi x}{a}\right) e^{ikz - i\omega t}$$

We used $\gamma^2 = \mu\epsilon\omega^2 - k^2 = \pi^2 / a^2$ and $E\perp = ik/\gamma^2\nabla\perp\Psi$ to get the second of these, and $H_\perp = \frac{k}{\omega\mu}(\hat{z}\times E_\perp)$ to get the last one.

There is a lot more one can study in Jackson associated with waveguides, but we must move on at this time to a brief look at resonant cavities (another important topic) and multipoles.

STRIP WAVEGUIDES

A strip waveguide is basically a strip of the guiding layer confined between cladding layers. The simplest case is a rectangular waveguide, which is formed when the guiding layer of the slab waveguide is restricted in both transverse directions rather than just one. Rectangular waveguides are used in integrated optical circuits, and in laser diodes. They are commonly used as the basis of such optical components as Mach-Zehnder interferometers and wavelength division multiplexers. The cavities of laser diodes are frequently constructed as rectangular optical waveguides. Optical waveguides with rectangular geometry are produced by a variety of means, usually by a planar process.

The field distribution in rectangular waveguide cannot be solved analytically, however approximate solution methods, such as Marcatili's method, are known.

RIB WAVEGUIDES

A rib waveguide is a waveguide in which the guiding layer basically consists of the slab with a strip (or several strips) superimposed onto it. Rib waveguides also provide confinement of the wave in two dimensions.

ARRAYED WAVEGUIDE GRATINGS

An arrayed waveguide grating is a (typically fibre-coupled) device which can separate or combine signals with different wavelengths. It is usually built as a planar lightwave circuit, where the light coming from an input fibre first enters a multimode waveguide section, then propagates through several single-mode waveguides to a second multimode section, and finally into the output fibres. Wavelength filtering is based on an interference effect and the different optical path lengths in the single-mode waveguides: any frequency component of the input propagates through all single-mode waveguides, and the output in any channel results from the superposition (interference) of all these contributions. The wavelength-dependent phase shifts lead to a wavelength-dependent overall throughput for any combination of an input port and an output port.

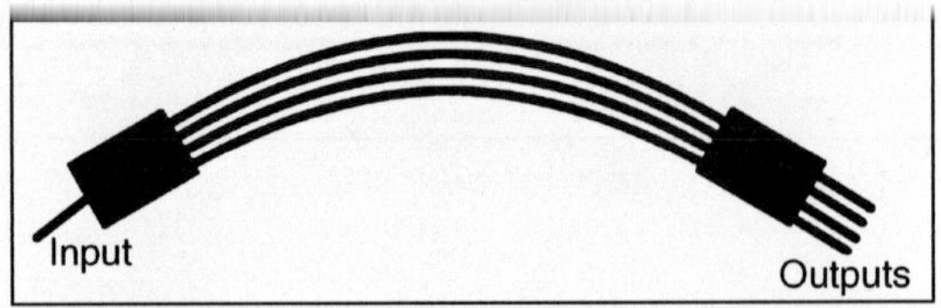

Figure : Structure of an Arrayed Waveguide Grating.

Particularly for AWGs with large numbers of channels, a high precision of the fabrication is required for achieving a low channel cross-talk.

AWGs can be realised with different material systems, *e.g.* based on silica (SiO_2), indium phosphide (InP), or silicon (Si).

Applications

Arrayed waveguide gratings are mainly applied in optical fibre communication systems, in particular in those based on multi-channel transmission with wavelength division multiplexing (WDM), where individual wavelength channels must be combined or separated. They can be part of more complex photonic integrated circuits, functioning *e.g.* as WDM data transmitters.

An arrayed waveguide grating may also be used for separating the lines in the optical spectrum of a supercontinuum source, or in a pulse shaper for ultrashort pulses.

Operation of AWG Devices

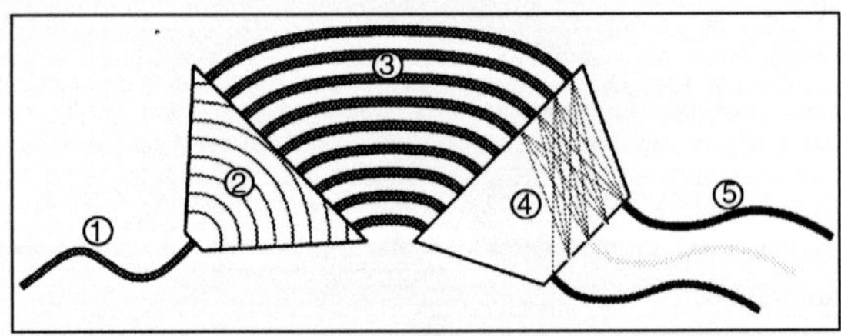

Figure: The Incoming Light (1) Traverses a Free Space (2) and Enters a Bundle of Optical Fibers or Channel Waveguides (3). The Fibers have Different Length and thus Apply a Different Phase Shift at the Exit of the Fibers (3) The Light Traverses another Free Space (4) and Interferes at the Entries of the Output Waveguides (5) in such a way that each Output Channel Receives only Light of a Certain Wavelength. The Orange Lines only Illustrate the Light Path. The Light Path from (1) to (5) is a Demultiplexer, from (5) to (1) a Multiplexer.

Conventional silica-based AWGs schematically are planar lightwave circuits fabricated by depositing doped and undoped layers of silica on a silicon substrate. The AWGs consist of a number of input *(1)*/output *(5)* couplers, a free space propagation region *(2)* and *(4)* and the grating waveguides *(3)*. The grating consists of a large number of waveguides with a constant length increment (ΔL). Light is coupled into the device via an optical fibre *(1)* connected to the input port. Light diffracting out of the input waveguide at the coupler/slab interface propagates through the free-space region *(2)* and illuminates the grating with a Gaussian distribution. Each wavelength of light coupled to the grating waveguides *(3)*, undergoes a constant change of phase attributed to the constant length increment in grating waveguides. Light diffracted from each waveguide of the grating interferes constructively and gets refocused at the output waveguides *(5)*, with the spatial position, the output channels, being wavelength dependent on the array phase shift.

REFERENCES

- Bhatia, A.S. : *Nanowires and Nanobelts*, Deep and Deep, Delhi, 2011.
- Geoffrey N.: *Introduction to Nonlinear Optics*, Cambridge University Press, Delhi, 2011.
- Herzberg, G. : *Molecular Spectra and Molecular Structure*, Van Nostrand Reinhold Company, New York, 1979.
- Imry, Y. : *Introduction to Mesoscopic Physics*. Oxford University Press, New York, 1997.
- Mark Fox: *Quantum Optics : An Introduction*, Oxford University Press, Delhi, 2009.
- Merkle R.: *"On Physics, Fundamentals, and Nanorobots."* Institute for Molecular Manufacturing, 2001.
- Sinha, D.N.: *Nanotubes and Nanowires*, Patna: Indian Academy of Sciences, 2005.

4

Applications to Optical Devices

The human eye is an important element for performing visual nondestructive tests. But in many cases the human eye is not sensitive enough or cannot access the test site.

As a part of maintenance programme and to ensure the safety and structural integrity of aircraft, 'Optically aided visual methods' are frequently used, where access is poor for other nondestructive tests. Such tests includes the magnifiers and borescopes.

Devices for viewing the interior of objects are called endoscopes, from the Greek words for ' inside view'. Industrial endoscopes are called borescopes because they are generally used in machined apertures and holes.

Probably in 1946, an ultraviolet light borescope was developed for fluorescent testing of the interior of hollow steel propeller blades of aircraft engine.

The 100W-viewing instrument detected interior surface discontinuities as glowing green lines. Later in 1958 the USAF B-47 Bomber fleet was grounded because of metal fatigue cracks resulting from low-level simulated bombing missions.

Visual testing with borescopes was proved the effective method for the first step to resolve the problem. In 1950s, a system was developed for automatic testing of helicopter blades.

The borescope supported by long bench could test the blades while the operator viewed results on a television screen.

In 1960s Mr. Lang developed the radiation optics in a project to keep functional the borescope system in high temperature environment. He also pioneers the use of closed circuit television with borescopes for testing the inner parts of jet engines and wings, hollow helicopter blades and nuclear reactors.

In 1965 a borescope was invented whose mirror could precisely controlled and this borescope could zoom to high magnification and could intensely illuminate the walls of chamber by means of a quartz incandescent lamp containing iodine vapour.

Since then various optical devices are using during aircraft maintenance.

INSPECTION DEVICES

Some optically aided visual test devices.

MAGNIFYING LENSES

An optical microscope is a combination of lenses used to magnify an image. The object is placed close to the lens in order to obtain as great a magnification as desired. The distance from lens to object is adjusted until the object is in the lens's depth of field and is in focus. The field of view is the area seen through the magnifier. With a simple magnifier, the diameter of the field of view is less than its focal length. So selection of a magnifier with the proper field of view is important.

Borescopes

Borescope is a precise optical instrument with built-in illumination. It can be used to visually check internal areas deep holes, bores and tubes.

Borescopes are available in rigid and flexible models from 2.00mm in diameter to 19mm in diameter and several feet in length.

Rigid Borescopes

- A rigid Borescope has a metal tube containing a series of lenses, which provide a view of the inspection area, and a lighting system that directs light to the inspection area.
- Rigid Borescopes are available in sizes ranging from approximately 2.0 mm [0.08 in] in diameter and a few centimeters in length, to approximately 19.0 mm [0.75 in.] in diameter and many centimeters in length.
- Borescopes can be supplied with optical systems, which will provide direct, right angle, rear and fore oblique directions of view.
- Many borescopes have an adaptor on the eyepiece, which allows the image of the inspection area to be recorded on photographic film, on video, or to be viewed with a Closed Circuit TV System (CCTV).
- Borescopes usually have high image resolution and a depth of field, ranging from a few millimetres (fractions of an inch) to infinity.
- Rigid borescopes are available with various magnification powers and adjustable focus controls.

Flexible Borescopes

- Flexible Borescopes are armoured plastic tubes, which can be manipulated into various curves and angles thus permitting access to areas, which are inaccessible to rigid borescopes. They are also available in various diameter/length combinations.

- These borescopes do not have an intermediate lens system within the flexible tube. At each end of the tube is a Lens system connected by two bundles of glass fibres. The outer bundle of fibres transmits light to the inspection area whilst the inner bundle transmits the image of the inspection area to the Lens in the eyepiece. The lighting is usually provided by an external, variable intensity, cold light source. At the eyepiece there is a focusing ring that permits adjustment of image clarity.
- Flexible borescopes may have replacement object lenses, which provide a change of the angle, or field, of view. Some borescopes have a knob, near the eyepiece, which controls a steerable tip (distal end). This control permits changes of angle of view and allows the distal end to be steered past obstructions.
- The image resolution of flexible borescopes is, generally, lower than that of rigid borescopes.
- Where access is difficult, it may be necessary to use guide tubes to enable the borescope to be directed to the inspection area. Details of guide tube requirements would be stated in the specific nondestructive testing manual for procedure.
- Adaptors are also available to enable the image of the inspection area to be recorded on photographic film, on video, or to be viewed with a CCTV system.

Microelectronic Video Borescopes

An electronic sensor embedded in the movable tip of the probe transmits signals to video processor, where the image is sent to monitor. The video borescope has a bright, high resolution colour image with no distortion or spots.

The device does not have an eyepiece like other borescopes. It has freeze frame feature that allows closer viewing of the image. The image can be electronically transferred for permanent documentation.

The image may be magnified for precise viewing.

The field of view is up to 90 degrees and the probe tip has four-way articulation.

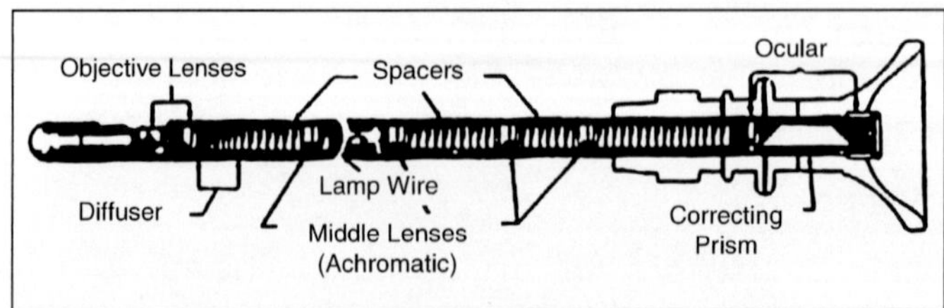

Figure: Sectional View of a Typical Borescope, Showing Relationship of Parts in its Optical System

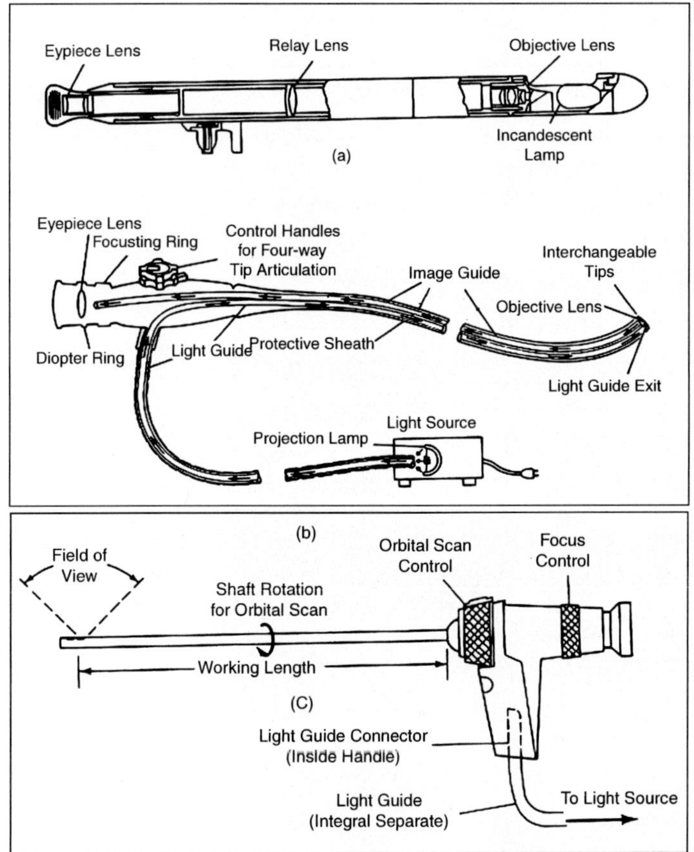

Figure: Three Typical Designs of Borescopes (a) A Rigid Borescope with a Lamp at the Distal end (b) A Flexible Fibrescope with a Light Source (c) A Rigid Borescope with a Light Guide Bundle in the Shaft.

Diffracted Light

A technique using diffracted light has been developed for visualizing surface distortions, depression, or protrusions as small as 10 ˜m (0.39mil). A real time technique particularly applicable to rapid inspection of large surfaces. The national research council of Canada suggested using this technique to inspect composite structures for barely visible impact damage. The optical setup for the diffracted light technique consists of a light source, a retroreflective screen, and the object being inspected. The surface being inspected must be reflective. Both flat and moderately curved surfaces can be inspected using this method.

Applications

Followings are some applications of 'optical devices' on aircraft structure and engines.

AIRCRAFT

Torsion Bar Core Corrosion Pitting

Stress corrosion cracks can cause failure in high strength steel spoiler torsion bars. Investigation revealed 'corrosion pitting' on the inside surface of the torsion bar cavity (bore) was found to have led to stress corrosion cracks and subsequent failure. For this inspection 70 or 90-degree borescopes are used.

Slat Drive Mechanism Bell Crank

The operator reported several slat bell crank failures. Investigation revealed that failures were caused by fatigue cracks that initiated at the bell crank-to-collar attachment holes. With the bell crank installed on the aircraft, access to the crack area is extremely poor, preventing the ultrasonic, eddy current or radiographic testing. This leaves the fibre optic borescope as the only option for inspecting the this component. A rigid right angle flexible borescope with defecting tip 5-mm diameter and about 1 meter in length is generally used for this inspection.

Spoiler Actuator Mechanism Lube Holes

Failures of the link or fitting assemblies of the slat drive mechanism were found to be caused by fatigue cracks generated at the inner surface of the lubrication holes in the link and fitting assemblies. After removing the lubrication fittings and grease visual inspection carried out with a 70-degree forward oblique borescope, 2.7mm in diameter by 180mm length. Crack indications are sought at the inboard and outboard sides of the lubrication bore.

Wing Rear Spar Doubler and Web under Trapezoidal Fitting

Fatigue cracks may occur in the wing rear spar cap web and doubler under trapezoidal panel attachment fitting. These fatigue cracks may occur in both the web and doubler or in each member separately. The cracks originate at the lower edge of both members. Direct access to the cracked areas of the web and doubler require removal of the trapezoidal fitting. Other methods of NDT cannot be used due to poor access caused by the fitting. The visual test can be done with optically aided O degree borescope along with a 70 to 90 degree borescope, 300 to 480 mm in length and 4 to 5mm in diameter.

Rudder Rib Flange

Cracks develop in the rib flanges of the rudder and investigation revealed that the said cracks resulted from acoustically induced vibration. It was also determined that installation of stiffeners to on the rudder ribs strengthens the rudder and minimizes the possibility of further crack development. A radiograph test is first conducted and if cracks are detected in the rib flanges at or adjacent to the skin attachment fastener holes, their lengths must be determined. Their lengths may be difficult to determine from radiographs alone. If cracks occur within or progress into the flange upper radius, their lengths must be determined by use of 3.2 mm (0.125 inch) diameter rigid borescope or flexible borescope of 'O' degree.

Main Landing Gear Truck Beam

Several instances of 'main landing gear truck beam assembly' failures were reported as resulting in major secondary damage to the aircraft. Investigation revealed that failure was a result of stress corrosion fracture that initiated at or immediately adjacent to the intersection of the lubrication hole and the pivot bore.

The stress corrosion fracture is result of severe pitting in the lubrication hole caused by inadequate lubrication.

If this condition is not corrected in time, the truck beam assemblies are vulnerable to failure. Removing corrosion or pitting from the surface of the four pivots bore lubrication holes and increasing the frequency of lubrication minimizes the possibility of failure and extends the service life of the beam assemblies.

The internal surface of each bore is inspected using 'O' degree 0.110-inch diameter and 0.110 inch diameter 70 degree borescopes.

Wing Front Spar Lower Cap

The purpose of this test is to check the lower spar cap forwarding fatigue cracks at the fastener locations. The area of interest is located at the wing pylons that support the jet engines. In addition to other NDT techniques borescopes inspection is also done to check the abnormalities of fasteners located under the fitting. Inspection of fasteners of inboard side fitting is done using a flexible borescope with minimum length of 2 feet. The area under the footstool fitting is inspected using flexible or rigid 90-degree borescope. With this inspection it has proven to detect small cracks at the forward and aft side of the fasteners and larger cracks that propagate to the leading edge [fwd] or vertical leg [aft] of the cap.

Engine

To make an effective airworthy inspection, different internal areas of engines are required on wing borescope inspection at certain interval.

Following areas of CF6-50 turbofan engines are periodically inspected.

Fan Section

Considering the Engine problem 1st stage & 2nd stages of fan rotors are generally visually inspected through the engine inlet.

As per CF6-50 NDTM borescope inspection frequently required to check at 3rd and 4th Stages fan rotor blades. During the inspection following defects are to be recorded.

- Cracks or Tears
- Nicks & Scratches
- Dents
- Erosion
- Tip Curl
- Pits
- Distortion leading or trailing edges
- Missing metal.

The primary Borescope used for fan blade inspection is the wide angle (60-65 degree) 'field of probe'.

Compressor Section

Borescope inspection carried out at compressor section [14 stage rotor blades] with either the rigid or flexible fibre optic system. The following defects are sought during inspection; (i) cracks at airfoil & tips, (ii) tears, (iii) leading and trailing edge damage, (iv) foreign object damage (FOD) domestic object damage cause random impact throughout the high-pressure compressor rotor stages (v) tip curl, (vi) missing metal [airfoil], (vii) airfoil surface defects, (viii) blade tip damage, (ix) corrosion and dirt (x) airfoil erosion, platform distortion, bowing and shingling, aluminum deposits.

Combustor Section

Combustor assembly inspection may be accomplished with either the focusing borescope system or the rigid optic fibre light type borescope system. The 60-degree field with 5.5mm to 6mm outer diameter probes is used for this inspection. The high intensity light source is used for this specific area inspection.

During on wing inspection following defects are sought discolouration, carbon accumulation, riveted joints defects, dome bands/plates defects, igniter tubes and ferrule defects, inner &outer liner assemblies defects, missing metals &burn- through, distortion or bowing of the liner assemblies.

High Pressure Turbine Section

Borescope inspection of the High pressure turbine nozzle inspection at stage-1 is done by rigid and flexible borescope to find nozzle radial & axial cracking, craze cracking, Nicks, scores, scratches/dents, cracks in airfoil fillet at platform, Metal splatter *etc.* and defects at turbine rotor section.

Turbine Mid Frame Area: Turbine mid frame section inspection required on condition engine maintenance plan. Both rigid and flexible borescopes used to check the said area.

Low-pressure Turbine: Low-pressure turbine rotor stages are inspecting by both rigid & flexible borescopes with greater magnification. In questionable circumstances regarding defect identification various angles and variable light levels probes are used.

NANOSTRUCTURE-BASED OPTICS

Nanooptics are optical thin-film structures that derive their physical properties by patterning materials on a subwavelength-scale roughly normal to the incident angle of the illuminating light. For ultraviolet, visible, and near-infrared wavelengths, the critical dimensions of these structures are on the nanometer scale. The structural attributes that determine the functional performance of a nano-optic device include:

- The nanopattern and its dimensions;
- The material used for the nanopatterned layer;
- The material used for the fill between the nanostructures;
- The formulation of the thin-film layers on either side of the nanostructure layer; and
- The substrate.

The critical dimensions of nanooptic structures for ultraviolet, visible, and near-infrared wavelength devices are on the order of tens to a few hundreds of nanometers, with a sub-ten-nanometer dimensional accuracy required in fabrication.

Physically, nanooptic devices derive their optical performance from the unique interactions of light with subwavelength-scale structures-including form birefringence and elimination of higher-order diffraction. At a macro level this is high optical performance in a grating layer often thinner than one micron. The optical function of nanooptic devices is the result of a convolution of material and structural properties, so the same optical function can be achieved using different material choices by making appropriate adjustments in the dimensions of the nanostructures.

Optical functionality that has been demonstrated based on nanostructures includes:

- Polarization effects such as polarizers and polarization beamsplitters or combiners;
- Phase retardation such as wave plates and retarders;
- Spectral selection; and
- Focal adjustment such as antireflection structures, diffusive layers, lenses, or mirrors.

In general, nanooptic devices can be designed to perform well in harsh environments by trading off material and structure to use the most robust possible material to achieve the desired function and to minimize the effect of thermal expansion or contraction.

FABRICATING NANOOPTICS

Nanooptic devices are fabricated using semiconductor-like deposition, lithography, etching, and coating processes. In general, a lithographic mask is prepared with the desired nanoscale features patterned on it. The original mask can be patterned using e-beam lithography, interference lithography, or by combining multiple partial mappings and exposures to create spatial variations or arrayed optics. Manufacturers can use this mask in a variety of methods to transfer the nanopattern to a target wafer; the specific topography of the mask depends on the method.

A fairly simple method is nanopattern transfer lithography, which is closely related to nanoimprint or nanoembossing. Nanopattern transfer uses physical contact to transfer a pattern to a resist layer. This approach allows for high fidelity in pattern replication, and avoids interactions between an energy beam and the polymer layer that can reduce the accuracy of the pattern transfer. The nanopattern transfer manufacturing process consists several steps.

First, one or more target wafers are prepared by selecting a substrate material, depositing appropriate thin-film boundary layers, and depositing an appropriate thickness of a target material. This target material is the layer into which the nanostructure is to be inscribed; usually it is a dielectric or a metal layer.

For each wafer, a thin, pliable, polymer resist layer is spun on top of the target layer, then a physical "mold" shaped with the complement of the desired pattern is brought into contact with the resist. The polymer resist layer is then set using ultraviolet or thermal curing, and the mold removed. This leaves a relief of the desired pattern in the now fixed polymer resist layer.

The fixed polymer resist layer then acts as an etching guide to transfer the pattern to the underlying target layer. After etching has been completed, various deposition processes fill the etched areas with a second material and add a thin-film boundary or surface layer. It is important to eliminate any voids and microchannels in this process.

Finally the wafer can be diced to size for application, or can be used as the base substrate for another deposition and etching cycle to create multilayer, multifunction, nanooptic devices.

NONLINEAR OPTICAL ABSORPTION

Nonlinear optical materials play a major role in the technology of photonics. To further advance the performance of optical devices, researchers have sought for materials with enhanced nonlinear optical properties, including inorganic semiconductors, polymeric systems and other organic molecules. The latter, specifically, are of great interest because they present high nonlinear optical properties combined with versatility of available routes of synthesis, used to alter and optimize molecular structure to maximize nonlinear responses and other properties. To select a material for a specific application, it is important to characterize its nonlinear optical absorption in a broad spectral range, in order to choose the optimized excitation wavelength. The resonant nonlinear optical absorption spectrum of the organic molecules Chlorophyll A, Indocyanine Green, Ytterbium Bisphthalocyanine and Cytochrome C, which are materials with potential applications in nonlinear optical devices. These organic molecules present distinct nonlinear optical absorption, such as saturable absorption (SA) and reverse saturable absorption (RSA), depending on the laser excitation wavelength and pulse duration. In order to investigate the excited state absorption and the population dynamics of the molecules, two distinct Z-scan methodologies were used, both based on measurements of the sample's transmittance under laser excitation. The main difference between the methods employed is the possibility to monitor nonlinear effects in distinct temporal regimes, and then observe effects related with singlet and triplet electronic transitions, which, in most cases, can only be discriminated by the temporal dynamic of the system.

While one technique uses a sequence of picoseconds pulses to induce and probe nonlinear absorption dynamics between singlet and triplet states, the other one employs femtosecond white-light supercontinuum pulses to map the spectrum of the excited state absorption in the visible and near infrared. By using both methods, one can determine and quantify nonlinear effects from femtoseconds up to hundreds of nanoseconds, including singlet excited state absorption cross-section spectrum and other spectroscopic features, such as decaying time (ranging from femtoseconds to a few nanoseconds) and triplet excited state cross-section.

Once one is able to temporally and spectrally map the nonlinearity of materials, it can be predicted how a given nonlinear optical device will act in distinct regimes (temporal, spectral and of intensity). Specifically, Chlorophyll A and Ytterbium Bisphthalocyanine present large RSA or SA depending on the excitation wavelength. Indocyanine Green presents high singlet and triplet excited state absorptions. Cytochrome C also presents saturable absorption, with short singlet-state lifetime and a relatively fast intersystem-crossing time. Therefore, information on the temporal dynamics combined with the excited state absorption cross-section and intensity regime can guide the development of applications based on the organic molecules presented here. Some notable applications of these materials in optical devices include all-optical switches, nonlinear optical limiters, and also bio-related applications, such as Photodynamic Therapy (PDT).

NONLINEAR OPTICS

Nonlinear optics processes are responsible for the nonlinearity of materials, phenomena which include nonlinear absorption, nonlinear refraction, and induced scattering, among others. The origin of such nonlinearities varies considerably. For instance, nonlinear absorption may be related to two-photon absorption, excited state absorption and free-carrier absorption. There are good reviews in the literature about two-photon absorption and free-carrier absorption.

Classification of optical nonlinearities usually falls in two main categories: instantaneous and accumulative (slow) nonlinearities. In the instantaneous case, the instantaneous polarization induced by the applied electromagnetic field in the medium can be expressed by an expansion in a Taylor series of the electric field amplitude, according to:

$$P^{\rightarrow} = \chi(1) \cdot E^{\rightarrow} + \chi(2) \cdot E^{\rightarrow} E^{\rightarrow} + \chi(3) \cdot E^{\rightarrow} E^{\rightarrow} E^{\rightarrow}$$

Options

The quadratic term with the electric field describes the first nonlinear effect. The coefficient

$$\chi(2)$$

is named complex susceptibility tensor of order 2. Its magnitude describes the intensity of the 2 order nonlinear effect, which, in turn, is related to second harmonic generation, optical rectification and sum and difference frequency mixing. Such effects are only observed in noncentrosymmetric materials.

The Coefficient

$$\chi(3)$$

is named complex susceptibility tensor of order 3 and describes the intensity of the 3 order nonlinear effect, such as self-focusing action, third harmonic generation and two-photon absorption. Materials presenting non-zero

$$\chi(3)$$

(*e.g.*, materials exhibiting two-photon absorption) are candidates as active medium for several technological applications, for which there are good reviews available in literature. However, such materials are out of the scope of this chapter, which will describe organic molecules' nonlinear effects arising only from resonant absorption. In the case of resonant absorption, the accumulative nonlinearities may occur in a time scale longer than the pulse duration. For instance, the polarization density generated by an applied field may develop or decay in a time scale longer than the excitation duration. Such interactions are normally dissipative, causing energy transfer from the absorbed field to the medium, which triggers the nonlinearity. Therefore, on the contrary to instantaneous nonlinearity, accumulative nonlinearities are strictly dependent on the amount of energy stored into the medium. Examples of such accumulative nonlinearities include resonant excited state absorption, such as saturation of absorption (SA) and reverse saturable absorption (RSA), which find applications in the construction of nonlinear optical devices.

Applications in Nonlinear Optical Devices

The integration of all-optical and electro-optical devices into the current technology has led to the development of a large number of schemes for controlling and manipulating the phase, direction, polarization and amplitude of optical beams. The ability to control light intensity is of prime relevance in optical-related applications, such as optical communications, optical computing and light-driven chemical reactions. The methods to control the amplitude of light can be generally divided into two major groups: dynamic and passive methods.

The first method is accomplished by a device that receives an external feedback, such as an iris or a power optical filter to control the light intensity incident on an optical system. Such method requires other devices to work along in the acquisition (*e.g.*, a sensor, a processor, etc.), leaving the process slower and more complex.

On the other hand, passive control is obtained by using a nonlinear optical material, in which sensing, processing and actuation are inherent, making the devices simpler and faster. Within the classification of passive devices to control the amplitude of an optical signal, two of great importance are all-optical switches and optical limiters. An ideal optical switch is a nonlinear optical device that is activated at a specific intensity or fluence threshold, becoming totally opaque. In contrast, an ideal optical limiter is a device that exhibits a linear transmittance below a specific intensity or fluence level but, above this level, its output intensity becomes constant. Optical limiters have been used in several applications, such as passive mode locking, pulse compression, and the most popular application: eye and sensor protection in optical system (*e.g.*, telescopes and night vision systems).

For instance, one can construct optical limiters with responses that are insensitive to the incident pulse duration. The resonant nature of the accumulative nonlinearities (such as SA and RSA), however, frequently results in a narrow bandwidth of operation for devices utilizing these mechanisms. By contrast, optical limiters that rely on instantaneous (nonresonant) nonlinearities can be broad band. These nonlinearities, however, require high intensities and typically operate effectively only for very short optical pulses. Therefore, by using an appropriate optical limiter in terms of nature and threshold value, one can extend the lifetime of a device and allow it to continue to operate under severe conditions.

NEAR-FIELD OPTICAL MICROSCOPY

Near-field scanning optical microscopy (NSOM/SNOM) is a microscopy technique for nanostructure investigation that breaks the far field resolution limit by exploiting the properties of evanescent waves. This is done by placing the detector very close (distance much smaller than wavelength λ) to the specimen surface.

This allows for the surface inspection with high spatial, spectral and temporal resolving power. With this technique, the resolution of the image is limited by the size of the detector aperture and not by the wavelength of the illuminating light.

In particular, lateral resolution of 20 nm and vertical resolution of 2–5 nm have been demonstrated. As in optical microscopy, the contrast mechanism can be easily adapted to study different properties, such as refractive index, chemical structure and local stress. Dynamic properties can also be studied at a sub-wavelength scale using this technique.

MODES OF OPERATION

Aperture and Apertureless Operation

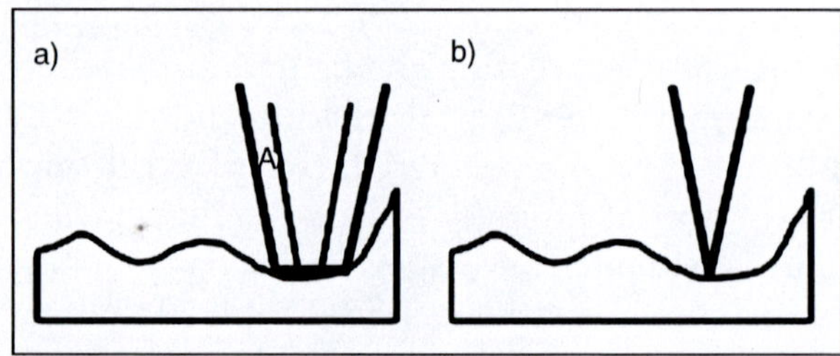

Figure: Sketch of a) Typical Metal-coated Tip, and b) Sharp Uncoated Tip.

There exist NSOM which can be operated in so-called aperture mode and NSOM for operation in a non-aperture mode. As illustrated, the tips used in the apertureless mode are very sharp and do not have a metal coating.

Though there are many issues associated with the apertured tips (heating, artifacts, contrast, sensitivity, topology and interference amongst others), aperture mode remains more popular.

This is primarily because apertureless mode is even more complex to set up and operate, and is not understood as well. There are five primary modes of apertured NSOM operation and four primary modes of apertureless NSOM operation.

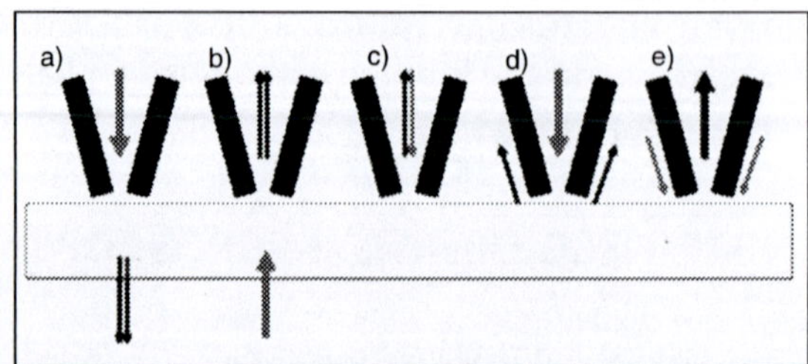

Figure: Apertured Modes of Operation: a) Illumination, b) Collection, c) Illumination Collection, d) Reflection and e) Reflection Collection.

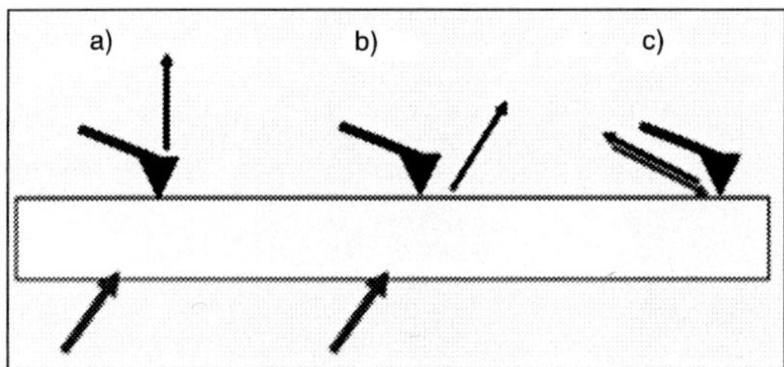

Figure: Apertureless Modes of Operation: a) Photon Tunneling (PSTM) by a Sharp Transparent Tip, b) PSTM by Sharp Opaque Tip on Smooth Surface, and c) Scanning Interferometric Apertureless Microscopy with Double Modulation.

Feedback Mechanisms

Feedback mechanisms are usually used to achieve high resolution and artifact free images since the detector must be positioned within a few nanometers of the surfaces.

Some of these mechanisms are:

- Constant force feedback: This mode is very similar to the feedback mechanism used in atomic force microscope (AFM). Experiments can be performed in contact, intermittent contact, and non-contact modes.
- Shear force feedback: In this mode, a tuning fork is mounted alongside the tip and made to oscillate at its resonance frequency. The amplitude is closely related to the tip-surface distance, and thus used as a feedback mechanism.

Contrast

It is possible to take advantage of the various contrast techniques available to optical microscopy though NSOM but with much higher resolution. By using the change in the polarization of light or the intensity of the light as a function of the incident wavelength, it is possible to make use of contrast enhancing techniques such as staining, fluorescence, phase contrast and differential interference contrast.

It is also possible to provide contrast using the change in refractive index, reflectivity, local stress and magnetic properties amongst others.

Instrumentation and Standard Setup

The primary components of an NSOM setup are the light source, feedback mechanism, the scanning tip, the detector and the piezoelectric sample stage. The light source is usually a laser focused into an optical fibre through a polarizer, a beam splitter and a coupler.

The polarizer and the beam splitter would serve to remove stray light from the returning reflected light.

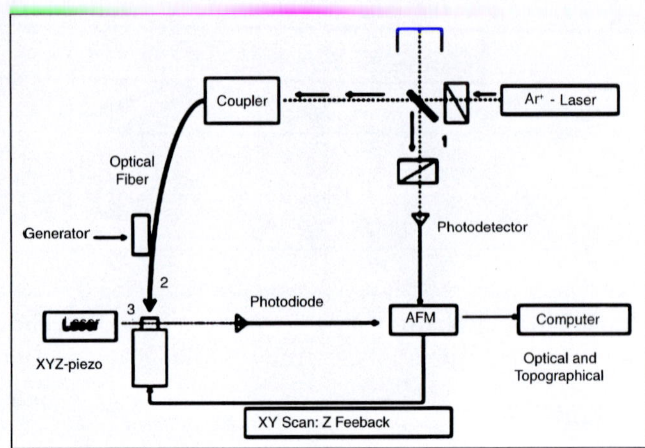

Figure: Block Diagram of An Apertureless Reflection-back-to-the-fibre NSOM Setup with Shear-force Distance Control and Cross-polarization; 1: beam Splitter and Crossed Polarizers; 2: Shear-force Arrangement; 3: Sample Mount on a Piezo Stage.

The scanning tip, depending upon the operation mode, is usually a pulled or stretched optical fibre coated with metal except at the tip or just a standard AFM cantilever with a hole in the centre of the pyramidal tip. Standard optical detectors, such as avalanche photodiode, photomultiplier tube (PMT) or CCD, can be used.

Highly specialized NSOM techniques, Raman NSOM for example, have much more stringent detector requirements.

NEAR-FIELD SPECTROSCOPY

As the name implies, information is collected by spectroscopic means instead of imaging in the near field regime. Through Near Field Spectroscopy (NFS), one can probe spectroscopically with subwavelength resolution. Raman SNOM and fluorescence SNOM are two of the most popular NFS techniques as they allow for the identification of nanosized features with chemical contrast. Some of the common near field spectroscopic techniques are:

- *Direct local Raman NSOM:* Aperture Raman NSOM is limited by very hot and blunt tips, and by long collection times. However, apertureless NSOM can be used to achieve high Raman scattering efficiency factors (around 40). Topological artifacts make it hard to implement this technique for rough surfaces.
- *Surface enhanced Raman spectroscopy (SERS) NSOM:* This technique can be used in an apertureless shear-force NSOM setup, or by using an AFM tip coated with gold. The Raman signal is found to be significantly enhanced under the AFM tip. This technique has been used to give local variations in the Raman spectra under a single-

walled nanotube. A highly sensitive optoacoustic spectrometer must be used for the detection of the Raman signal.

- *Fluorescence NSOM:* This highly popular and sensitive technique makes use of the fluorescence for near field imaging, and is especially suited for biological applications. The technique of choice here is the apertureless back to the fibre emission in constant shear force mode. This technique uses merocyanine based dyes embedded in an appropriate resin. Edge filters are used for removal of all primary laser light. Resolution as low as 10 nm can be achieved using this technique.
- Near field infrared spectrometry and near field dielectric microscopy

FIBRE PROBES

FABRICATION

The first tarred fibre robes which have been used in near-field optical microscopy were sharpened using etching in hydrofluoric acid (HF). The etching process is influenced by a number of parameters, *e.g.* the etchant dilution, the oil-like layer on to of the etchant, the etching time and the type of fibre or do ant. As different do ants and/or do ant-concentrations in the fibre are used to create the refractive index profile between the cladding and the core of the fibre, the etch rates of cladding and core are different, which allows to etch fibres with very high a ex angles up to 40 degrees at tip sizes around 30 nm.

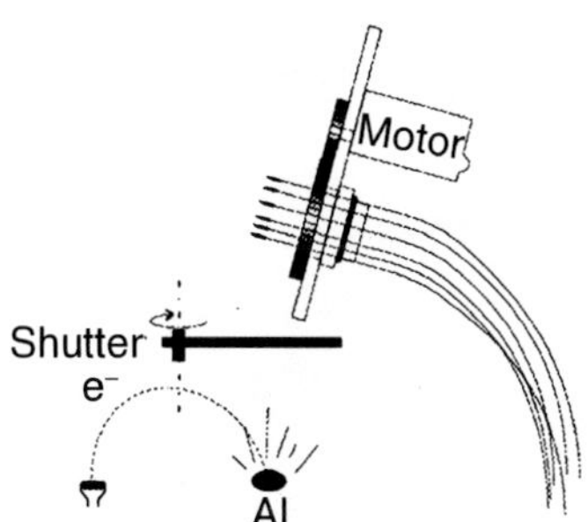

Figure: Aluminum Evaporation using an E-beam Evaporator.

The major disadvantage of the etching process is the very irregular surface of the fibre after etching, with roughness features of few tens of nanometers, which will lead to pinholes in the metal coating after evaporation. A way of generating fibres with a very smooth surface is pulling.

For this a commercial pipette puller was used. Here, the fibre is locally heated by a CO_2-laser, while a controlled force is pulling at the fibre. When a re-set pulling velocity is reached, the laser is switched off and after an adjustable delay a hard- pull is exerted on the fibre, pulling both arts a art. Aimed is for a fibre with a high a ex angle in order to minimize the cut-off region, which is the region where the size of the fibre is too small to contain a light propagating mode. In this way the optical losses in the fibre are kept as low as possible.

To have a clear circular aperture the very end of the fibre should be flat with a diameter between 50 to 100 nm. The fibre is subsequently coated with approximately 200 nm of aluminum using the configuration.

Aluminum is chosen because of the high extinction coefficient, resulting in a penetration depth of about 7 nm at a wavelength of 500 nm. Evaporation is performed under an angle to the fibre in order to leave the aperture free from aluminum. To minimize the formation of aluminum grains in the coating the evaporation rate should be as high as possible. Therefore, in the evaporation set-up the distance between the evaporation source and the tips was minimized from its initial distance of 34 cm to approximately 15 cm.

Several groups have studied the fibre pulling process. Especially Valaskovic *et al.* have performed an extensive study on the optimization of the pulling parameters of the fibre probes. They measured transmission efficiencies between 10^{-7} and 5×10^{-5} for tip sizes ranging from 60 to 100 nm.

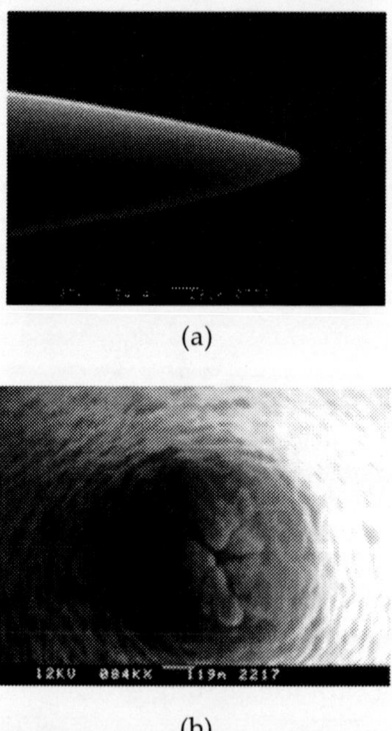

(a)

(b)

Figure: SEM Images: (a) An Uncoated Fibre Probe (Sideview), (b) Front View of an Aluminum Coated Fibre.

A scanning electron microscope (SEM) image of the side view an uncoated fibre tip. An example of the front view of a coated tip. This image clearly shows the aluminum grain around the aperture of the probe, partially due to the relatively low evaporation rate which was used as at that time, as the distance between the evaporation source and the tip was not minimized yet.

ELECTRIC-MAGNETIC FIELD PROPERTIES OF SMALL APERTURES

The electric field at the tip aperture is highly non-homogeneous due to the metal coating and the polarization of the light: the electric field is zero at position where the polarization is parallel to the metal coating. Bethe derived a theory which describes the field at a small circular aperture in an infinitely thin perfectly conducting plane.

Later Bouwkamp corrected for an error he discovered in Bethe's theory. In the first single molecule experiment by Betzig a remarkably good correspondence between this theory and the field from a fibre probe was found. By scanning the relatively large probe over the single molecule, the molecule effectively probes the field profile at the aperture. The emitted single molecular fluorescence is proportional to overlap between the absorption dipole of the molecule and the electric-field.

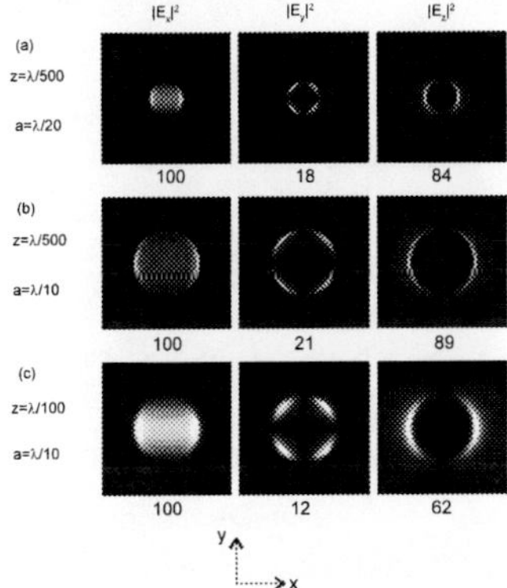

Figure: Electric-field Profiles According to the Bethe/Bouwkamp Theory, the Image width is l/2.5. The Intensities Relative to the Intensity in the x-direction are Denoted below the Images (in %).

The electric field, using the Bethe/Bouwkamp theory, at a distance z of a perfectly conducting infinite metal plate with an aperture of radius a, with the incident light polarized along the horizontal x-direction. The field-profiles show a maximum intensity along the incident polarization and a relatively large electricfield component in the z-direction. At the rim of the aperture the field vanishes when its polarization is parallel to the metal. In the direct neighbourhood of the aperture the field decays exponentially in the z-direction. The total throughput of the aperture is found to be proportional to the sixth power of the aperture radius ($\sim a^6$).

The throughput of the fibres which were used in the experiments was approximately 10^{-6} for aperture between 50 and 100 nm, resulting in an optical output power of the fibre between 100 pW and 10 nW.

SCANNING NEAR-FIELD OPTICAL MICROSCOPY

Scanning near-field optical microscopy (SNOM) is an imaging technique that can work beyond the diffraction limit. It can do this by collecting the non-propagating evanescent field, which exists up to a few nanometres away from a sample's surface. This is achieved by positioning a very small aperture (usually a glass fibre heated and pulled down to form a sharp tip) just a few nanometres from a sample's surface.

Surpassing the resolution of standard optical microscopes, SNOM can investigate the local electric field properties of devices that have features much smaller than the wavelength of light under study. The resolution of SNOM is limited by the aperture size of the collector, rather than the wavelength of light used to image.

Our group uses SNOM's unique ability of collecting the evanescent wave to investigate the properties of waveguides. We have studied the electric field behaviour inside fibre Bragg gratings, the mode profiles in micro-structured fibres, supercontinuum generation in highly nonlinear waveguides, and the inter-modal delay of pulses inside waveguides. Within the group, we have the unusual ability to combine ultra-fast science with the high-spatial resolution of SNOM. This ability enables us to investigate phenomena with femtosecond time resolutions and nanometre spatial resolutions.

AN EXPERIMENT USING SCANNING NEAR-FIELD OPTICAL MICROSCOPY

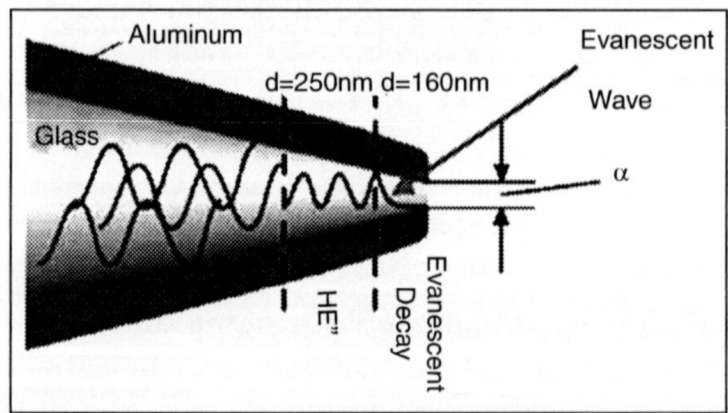

Figure : Scheme of a Pulled Optical Fibre with an Aluminium Coating.

In order to make an NSOM/SNOM experiment, a point light source must be brought near the surface that will be imaged (within nanometers). The point light source must then be scanned over the surface, without touching it, and the optical signal from the surface must be collected and detected. There are a few different ways to obtain a point light source: One can use pulled or etched optical fibres (tapered optical fibres) that are coated with a metal except for at an aperture at the fibre's tip. The light is coupled into the fibre and is then emitted at the sub-wavelength (50 nm or larger) aperture of the fibre. Or, one can use a standard AFM cantilever with a hole in the centre of the pyramidal tip. A laser is focused onto this hole, which is of sub-wavelength dimensions.

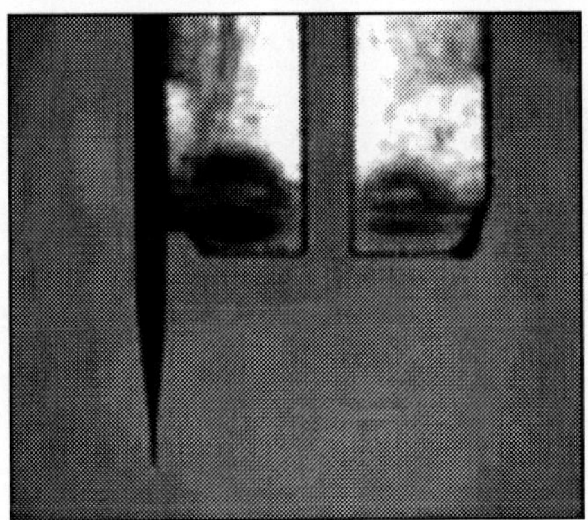

Figure : This Picture Shows an Optical Fibre Mounted on a Tuning Fork.

Figure : Electron Microscopy Image of the Hollow Aluminium Tip. The Standard Aperture Size is 100 nm.

Shear Force Feedback and Normal Force Feedback

The resolution of an NSOM/SNOM measurement is defined by the size of the point light source used (typically 50-100 nm). The distance between the point light source and the sample surface is usually controlled through a feedback mechanism that is unrelated to the NSOM/SNOM signal. Currently, most instruments use one of the following two types of feedback: shear force or normal force. In the shear force feedback, or tuning fork feedback, the straight tip is mounted to a tuning fork, which is then oscillated at its resonance frequency. The amplitude of this oscillation is strongly dependent on the tip-surface distance, and it can be effectively used as a feedback signal.

Atomic Force Microscopes (AFMs)

In the normal force feedback (the standard feedback mode used in AFM), it is possible to perform experiments in contact and in intermittent contact mode. This feedback mechanism is only possible with cantilevered, tapered optical fibres and with AFM cantilevers with holes.

Different Operating Methods

Scanning near-field optical microscopy can be performed in many different ways of operation. Most common today is the use of aperture probes for transmission microscopy, either in illumination (a) or in collection (b). However, many samples or substrates are opaque, so that working in reflection is necessary (c). The reflected light can be collected by optics close to the tip, or by the fibre probe itself, in which case often uncoated fibre tips are used.

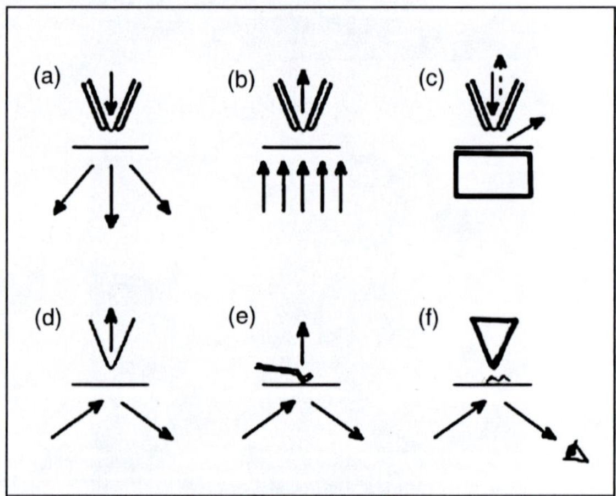

Figure : Diagram Showing Different Methods of Operation for Scanning Near-Field Optical Microscopy (SNOM).

Photon Scanning Tunneling Microscope (PSTM)

A different approach is taken by the Photon Scanning Tunneling Microscope (PSTM), where evanescent waves are created at the sample surface by oblique far-field illumination (d). The probe tip acts as a scatterer of the evanescent field, leading to homogeneous waves which can be easily detected. Easy to operate, this mode suffers somewhat from difficulties in data interpretation. Of high interest is this arrangement with inverted light path, the i-PSTM or Tunnel Near-Field Optical Microscope (TNOM) or forbidden light near-field optical microscope.

Plasmon Near-Field Microscope

Similarly to the PSTM, light can be scattered from the evanescent field by other probe tips, such as a force microscope tip on a cantilever (e). In the

Plasmon Near-Field Microscope, surface plasmons are generated at the surface of a sample on a thin film metallic substrate, and scattered by a probe tip (f). The SNOM system at the Materials and Microsystems Laboratory is an AlphaSNOM by Witec, where the optical probe is a silicon cantilever with a hole in the centre of a metallic tip.

Using Scanning Near-Field Optical Microscopy (SNOM) in Lithography

SNOM can be fruitfully used for direct writing optical nanolithography. In fact, topography and physical properties of photosensitive surfaces such as photoresists may be changed by laser pulses emitted from the SNOM probe, offering resolutions unachievable by conventional optical and laser systems (going well beyond the diffraction limit).

The first results obtained in several labs dealt with nanometric strips on standard positive photoresist and on self-assembled monolayers of alkanethiols (with transfer to metallic substrates) by using laser sources in the visible-UV range.

OPTICAL PROBES OF NEAR-FIELD

Optical near-fields are created near material boundaries. More general, the concentration of optical fields to dimensions smaller than the diffraction limit relies on inhomogeneities in an otherwise empty space. The understanding, construction, and optimization of light confining structures is one of the key topics in Nanooptics. It is desirable that the actual light confining structure, the optical probe, can be positioned and manipulated. This requires some kind of mechanical support. Because of the evanescent nature of optical near-fields, strong light confinement can only be guaranteed near the material boundaries of the probe. With increasing distance from the probe the fields spread out and the field strength decays rapidly. Near-field optical probes, such as laser-irradiated metal tips, are the key components of near-field optical microscopes. No matter whether the probe is used as a local illuminator, a local collector, or both, the optical spatial resolution solely depends on the confinement of the optical energy at the apex of the probe. This chapter discusses light propagation and light confinement in different probes used in near-field optical microscopy. Where applicable we study fundamental properties using electromagnetic simulations and provide an overview of current methods used for the fabrication of optical probes. We hope to provide the basic knowledge to develop a clear sense of the potentials and the technical limitations of such probes. The most common optical probes are;

- Uncoated fibre probes,
- Aperture probes,

- Pointed metal and semiconductor probes, and
- Nano emitters, such as single molecules or nano crystals.

The reciprocity theorem of electromagnetism states that a signal remains unchanged upon exchange of source and detector. Therefore, it suffices to investigate a given probe only in one mode of operation. In the majority of applications it is undesirable to expose the sample surface on a large scale due to the risk of photo damage or long-range interference effects complicating image-reconstruction. Therefore, we will preferentially consider the local illumination configuration.

DIELECTRIC PROBES

Dielectric, *i.e.* transparent tips are an important class of near-field optical probes and in addition are the key components for the fabrication of more complicated probes. *e.g.* aperture probes. Transparent tips can be produced by tapering of optical fibres yielding conical tips, by suitable breaking of glass slides to produce tetrahedral tips, by polymer moulding processes, or by silicon (nitride or oxide) microfabrication techniques. Tips at the end of glass fibres have the distinct advantage that the coupling of light into the taper region can be done easily by exciting the guided modes in the fibre at the far fibre end. Microfabricated or moulded tips can be mounted at the end of cleaved fibres.

NEAR-FIELD SCANNING OPTICAL MICROSCOPE

Certus NSOM - is a Scanning Probe Microscope (SPM), equipped with special probe holder and optical equipment to conduct studies using near-field effects.

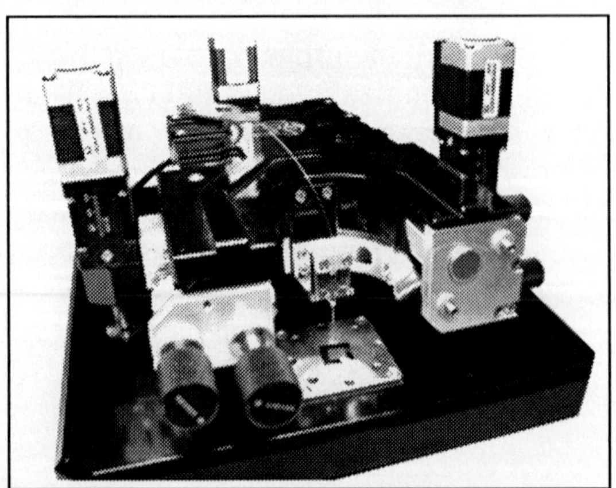

Near-field optical microscopy is based on the properties of near (evanescent) field, which allows to overcom a diffraction limit of classical optical microscopy.

Each near-field optical microscope includes several basic elements:

- The probe;
- Scanning system to move a probe towards the sample surface for raster scan of specimen, line by line;
- Detecting system;
- Optical excitation system.

Optical probe is the main element of any near-field optical microscope. The design of such probe allows to localize electromagnetic field in a region of space with size less than wavelength of applied radiation.

There are two methods of localization of electromagnetic field: aperture and apertureless.

- To illuminate the object and/or to detect the signal the aperture is used, whose size can be substantially smaller than the wavelength of light ($d \ll \lambda$). Usually the probes based on metal coated optical fibres are used with an aperture at the probe tip. In this case the aperture should be located at a distance which is smaller than the wavelength ($h \ll \lambda$) above the surface.
- To radiation localize used by probe tip, approached to the illuminated surface at a distance less than the wavelength. In this mode, the tip scatters (turns to far-field light) the near-field, localized near sample surface.

NEAR-FIELD OPTICAL MICROSCOPY WITH APERTURE PROBE

In the most common case, the probe for aperture near-field microscopy is metal coated optical fibre with truncated cone on the end and aperture on top of the cone. Aperture size and coating material varies depending on the technique and the required resolution. Typically, the aperture diameter is in the range of 50-100 nm, and coating is done with aluminum or silver.

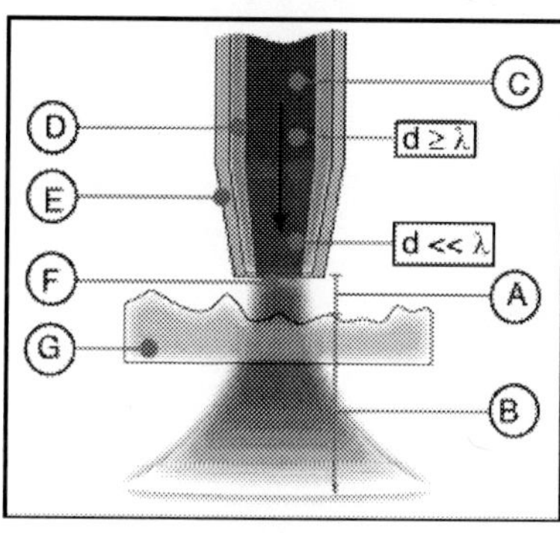

Use probe with an aperture for near-field optical image at point. Model.
A - near-field area ($h \ll \lambda$)
B - far-field area ($h \geq \lambda$)
C - laser beam

D - optical fibre
E - metallic coating of optical fibre
F - aperture (d << λ)
G – sample.
Exist for some basic techniques of lighting and/or signal detection.

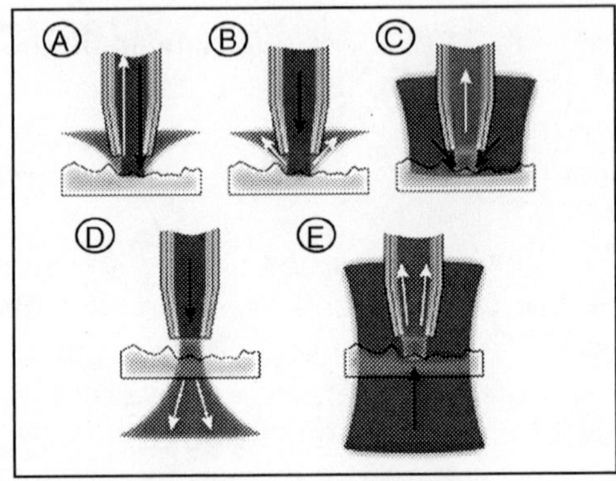

Figure: Different Methods of Near-field Optical Microscopy with Aperture.

A - for illuminate and collect reflected radiation use the same optical fibre.

B - for illuminate use optical fibre probe, collect the reflected light from the sample surface.

C - for illuminate sample use an external source of laser radiation, to collect reflected radiation use probe.

D - for illuminate the sample use a probe, collect transmitted light. The technique is applicable only to transparent samples.

E - for illuminate sample use an external source of laser radiation, to collect transmitted light use probe. The technique is applicable only to transparent samples.

Raster Scanning System

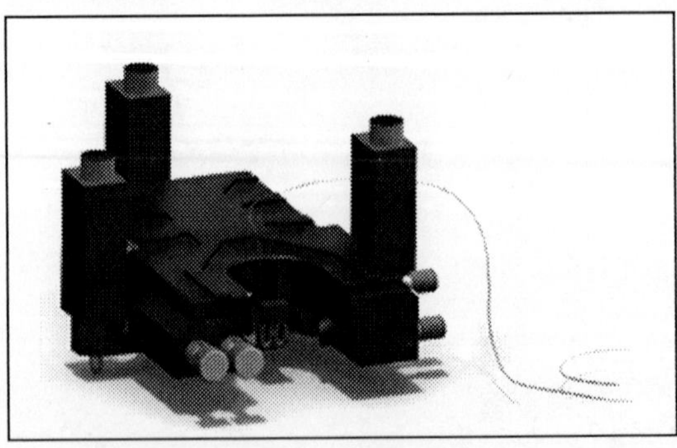

Figure: Certus NSOM (model).

Regardless of optical probe can collect the signal only at one point. To construct a raster image requires a system of moving the probe relative to surface in XY or XYZ coordinates (raster scanning system).

For scanning systems use scanners, similar to scanners tunneling and atomic force microscopes. That is scanners with piezo-tube or plane-parallel scanners.

At Certus NSOM scanning head Certus and XY-scanning stage Ratis allows you to use external large aperture objectives, illuminators, microscopes condensers, *etc.* to illuminate the work area, monitor the sample and the probe position, to take stock of radiation in the point of contact of the probe and the sample.

Which allows the use of all basic techniques of near-field microscopy.

Signal Detection System

To register change distance between probe and sample surface (Z) use same methods as in atomic force microscopy. The most widely used method "shear-force" and registration of probe position by deflectometer.

"Shear-force" method is based on detection changes oscillation frequency of probe caused by tangential component of interaction forces between probe and surface.

To determine changes in oscillation frequency use tuning fork resonators, oscillation which are detected by direct piezoelectric effect of quartz crystal.

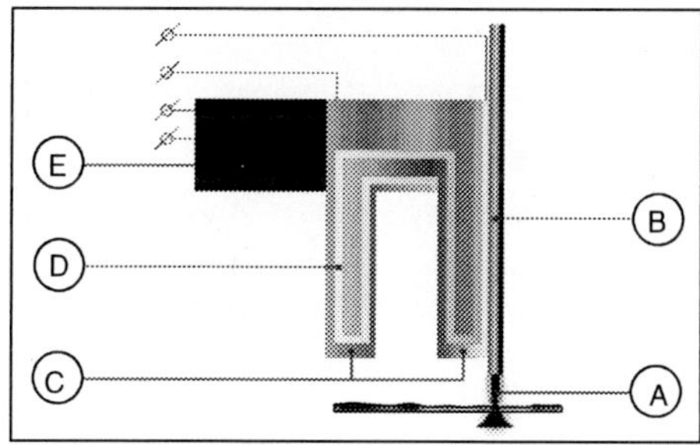

Figure: The Quartz Tuning Fork Resonator, used for Feedback. Model.

A - probe.

B - optical fibre.

C - electrodes used to measure voltage changes occurring on sides of a quartz crystal (direct piezoelectric effect) due to changes in frequency during interaction with surface. Information about voltage change to determine current frequency of resonator.

D - quartz crystal (piezoelectric).

E - dither piezo used to create a forced oscillation of resonator (setting initial frequency of resonator).

Figure: Certus NSOM

A - quartz tuning fork resonator with probe.

B - probe holder.

C - optical fibre.

D - SPM head Certus.

Optical Unit

For excitation and detection of optical signal use optical unit, composed of laser light source, polarizers and signal detectors (avalanche photodiodes, PMT or CCD). Optical unit configuration collect for task and budget research laboratories.

A company has developed three versions of the microscope Certus NSOM:

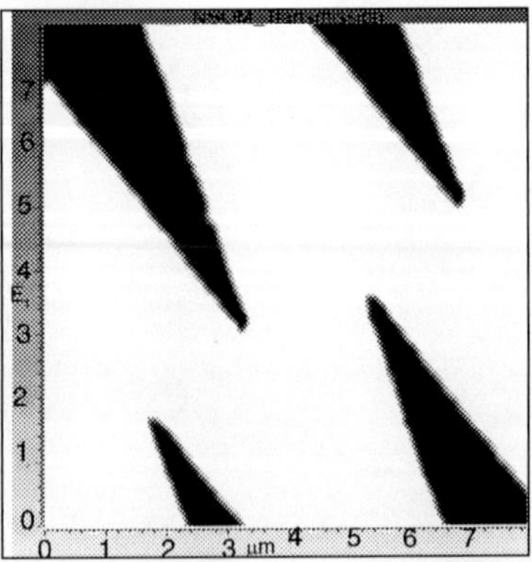

Figure: Vanadium Test Grating. NSOM Image.

Certus NSOM S - Modification of Scanning Probe Microscope Certus Standard

Certus NSOM S includes:

- SPM scanning head Certus with probe holders for standard cantilever probes;
- Probe holders for the tuning fork with a vertical or horizontal arrangement;
- Video microscope with USB camera;
- Optical unit;
- Integrated mechanical XY-stage for sample adjustment;
- SPM controller EG-3000;
- NSpec software package.

Microscope Certus NSOM S allows the use of near-field optical microscopy techniques based on light reflection from the surface or probes. XYZ probe scanning with head scanner.

Certus NSOM O - Modification of Scanning Probe Microscope Certus Optic

Certus NSOM O includes:

- SPM scanning head Certus with probe holders for standard cantilever probes;
- Probe holders for the tuning fork with a vertical or horizontal arrangement;
- XY-scanning stage Ratis;
- Optical microscope (upright or inverted);
- Z-axis stage for objective;
- Optical unit;
- Integrated mechanical XY-stage for sample adjustment;
- SPM controller EG-3000;
- NSpec software package.

Complex Certus NSOM O enables use near-field optical microscopy techniques based on both reflection of light from the surface and transmittion light through the sample. Two scanning modes: XY sample scanning with Ratis stage and Z scanning with head scanner Certus, or XYZ probe scanning with head scanner Certus.

In addition, when XY sample scanning Certus NSOM O operates as the laser confocal microscope.

Certus NSOM Modification for AFM/Raman/Confocal System Centaur (HR)

For complexes Centaur (HR) based on SPM Certus NSOM as an additional option implemented techniques of near-field optical microscopy with aperture. Apertureless techniques implemented in these complexes by default.

SUPER-RESOLUTION OPTICS AND SUPER-LENSES

SUPERRESOLUTION

Nanoscale optical elements that can mold the flow of light offer the potential of entirely new modalities of superresolution imaging. With a vision to build superlenses capable of resolving subwavelength details, in the last decade the photonics community has focused on fabricating tiny metallic inclusions (or, closely related, plasmonic architectures) characterized by a negative index of refraction. Unfortunately, such metamaterials suffer from material losses enhanced by resonances, leading to a degradation of functionality. Instead, a group has demonstrated successful performance of highly functional nanostructured metamaterials in 'low-loss' dielectric platforms such as metal-dielectric composites and III-V semiconductors (composed of indium phosphide/indium gallium arsenide phosphide: InP/InGaAsP).

They recently reported the experimental realization of superresolution imaging with a low-loss 3D nanolens, comprised of gold nanowires embedded in nanoporous alumina. This 3D nanolens, manufactured using a combination of bottom-up self-assembly and electrochemical process, transmits subwavelength details down to $\lambda/4$ (λ: Wavelength) and over a significant distance of more than 6λ. They show an example of the superresolution. The nanolens guided modes do not penetrate the nanowires, so that most of the energy propagates between them. Consequently, image reconstruction can be done with low loss and over a broad spectral range. The nanolens possesses a figure of merit (a metric for functionality) that is four times higher than the best fabricated metallic-based metamaterial at telecommunications wavelengths. Manufacturable on large scales, the nanolens has immediate applications in contact-lithography techniques or optical projection printing.

OPTICAL SUPERRESOLUTION

Optical superresolution is a term for techniques in which object details beyond the normal diffraction limit are transmitted to the image of an optical device.

Geometrical superresolution is the term applied to image processing techniques for improving information retrieval of details contained in the optical image *within* the diffraction limit. A prominent example of a technique categorized as geometrical superresolution is noise reduction by averaging multiple exposures.

Optical Imaging Resolves beyond the Diffraction Limit

To overcome this limitation, information theory provides the necessary theoretical background by establishing an invariance theorem that states the information capacity (given by the number of degrees of freedom) of an imaging system remains constant.

Thus, provided that the input object belongs to a restricted class, it is possible to extend the spatial bandwidth (or any other desired parameter) by encoding-decoding additional spatial-frequency information onto the independent (unused) parameter(s) of the imaging system in such a way that it will then pass through the limited system aperture. This is the fundamental principle underlying all superresolution schemes.

An example of such object classification using a priori information is an object that is approximately independent of one variable, such as time, polarization, wavelength, dimensions, or dynamic range. For instance, because of the finite size of the system's aperture, some of the spatial-frequency signal information will be lost. If it is known that the signal information is monochromatic, it is possible to convert part of the spatial information into wavelength information in such a way that the aperture of the system is synthetically expanded, which ultimately improves resolution.

Motivation

In this context, superresolution techniques aim to improve spatial resolution without changing the physical properties of the optical system. These techniques can be understood as the generation of a synthetic aperture (SA) that increases the frequency coverage of the optical system beyond the limit defined by its cut-off frequency; that is, beyond its physical limited aperture.

But the question is: which techniques provide synthetic-aperture superresolution? By definition, such methods imply the generation of a synthetic aperture that permits high-resolution images using low-NA lenses. In microscopy, low-NA microscope objectives have attractive advantages such as long working distance, large field of view, large depth of focus, and relatively low cost compared to higher-NA microscopes objectives. So, besides the fact that high-NA lenses are not useful for some applications, low-NA microscope lenses obtain only questionable benefit from the superresolution techniques.

For those applications that present conflicting requirements for high-resolution imaging over large fields of view at a large working distance from the sample, synthetic-aperture superresolution techniques can be the key to success.

Mask inspection in the semiconductor industry and microscopic biological applications are two potential applications for synthetic-aperture superresolution techniques. In the semiconductor industry, the decreasing dimensions of mask features require higher-resolution optics, while the need for increasing inspection speed demands that the imaging optics must maintain a long working distance from the mask to provide a large field of view. In biology, many samples have protective pellicles mounted above them or are hard to access. In these cases, it is essential to have a long working distance to image the sample, which decreases resolution in the final image. Once again, synthetic-aperture superresolution techniques provide a solution.

Many approaches have been proposed to achieve superresolution effects. In particular, far-field-imaging superresolution techniques are attractive in applied-optics disciplines such as digital microscopy, optical lithography, and information transmission. However, samples that are temporally restricted require exploitation of the temporal degree of freedom.

TIME MULTIPLEXING

In general, superresolution effects are achieved in at least two basic steps. First, we encode the object's spectrum to divert the higher spatial-frequency content through the imaging-lens aperture. The second stage is to recover and correctly reassemble the frequency content transmitted in the encoding stage. In some cases, the decoding procedure may not enable restoration of new bandpass frequency content to its original position in the spectrum. If so, a third post-processing stage allows final superresolved imaging. Because digital sensors are a relatively modern development, historically superresolution involves only the first two stages.

The encoding stage typically uses external elements such as diffraction gratings, prisms, or special illumination methods that allow the transmission of a wider object's spectrum through the limited system aperture. The classical way to accomplish this encoding is to use diffraction gratings as the encoding mask.

Other methods involve optically generated gratings that can increase the lateral resolution beyond the diffraction limit in conventional microscopy imaging. Optically generated gratings are also used in fluorescence microscopy, using both structured illumination and optical nonlinearity in the pattern excitation processes.

All of these methods have the same underlying purpose: the diffraction grating diverts the on-axis high-frequency-diffracted bands of the object's spectrum, and passes spatial-frequency information through the system aperture. The rest of the information remains to be recovered. This is the aim of the decoding stage, recovering each frequency band and properly placing it in its original position in the object's spectrum

Another way to achieve time-multiplexing superresolution is the use of speckle patterns instead of gratings. Speckle patterns can be understood as a continuous case compared to the discrete case of a diffraction grating. Simply coding and decoding the input and the output, respectively, with a given speckle pattern enables superresolution in two dimensions. A speckle-pattern projection method greatly reduces complexity compared to the projection systems used in structured illumination.

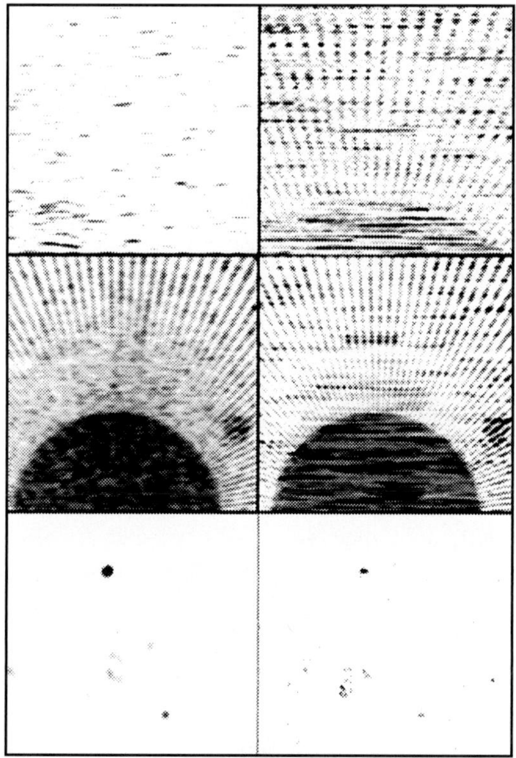

Figure: Images of a Spoke Target with Coherent Illumination (top), Incoherent Illumination (centre), and 1-μm-diameter Fluorescent Beads (bottom) are Low Resolution without Superresolution Techniques (left) and Superresolved (right).

Coherence coding also facilitates time-multiplexing superresolution in both spatial and temporal ways. In spatial coherence coding, the mutual intensity function of the illumination beam is used to code spatial information in an analogous way to time multiplexing. The information is multiplexed in time slots separated by an interval of at least the coherence time of the illumination beam.

Instead of generating diverse distributions, each one with different transverse coherence functions, the longitudinal axis is used to code the transverse spatial information in the temporal coherence coding by using incoherent light interferometry. After that, interferometric image-plane recording and a digital post-processing stage allow the needed decoding to achieve the superresolution effect.

An analogy of tilted beams helps explain coherence coding. The key idea is the equivalence between a diffraction grating and a set of tilted beams (on-axis and off-axis beams) with the same propagation angles as those diffracted by the grating. Thus, the off-axis illumination is used to downshift the high-frequency components of the object's spectrum to low-frequency ones, allowing its passage through the system aperture. In that sense, the encoding stage is defined as a set of illumination beams having different incidence angles (on-axis and off-axis). When these tilted beams impinge upon the input object, each of them produces a frequency shift in the object's spectrum according to the beam propagation angle. Once again, a decoding stage and in most cases a digital post-processing stage restore each diffracted band according to its original diffraction angle.

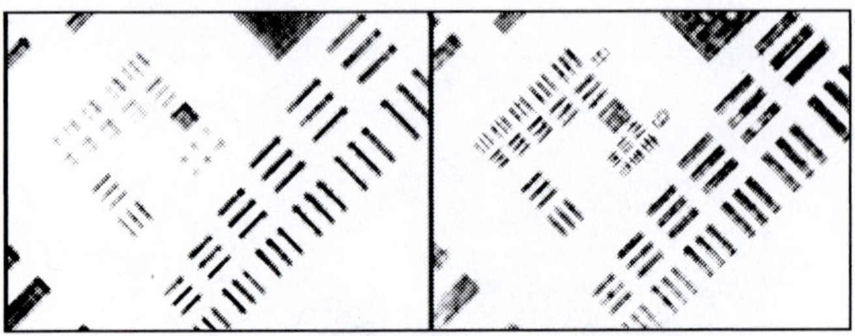

Figure: Images of 1951-USAF Test Targets are Low Resolution without off-axis Illumination (left) and Superresolved with off-axis Illumination (right).

Over the last few years, time multiplexing has been combined with digital holography to recover the complex amplitude distribution of the sample under test.

In particular, one interesting application for this interferometric imaging is microscopy, in which synthetic-aperture generation permits the extraction of medium-and high-resolution images using low-NA microscope lenses. Much like the grating and speckle-pattern processes, this technique also directs off-axis illumination onto the object and enables the transmission of high-frequency bands through a limited system aperture. Using holographic image-plane recording, a digital filtering and relocation process generates a synthetic aperture and restores each transmitted frequency band to its original position in the object's spectrum. Finally, a simple Fourier transformation of the information contained in the sythetic aperture creates a superresolved image.

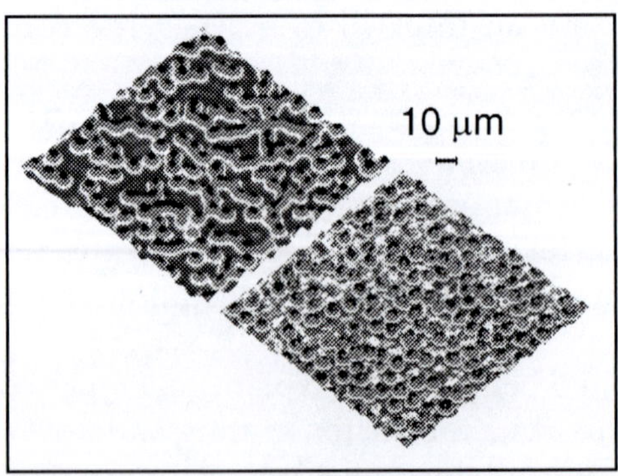

Figure: False-colour Three-dimensional Images of Human Red Blood Cells are Low Resolution without Off-axis Illumination (top) and Superresolved (bottom).

SUPERLENS: PERFECT LENS

A superlens, super lens or perfect lens is a lens which uses metamaterials to go beyond the diffraction limit. The diffraction limit is an inherent limitation in conventional optical devices or lenses. In 2000, a type of lens was proposed, consisting of a metamaterial that compensates for wave decay and reconstructs images in the near field. In addition, both propagating and evanescent waves contribute to the resolution of the image. Theory and simulations show that the superlens and hyperlens can work but engineering obstacles need to be overcome.

The first superlens with a negative refractive index provided resolution three times better than the diffraction limit and was demonstrated at microwave frequencies. In 2005, the first near field superlens (which exceeded the diffraction limit) was demonstrated by N.Fang *et al.*, but the lens did not rely on negative refraction. Instead, a thin silver film was used to enhance the evanescent modes through surface plasmon coupling. Almost at the same time Melville and Blaikie succeeded with the near field superlens. Other groups followed. Two developments in superlens research were reported in 2008. In the second case, a metamaterial was formed from silver nanowires which were electrochemically deposited in porous aluminium oxide. The material exhibited negative refraction.

The superlens has not yet been demonstrated in the visible or near-infrared domains. Furthermore as dispersive materials these are limited to functioning at a single wavelength. Proposed solutions are metal–dielectric composites (MDCs) and multilayer lens structures. The multi-layer superlens appears to have better subwavelength resolution than the single layer superlens. Losses are less of a concern with the multi-layer system, but so far it appears to be impractical because of impedance mis-match.

PERFECT LENS

When the world is observed through conventional lenses, the sharpness of the image is determined by and limited to the wavelength of light. Around the year 2000, a slab of negative index metamaterial was theorized to create a lens with capabilities beyond conventional (positive index) lenses. Sir John Pendry, a British physicist, proposed that a thin slab of negative refractive metamaterial might overcome known problems with common lenses to achieve a "perfect" lens that would focus the entire spectrum, both the propagating as well as the evanescent spectra.

A slab of silver was proposed as the metamaterial. As light moves away (propagates) from the source, it acquires an arbitrary phase. Through a conventional lens the phase remains consistent, but the evanescent waves decay exponentially. In the flat metamaterial DNG slab, normally decaying evanescent waves are contrarily amplified. Furthermore, as the evanescent waves are now amplified, the phase is reversed. Therefore, a type of lens was proposed, consisting of a metal film metamaterial. When illuminated near its plasma frequency, the lens could be used for superresolution imaging that compensates for wave decay and reconstructs images in the near-field. In addition, both propagating and evanescent waves contribute to the resolution of the image.

Pendry suggested that left-handed slabs allow "perfect imaging" if they are completely lossless, impedance matched, and their refractive index is "1 relative to the surrounding medium. Theoretically, this would be a breakthrough in that the optical version resolves objects as minuscule as nanometers across. Pendry predicted that Double negative metamaterials (DNG) with a refractive index of *n* = "*1*, can act, at least in principle, as a "perfect lens" allowing imaging resolution which is limited not by the wavelength, but rather by material quality.

OTHER STUDIES CONCERNING THE PERFECT LENS

Further research demonstrated that Pendry's theory behind the perfect lens was not exactly correct. The analysis of the focusing of the evanescent spectrum was flawed. In addition, this applies to only one (theoretical) instance, and that is one particular medium that is lossless, nondispersive and the constituent parameters are defined as:

$\varepsilon(\omega)/\varepsilon_0 = \mu(\omega)/\mu_0 = -1$, which in turn results in a negative refraction of $n = -1$

However, the final intuitive result of this theory that both the propagating and evanescent waves are focused, resulting in a converging focal point within the slab and another convergence (focal point) beyond the slab turned out to be correct.

If the DNG metamaterial medium has a large negative index or becomes lossy or dispersive, Pendry's perfect lens effect cannot be realised. As a result, the perfect lens effect does not exist in general. According to FDTD simulations at the time, the DNG slab acts like a converter from a pulsed cylindrical wave to a pulsed beam. Furthermore, in reality (in practice), a DNG medium must be and is dispersive and lossy, which can have either desirable or undesirable effects, depending on the research or application. Consequently, Pendry's perfect lens effect is inaccessible with any metamaterial designed to be a DNG medium.

Another analysis, in 2002, of the perfect lens concept showed it to be in error while using the lossless, dispersionless DNG as the subject. This analysis mathematically demonstrated that subtleties of evanescent waves, restriction to a finite slab and absorption had led to inconsistencies and divergencies that contradict the basic mathematical properties of scattered wave fields. For example, this analysis stated that absorption, which is linked to dispersion, is always present in practice, and absorption tends to transform amplified waves into decaying ones inside this medium (DNG).

A third analysis of Pendry's perfect lens concept, published in 2003, used the recent demonstration of negative refraction at microwave frequencies as confirming the viability of the fundamental concept of the perfect lens. In addition, this demonstration was thought to be experimental evidence that a planar DNG metamaterial would refocus the far field radiation of a point source. However, the perfect lens would require significantly different values for permittivity, permeability, and spatial periodicity than the demonstrated negative refractive sample.

This study agrees that any deviation from conditions where $\varepsilon = \mu = -1$ results in the normal, conventional, imperfect image that degrades exponentially *i.e.*, the diffraction limit. The perfect lens solution in the absence of losses is again, not practical, and can lead to paradoxical interpretations.

It was determined that although resonant surface plasmons are undesirable for imaging, these turn out to be essential for recovery of decaying evanescent waves. This analysis discovered that metamaterial periodicity has a significant effect on the recovery of types of evanescent components. In addition, achieving subwavelength resolution is possible with current technologies.

Negative refractive indices have been demonstrated in structured metamaterials. Such materials can be engineered to have tunable material parameters, and so achieve the optimal conditions. Losses can be minimized in structures utilizing superconducting elements. Furthermore, consideration of alternate structures may lead to configurations of left-handed materials that can achieve subwavelength focusing. Such structures were being studied at the time.

IMAGE FORMATION

An image of an object can be defined as a tangible or visible representation of the features of that object. A requirement for image formation is interaction with radiation fields.

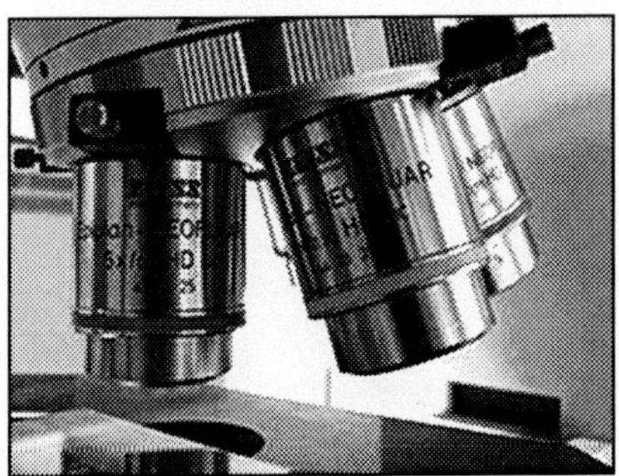

Figure: The Binocular Microscope is a Conventional Optical System. Spatial Resolution is Confined by a Diffraction Limit that is a Little above 200 Nanometers.

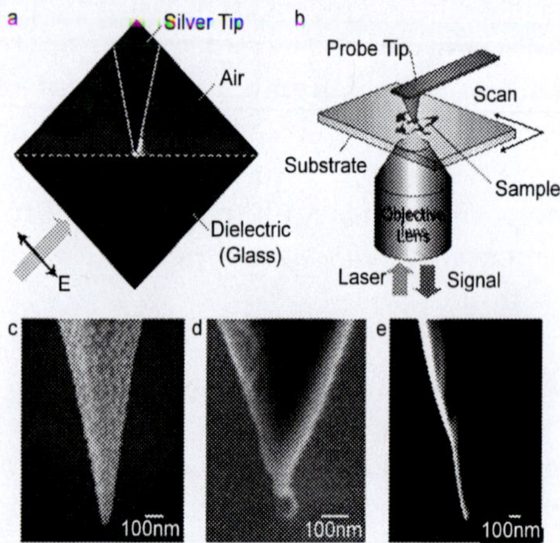

Figure: Schematic Depictions and Images of Commonly used Metallic Nanoprobes that can be used to see a Sample in Nanometer Resolution.

The level of feature detail, or image resolution, depends on the wavelength of radiation. For example, with optical microscopy, image production and resolution depends on the wavelength of visible light. And visible light is a small part of the electromagnetic spectrum. However, with a superlens, this limitation on the wavelength of the field may be removed, and a new class of image generated.

Electron beam lithography can overcome this resolution limit. Optical microscopy, on the other hand, has been traditionally limited to some value just above 200 nanometers. However, new technologies combined with optical microscopy are beginning to allow for increased feature resolution.

One definition of being constrained by the diffraction barrier, is a resolution cut off at half the wavelength of light. The visible spectrum has a range that extends from 390 nanometers to 750 nanometers. Green light, half way in between, is around 500 nanometers. Microscopy takes into account parameters such as lens aperture, distance to the object from the lens, and the refractive index of the observed material. This combination defines the resolution cutoff, or optical limit, which tabulates to 200 nanometers. Therefore, conventional lenses, which literally construct an image of an object by using "ordinary" light waves, discard information that produce very fine, and minuscule details of the object that are contained in evanescent waves. These dimensions are less than 200 nanometers. For this reason, conventional optical systems, such as microscopes, have been unable to accurately image very small, nanometer-sized structures or nanometer-sized organisms in vivo, such as individual viruses, or DNA molecules.

The limitations of standard optical microscopy (bright field microscopy) lie in three areas:

- The technique can only image dark or strongly refracting objects effectively.

- Diffraction limits the object, or cell's, resolution to approximately 200 nanometers.
- Out of focus light from points outside the focal plane reduces image clarity.

Live biological cells in particular generally lack sufficient contrast to be studied successfully, because the internal structures of the cell are colourless and transparent. The most common way to increase contrast is to stain the different structures with selective dyes, but this involves killing and fixing the sample. Staining may also introduce artifacts, apparent structural details that are caused by the processing of the specimen and are thus not a legitimate feature of the specimen.

CONVENTIONAL LENS

The conventional glass lens is pervasive throughout our society, and the sciences. It is one of the fundamental tools of optics. However, the wavelength of light is like the width of a pencil used to draw the ordinary images. The limit becomes noticeable, for example, where the laser used in a digital video system can only detect and deliver details based on the wavelength of light from a DVD. The image cannot be rendered any sharper beyond this limitation.

Figure: DVD – A DVD is a Large, Fast Compact Digital Disk that can Hold Video, Audio, and/or Computer Data. A Laser is Employed for Data Transfer.

When an object emits or reflects light there are two types of electromagnetic radiation associated with this phenomenon. These are the near field radiation and the far field radiation. As implied by its description, the far field escapes beyond the object. It is then easily captured and manipulated by a conventional glass lens. However, useful (nanometer-sized) resolution details are not observed, because they are hidden in the near field. They remain localized, staying much closer to the light emitting object, unable to travel, and be captured by the conventional lens.

Controlling the near field radiation, for high resolution, can be accomplished with a new class of materials not found in nature. These are unlike familiar solids, such as crystals, which derive their properties from atomic and molecular units. The new material class, termed metamaterials, obtains its properties from its artificially larger structure. This has resulted in novel properties, and novel responses, which allow for details of images that surpass the limitations imposed by the wavelength of light.

Subwavelength Imaging

Figure: The "Electrocomposeur" was an Electron-beam Lithography Machine (Electron Microscope) Designed for Mask Writing. It was Developed in the early 1970s and Deployed in the Mid 1970s.

This has led to the desire to view live biological cell interactions in a real time, natural environment, and the need for *subwavelength imaging*. Subwavelength imaging can be defined as optical microscopy with the ability of an object or organism below the wavelength of visible light. In other words, to have capability to observe, in real time, below 200 nanometers. Optical microscopy is a non-invasive technique and technology, because everyday light is the transmission medium. Imaging below the optical limit in optical microscopy (subwavelength) can be engineered for the cellular level, and nanometer level.

For example, in 2007 a technique was demonstrated where a metamaterials-based lens coupled with a conventional optical lens could manipulate visible light to see (nanoscale) patterns that were too small to be observed with an ordinary optical microscope.

This has potential applications not only for observing a whole living cell, or for observing cellular processes, such as how proteins and fats move in and out of cells. In the technology domain, it could be used to improve the first steps of photolithography and nanolithography, essential for manufacturing ever smaller computer chips.

Focusing at subwavelength has become a unique imaging technique which allows visualization of features on the viewed object which are smaller than the wavelength of the photons in use.

A photon is the minimum unit of light. While previously thought to be physically impossible, subwavelength imaging has been made possible through the development of metamaterials. This is generally accomplished using a layer of metal such as gold or silver a few atoms thick, which acts as a superlens, or by means of 1D and 2D photonic crystals. There is a subtle interplay between propagating waves, evanescent waves, near field imaging and far field imaging.

Early Subwavelength Imaging

Metamaterial lenses (Superlens) are able to compensate for the exponential evanescent wave decay via negative refractive index, and in essence reconstruct the image. Prior to metamaterials, proposals were advanced in the 1970s to avoid this evanescent decay.

For example, in 1974 proposals for two-dimensional, fabrication techniques were presented. These proposals included contact imaging to create a pattern in relief, photolithography, electron lithography, X-ray lithography, or ion bombardment, on an appropriate planar substrate.

The shared technological goals of the metamaterial lens and the variety of lithography aim to optically resolve features having dimensions much smaller than that of the vacuum wavelength of the exposing light.

In 1981 two different techniques of contact imaging of planar (flat) submicroscopic metal patterns with blue light (400 nm) were demonstrated. One demonstration resulted in an image resolution of 100 nm and the other a resolution of 50 to 70 nm.

Since at least 1998 near field optical lithography was designed to create nanometer-scale features. Research on this technology continued as the first experimentally demonstrated negative index metamaterial came into existence in 2000–2001.

The effectiveness of electron-beam lithography was also being researched at the beginning of the new millennium for nanometer-scale applications. Imprint lithography was shown to have desirable advantages for nanometer-scaled research and technology.

Advanced deep UV photolithography can now offer sub-100 nm resolution, yet the minimum feature size and spacing between patterns are determined by the diffraction limit of light. Its derivative technologies such as evanescent near-field lithography, near-field interference lithography, and phase-shifting mask lithography were developed to overcome the diffraction limit.

CYLINDRICAL SUPERLENS VIA COORDINATE TRANSFORMATION

This began with a proposal by Sir John Pendry, in 2003. Magnifying the image required a new design concept in which the surface of the negatively refracting lens is curved. One cylinder touches another cylinder, resulting in a curved cylindrical lens which reproduced the contents of the smaller cylinder in magnified but undistorted form outside the larger cylinder. Coordinate transformations are required to curve the original perfect lens into the cylindrical, lens structure.

This was followed by a 36 page conceptual and mathematical proof in 2005, that the cylindrical superlens works in the quasistatic regime.

In 2007, a superlens utilizing coordinate transformation was again the subject. However, in addition to image transfer other useful operations were discussed; translation, rotation, mirroring and inversion as well as the superlens effect. Furthermore, elements that perform magnification are described, which are free from geometric aberrations, on both the input and output sides while utilizing free space sourcing (rather than waveguide). These magnifying elements also operate in the near and far field, transferring the image from near field to far field.

The cylindrical magnifying superlens was experimentally demonstrated in 2007 by two groups, Liu *et al.* and Smolyaninov *et al.*

NANOHOLE ARRAY SUBWAVELENGTH IMAGING

Nanohole Array as a Lens

A recent prior work (2007) demonstrated that a quasi-periodic array of nanoholes, in a metal screen, were able to focus the optical energy of a plane wave to form subwavelength spots (hot spots). The distances for the spots was a few tens of wavelengths on the other side of the wiktionary:array, or, in other words, opposite the side of the incident plane wave. The quasi-periodic array of nanoholes functioned as a light concentrator.

In June 2008, this was followed by the demonstrated capability of an array of quasi-crystal nanoholes in a metal screen. More than concentrating hot spots, an image of the point source is displayed a few tens of wavelengths from the array, on the other side of the array (the image plane). Also this type of array exhibited a 1 to 1 linear displacement, – from the location of the point source to its respective, parallel, location on the image plane. In other words from x to x + δx.

For example, other point sources were similarly displaced from x′ to x′ + δx′, from x^ to x^ + δx^, and from x^^ to x^^ + δx^^, and so on. Instead of functioning as a light concentrator, this performs the function of conventional lens imaging with a 1 to 1 correspondence, albeit with a point source.

However, resolution of more complicated structures can be achieved as constructions of multiple point sources. The fine details, and brighter image, that are normally associated with the high numerical apertures of conventional lenses can be reliably produced. Notable applications for this technology arise when conventional optics is not suitable for the task at hand. For example, this technology is better suited for X-ray imaging, or nano-optical circuits, and so forth.

Nanolens

In 2010, a nano-wire array prototype, described as a three-dimensional (3D) metamaterial-nanolens, consisting of bulk nanowires deposited in a dielectric substrate has been fabricated and tested. The metamaterial nanolens was constructed of millions of nanowires at 20 nanometers in diameter. These were precisely aligned, and a packaged configuration was applied. The lens is able to depict a clear, high-resolution image of nano-sized objects because it uses both the normal propagating EM radiation, and evanescent waves to construct the image. Super-resolution imaging was demonstrated over a distance of 6 times the wavelength (λ), in the far-field, with a resolution of at least $\lambda/4$. This is a significant improvement over previous research and demonstration of other near field and far field imaging.

Light Transmission Properties of Holey Metal Films

The light transmission properties of holey metal films in the metamaterial limit, where the unit length of the periodic structures is much smaller than the operating wavelength, are analysed theoretically.

Transporting an Image Through a Subwavelength Hole

Theoretically it appears possible to transport a complex electromagnetic image through a tiny subwavelength hole with diameter considerably smaller than the diameter of the image, without losing the subwavelength details.

Nanoparticle Imaging – Quantum Dots

When observing the complex processes in a living cell, significant processes (changes) or details are easy to overlook. This can more easily occur when watching changes that take a long time to unfold and require high-spatial-resolution imaging. However, recent research offers a solution to scrutinize activities that occur over hours or even days inside cells, potentially solving many of the mysteries associated with molecular-scale events occurring in these tiny organisms.

A joint research team, working at the National Institute of Standards and Technology (NIST) and the National Institute of Allergy and Infectious Diseases (NIAID), has discovered a method of using nanoparticles to illuminate the cellular interior to reveal these slow processes. Nanoparticles, thousands of times smaller than a cell, have a variety of applications. One type of nanoparticle called a quantum dot glows when exposed to light. These semiconductor particles can be coated with organic materials, which are tailored to be attracted to specific proteins within the part of a cell a scientist wishes to examine. Notably, quantum dots last longer than many organic dyes and fluorescent proteins that were previously used to illuminate the interiors of cells. They also have the advantage of monitoring changes in cellular processes while most high-resolution techniques like electron microscopy only provide images of cellular processes frozen at one moment. Using quantum dots, cellular processes involving the dynamic motions of proteins, are observable (elucidated).

The research focused primarily on characterizing quantum dot properties, contrasting them with other imaging techniques. In one example, quantum dots were designed to target a specific type of human red blood cell protein that forms part of a network structure in the cell's inner membrane. When these proteins cluster together in a healthy cell, the network provides mechanical flexibility to the cell so it can squeeze through narrow capillaries and other tight spaces. But when the cell gets infected with the malaria parasite, the structure of the network protein changes.

Because the clustering mechanism is not well understood, it was decided to examine it with the quantum dots. If a technique could be developed to visualize the clustering, then the progress of a malaria infection could be understood, which has several distinct developmental stages.

Research efforts revealed that as the membrane proteins bunch up, the quantum dots attached to them are induced to cluster themselves and glow more brightly, permitting real time observation as the clustering of proteins progresses. More broadly, the research discovered that when quantum dots attach themselves to other nanomaterials, the dots' optical properties change in unique ways in each case. Furthermore, evidence was discovered that quantum dot optical properties are altered as the nanoscale environment changes, offering greater possibility of using quantum dots to sense the local biochemical environment inside cells.

Some concerns remain over toxicity and other properties. However, the overall findings indicate that quantum dots could be a valuable tool to investigate dynamic cellular processes.

OPTICAL NANO-ANTENNAS

Optical nano-antennas are the visible spectrum counterpart of microwave and radio frequency (RF) antennas. The physics of these devices differs from RF antennas in two important aspects: first, due to high losses at optical frequencies, the assumption of perfect electrical conductor is no longer valid; and second, due to unique phenomena at nanoscale, *e.g.*, surface plasmon polaritons (SPP), the response of these structures can be quite different, the most notable of which is the subwavelength field confinement. Therefore, serious efforts are being devoted to extend current understanding from radio frequency antennas to their optical counterparts.

Optical nano-antennas have been used for several unique applications including cloaking, higher harmonic generation, biological and chemical sensing, sub-diffraction lithography, optical imaging, energy harvesting and various other visible and near-infrared applications. An important aspect of optical antennas is field enhancement in sub-wavelength regions that finds applications in fluorescence and Raman spectroscopy with capabilities of single molecule sensing. By placing dimers or multiple particles a few nanometers apart, unique phenomena can be observed, including dramatically high electromagnetic fields, polarization independent sensors, and colour sorters. A wide range of techniques have been utilized for characterizing the near-field as well as far-field behaviour of optical antennas.

OPTICAL NANO-ANTENNAS AND NANO-FABRICATION

Nanooptics and nano-fabrication for the development of optical antennas. In a three-step approach, our structures are first modelled, designed and simulated using Microwave Studio, Finite Difference Time Domain (FDTD), and Discrete Dipole Approximation (DDA). Extensive studies of field distribution, extinction efficiency and their dependence on geometry are carried out prior to physical design.

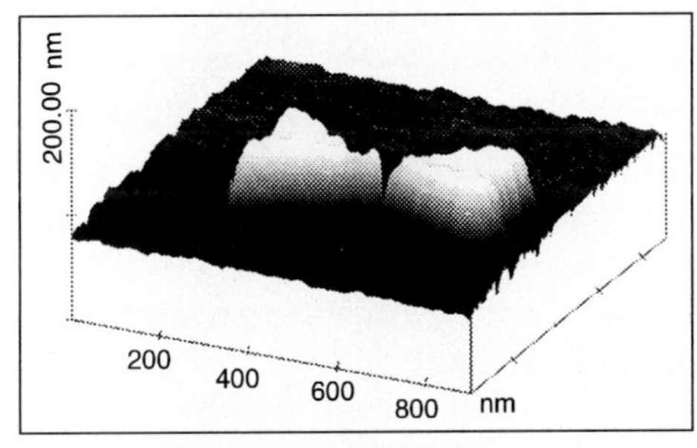

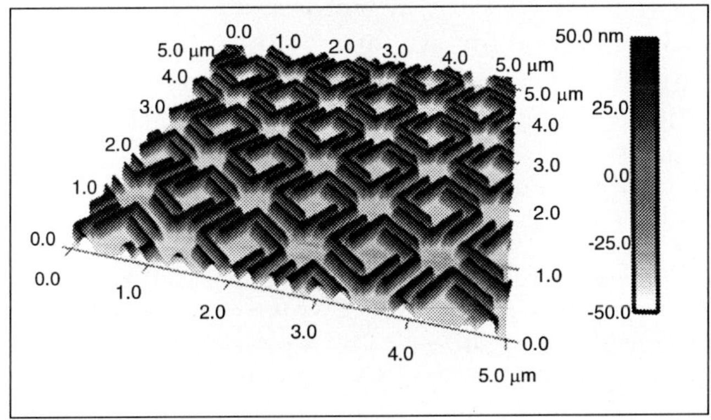

The nanostructures are fabricated using electron-beam lithography and a novel nanofabrication technique developed by our group called Solid-state superionic stamping. We focus on improving the capabilities of current fabrication capabilities beyond 20 nm feature size with added advantages of low cost, high throughput, ambient conditions, and large area patterning.

We apply our understanding of plasmonics and antenna theory to optical wavelengths and explore ways to utilize these sub-wavelength resonant structures for molecular imaging, nanolithography, solar cell efficiency improvement, data storage and optical communication.

Some other areas of my interest are: biomimetics, thin films (especially noble metal films), and biosensing.

PLASMONICS NANO-DEVICES

Plasmonic devices are naturally suited for electrical control because these nanometallic structures can serve as electrodes for current injection or application of electric fields. Thus far, demonstrations of electrical manipulations in nanoplasmonics are relatively sparse in the literature. A mainstream practice of electrically switchable plasmonic responses is to interface the metal with electrorefractive media such as liquid crystals, ferroelectric materials, or carrier-injected semiconductors. Sometimes nanoscale cavities are formed in the metallic architecture to enhance the light-matter interaction and the resultant modulation depth. Use of semiconductor gain media has led to a number of notable demonstrations, such as electrical amplification of surface plasmon propagation and electrically pumped plasmonic nanolasers.

We observed multiple active plasmonic devices based on a nanocavity consisting of a slit carved in a metal film surrounded by a periodic grating. This structure brings several distinct advantages to combined electrical/photonic applications. The slit that defines the plasmonic cavity is less than 100nm wide. Low voltages across the slit result in very high electric fields. A properly designed nanogroove array works as an optical antenna that enables either the focusing of impinging waves or the beaming of emerging light from the nanoslit region, depending on the experiment's configuration. The property of emitted light is further modified with an increased radiative decay rate through the Purcell effect, thanks to the substantially enhanced optical density of states within the plasmonic cavity.

When we place fluorescent emitters within the metallic nanocavity, their emission properties, including the spectral peak location and radiative strength, can be electrically tuned by an externally applied voltage. We saw this phenomenon by filling the cavity with colloidal quantum dots. The voltage-dependent emission spectra from the quantum dots exhibits a noticeable spectral shift toward the red along with a decreased intensity magnitude when the control field increases. We attributed these observations to the combined contributions from the quantum-confined Stark effect and luminescence quenching. As expected for the Stark effect, the spectral red shift scales quadratically with the applied voltage until saturation occurs. To test the dynamic response of the electrically controlled spectral shift of the emitters, we applied a 100kHz alternating voltage signal across the plasmonic device and obtained time-resolved traces. We expect the maximum modulation speed to be orders of magnitude higher than this test value, considering the fast response of the quantum-confined Stark effect.

Electrical manipulations of plasmonic nanophotonics have progressed beyond the scope of tuning the refractive index, the absorption/gain coefficient, and emission behaviour. They have enabled an intriguing opportunity for active nonlinear optics at the deep-subwavelength scale. We recently checked electrically controlled harmonic generation of light in a plasmonic cavity similar to the one used previously, where electrical signals can be encoded in the frequency-doubled light originating from a third-order nonlinear susceptibility $\chi^{(3)}$ or 2ω; ω; ω,0. Unlike conventional second harmonic generation, which imposes a rigorous requirement on the nonlinear medium's lattice asymmetry, third-order nonlinear responses are ubiquitous in all materials. The micrometer-scale footprint of the proposed device also eases the strict phase-matching condition and makes it suitable for chip-scale applications where dense integration is necessary.

VISION FOR THE NEXT DECADE

Both nanophotonics and plasmonics concern investigations into building, manipulating, and characterizing optically active nanostructures with a view to creating new capabilities in instrumentation for the nanoscale, chemical and biomedical sensing, information and communications technologies, enhanced solar cells and lighting, disease treatment, environmental remediation, and many other applications. Photonics and plasmonics share the characteristic that at least some of their basic concepts have been known for 40–50 years, but they have come into their own only in the last ten years, based on recent discoveries in nanoscience. Photonic materials and devices have played a pervasive role in communications, energy conversion, and sensing since the 1960s and 1970s. Photonics at the nanoscale, or *nanophotonics* might be defined as "the science and engineering of light-matter interactions that take place on wavelength and subwavelength scales where the physical, chemical, or structural nature of natural or artificial nanostructure matter controls the interactions". Broadly speaking, over the next ten years nanophotonic structures and devices promise dramatic reductions in energies of device operation, densely integrated information systems with lower power dissipation, enhanced spatial resolution for imaging and patterning, and new sensors of increased sensitivity and specificity.

Plasmonics aims to exploit the unique optical properties of metallic nanostructures to enable routing and active manipulation of light at the nanoscale. It has only been in the past 10 years that the young field of plasmonics has rapidly gained momentum, enabling exciting new fundamental science as well as groundbreaking real-life applications in the coming 10 years in terms of targeted medical therapy, ultrahigh-resolution imaging and patterning, and control of optical processes with extraordinary spatial and frequency precision. In addition, because plasmonics offers a natural integration compatibility with electronics and the speed of photonics, circuits and systems formed of plasmonic and electronic devices hold promise for next-generation systems that will incorporate the best qualities of both photonics and electronics for computation and communication at high speed, broad bandwidth, and low power dissipation.

Changes in the Vision over the Last Ten Years

Despite their relevance to many key technologies going forward, neither nanophotonic nor plasmonics devices were explicitly considered in the 1999 report *Nanotechnology research directions: Vision for nanotechnology R&D in the next decade*. In part, this may have resulted from the fact that one-dimensional nanoscale structures, such as antireflection coatings and distributed Bragg mirrors, have long been a part of optical design and engineering. The first quantum well laser, operating at room temperature with an active layer 20 nm thick, was reported in 1978; today quantum well lasers are the standard for room-temperature solid-state semiconductor devices; they have high efficiency and can be manufactured at relatively low cost.

Many of the critical concepts for today's rapidly growing and diverse field of nanophotonics were established in past decades, but recent progress in the ability to control materials at the nanoscale in multiple dimensions has allowed the validation of those concepts and the anticipation of yet more intriguing and powerful photonic behaviours. For example, an early paper by Yablonovitch (1987) discussed dielectric materials with spatial variations in index of refraction on the order of a wavelength of light. He anticipated that these *photonic crystals* would have a dramatic effect on the spontaneous emission of light within these structures. Negative index materials, a component of a class of *metamaterials,* were anticipated as early as the 1960s by Veselago (1968). *Surface plasmons* have played an important role in surface-enhanced Raman spectroscopy (SERS), an active area of research since the late 1970s. Today's research in plasmonics encompasses an even broader range of structures and applications, as is the case for nanophotonics.

Vision for the Next Ten Years

Advances in the fabrication of optical structures at the nanoscale and improved control of materials properties have allowed researchers to demonstrate and realise the potential of nanophotonics and plasmonics, and they provide strong impetus for further investments in these fields. In the next ten years, we envision that nanophotonics and plasmonics will have dramatic enabling capabilities for new medical therapies; low-power, high-bandwidth, and high-density computation and communications; high-spatial-resolution imaging and sensing with high spectral and spatial precision; efficient optical sources and detectors; and a host of profound scientific discoveries about the nature of light–matter interactions.

Exploit Synergies between Plasmonics, Photonics, and Electronics

Over the last decade, it has gradually become clear what role plasmonics can play in future device technologies and how it can complement electronics and conventional photonics. Each of these device technologies can perform unique functions that play to the strength of the key materials. The electrical properties of semiconductors enable the realization of truly nanoscale elements for computation and information storage; the high transparency of dielectrics (*e.g.*, glass) facilitates information transport over long distances and at very high data rates. Unfortunately, semiconductor electronics is limited in speed by interconnect (RC) delay-time issues, and photonics is limited in size by the fundamental laws of diffraction. Plasmonics offers precisely what electronics and photonics do not have: the size of electronics and the speed of photonics. Plasmonic devices might therefore naturally interface with similar-speed photonic devices and with similar-size electronic components, increasing the synergy between these technologies.

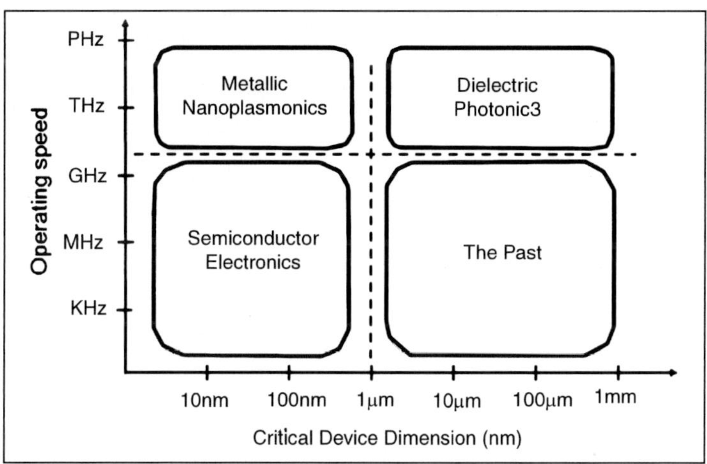

Figure : Graph of the Operating Speeds and Critical Dimensions of Different Chip-scale Device Technologies.

The semiconductor and photonics industries have continued to rapidly develop, and it will be exciting to see what the next decade will bring for plasmonics in these industries. In order to continue and even accelerate the advances in nanophotonics and plasmonics, a number of barriers will have to be overcome. The barriers relate to the enabling resources in the scientific and technological infrastructure.

SCIENTIFIC AND TECHNOLOGICAL INFRASTRUCTURE NEEDS

Develop Simple Design Rules

Although advances in electromagnetic and electronic device simulation tools have been considerable, new challenges have also emerged. As the complexity of integrated photonic systems increases, the development of simple design rules for nanophotonics and plasmonics components is absolutely essential.

The power of good design rules lies in their ability to hide much of the complexity within an individual device; the aim is to capture the essence of the device function and focus on its interactions with other devices. Such simplifications then enable the construction of system-level theories and simulators that can predict the behaviour of larger circuits. Theoretical frameworks should continue to be developed to address functions of nanostructured optical, or metactronic, circuits.

To properly predict the behaviour of optoelectronic circuits, simulation tools are required that can simultaneously capture the flow of electrons and photons and their impact on each other. Currently, these types of simulations tend to be performed separately for the electrons and photons and then combined in an *ad hoc* fashion. The coupling to other degrees of freedom (including mechanical, thermal, magnetic, biological) will become important as well. Coupling of optical simulations to quantum mechanical simulation tools is also needed for an increasing number of applications. In many optical simulation tools, the materials properties are taken into account by utilizing macroscopic properties, such as the real and imaginary part of the dielectric constant. This is inadequate for many nanostructured active and passive systems, and we will need to return to more basic theories.

Expanded Fabrication Tools and Facilities

Over the last decade, the field of nanotechnology has provided a wide range of new nanofabrication and synthesis techniques. With the rapid advances in CMOS and CMOS-compatible processing techniques, it comes as no surprise that silicon-based devices have started to gain popularity (despite the indirect gap nature of Si). This trend is further augmented by the fact that more materials are becoming allowed in Si fabs, and new techniques such as wafer bonding enable convenient ways to add materials with more desirable optical properties for active devices. It will be of value to educate and train the next generation of optical engineers to take advantage of Si fabs and provide easy access to such facilities. It also will be important to build new infrastructure to facilitate the anticipated increase in demand.

Other important fabrication or prototyping techniques for nanophotonics/ nanoplasmonic circuits and devices are state-of-the-art electron beam lithography and focused ion beam milling tools. These tools are extremely powerful and enable the realization of feature sizes below the 10 nm length scale. The cost of such tools will need to come down in order to provide access to an increasing number of users. The increasing divergence between capital equipment costs and equipments grants in the United States is making it harder and harder to provide critical infrastructure to research groups in academia, and industry is facing similar challenges. For this reason, it is also of the utmost importance to further develop less expensive printing technologies (*e.g.*, nanoimprint lithography), patterning techniques (*i.e.*, nanosphere lithography), and materials chemistry approaches (*e.g.*, directed self-assembly, nanoparticles and nanowire growth, and catalysis). Many of these provide access to similar (10 nm) length-scales, although cleanliness, reproducibility, and versatility need to be further improved.

All nanophotonics fabrication technologies will require (1) higher precision tools to control structure size, shape and placement, (2) larger reproducibility, (3) improved control over the material's purity, and (4) new ways to integrate and interface with electronics and conventional optics. Chip-scale devices face additional challenges in terms of cleanliness requirements and materials compatibilities. We must also give thought to newly emerging health and safety issues associated with the use of photonic and plasmonic (and all) nanostructures.

Nanophotonic and Plasmonic Materials and Devices

While the research community has benefited greatly from access to optical nanocharaterization tools, quantitative interpretation of SNOM images is still a challenge and requires significant expertise, the development of better SNOM tips, and better software. There is an increasing need for characterization tools that can correlate the structural and materials properties at the atomic-scale to the optical properties of nanophotonic devices and nanostructured materials. Emerging optical characterization tools in the fields of nanophotonics and plasmonics include cathodoluminescence, optical scanning tunneling microscopy, and electron energy loss spectroscopy (EELS) in a transmission electron microscope; these techniques have already enabled studies of plasmonic modes with nanometer resolution (1-2 orders of magnitude better than SNOM). Equipment like this will help realise simulation tools that properly take into account the optical, electronic, and quantum mechanical effects that will play an increasingly important role in nanophotonic and nanoplasmonic devices.

Researchers will also continue to rely on the availability of and access to state-of-the-art structural characterization tools, as their performance parameters keep on improving every year.

Build New Educational Systems

As the fields of nanophotonics and plasmonics expand rapidly into many applications, ranging from information transport to biology, it will also be important to restructure and further develop our education system in the right way.

The interdisciplinary nature of these fields requires the development of a common language and new educational methods and tools. In such a rapidly-developing field that has widespread applications, the background training of participants in the field will rapidly diversify. For example, just 10 years ago a symposium on plasmonics was rare, and the audience would primarily consist of hardcore theoreticians and a few scattered experimentalists.

Currently, there are well over 20 plasmonics symposia/conferences around the globe annually, filled with people from academia and industry with backgrounds in physics, chemistry, materials science, electrical engineering, biology, chemistry, and medicine. In order to take advantage of the many views and uses of optics, a diverse educational programme should be constructed to reduce misunderstandings due to the differences in vocabulary used to identify common concepts and enhance new and valuable collaborations.

QUANTUM DOTS LASERS AND DEVICES

QUANTUM DOT LASERS

Semiconductor lasers are key components in a host of widely used technological products, including compact disk players and laser printers, and they will play critical roles in optical communication schemes. The basis of laser operation depends on the creation of nonequilibrium populations of electrons and holes, and coupling of electrons and holes to an optical field, which will stimulate radiative emission.

Calculations carried out in the early 1970s by C. Henry predicted the advantages of using quantum wells as the active layer in such lasers: the carrier confinement and nature of the electronic density of states should result in more efficient devices operating at lower threshold currents than lasers with "bulk" active layers.

In addition, the use of a quantum well, with discrete transition energy levels dependent on the quantum well dimensions (thickness), provides a means of "tuning" the resulting wavelength of the material. The critical feature size-in this case, the thickness of the quantum well-depends on the desired spacing between energy levels. For energy levels of greater than a few tens of millielectron volts (meV, to be compared with room temperature thermal energy of 25 meV), the critical dimension is approximately a few hundred angstroms. Although the first quantum well laser, demonstrated in 1975, was many times less efficient than a conventional laser, the situation was reversed by 1981 through the use of new materials growth capabilities (molecular beam epitaxy), and optimization of the heterostructure laser design.

Even greater benefits have been predicted for lasers with quantum dot active layers. Arakawa and Sakaki (1982) predicted in the early 1980s that quantum dot lasers should exhibit performance that is less temperature-dependent than existing semiconductor lasers, and that will in particular not degrade at elevated temperatures.

Other benefits of quantum dot active layers include further reduction in threshold currents and an increase in differential gain-that is, more efficient laser operation. Some of the key concepts in the laser operation. Stimulated recombination of electron-hole pairs takes place in the GaAs quantum well region, where the confinement of carriers and of the optical mode enhance the interaction between carriers and radiation. In particular, note the change in the electronic density of states, as a function of the "dimensionality" of the active layer. The population inversion (creation of electrons and holes) necessary for lasing occurs more efficiently as the active layer material is scaled down from bulk (3-dimensional) to quantum dots (0-dimensional). However, the advantages in operation depend not only on the absolute size of the nanostructures in the active region, but also on the uniformity of size. A broad distribution of sizes "smears" the density of states, producing behaviour similar to that of bulk material.

Thus, the challenge in realizing quantum dot lasers with operation superior to that shown by quantum well lasers is that of forming high quality, uniform quantum dots in the active layer. Initially, the most widely followed approach to forming quantum dots was through electron beam lithography of suitably small featured patterns (~300 Å) and subsequent dry-etch transfer of dots into the substrate material. The problem that plagued these quantum dot arrays was their exceedingly low optical efficiency: high surface-to-volume ratios of these nanostructures and associated high surface recombination rates, together with damage introduced during the fabrication itself, precluded the successful formation of a quantum dot laser.

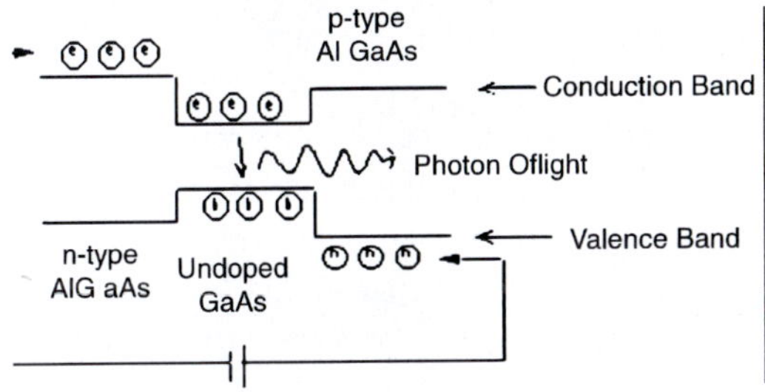

Figure : Schematic of a Semiconductor Laser.

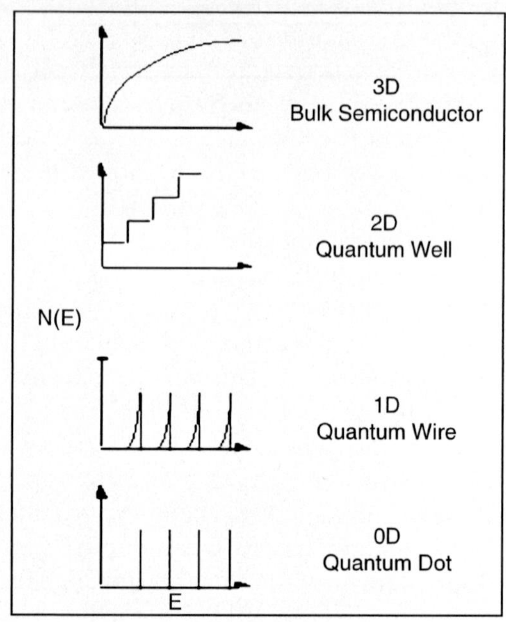

Figure: Density of Electronic States as a Function of Structure Size.

With the demonstration of the high optical efficiency self-assembled formation of quantum dots, formed without need of external processing and having the natural overgrowth of cladding material (which addressed issues of surface recombination), there ensued a marked increase in quantum dot laser research. The first demonstration of a quantum dot laser with high threshold density was reported by Ledentsov and colleagues in 1994. Bimberg *et al.* (1996) achieved improved operation by increasing the density of the quantum dot structures, stacking successive, strain-aligned rows of quantum dots and therefore achieving vertical as well as lateral coupling of the quantum dots. In addition to utilizing their quantum size effects in edge-emitting lasers, self-assembled quantum dots have also been incorporated within vertical cavity surface-emitting lasers.

As with the demonstration of the advantages of the quantum well laser that preceded it, the full promise of the quantum dot laser must await advances in the understanding of the materials growth and optimization of the laser structure. Although the self-assembled dots have provided an enormous stimulus to work in this field, there remain a number of critical issues involving their growth and formation: greater uniformity of size, controllable achievement of higher quantum dot density, and closer dot-to-dot interaction range will further improve laser performance. Better understanding of carrier confinement dynamics and capture times, and better evaluation of loss mechanisms, will further improve device characteristics. It should be noted that the spatial localization of carriers brought about by the quantum dot confinement may play a role in the "anomalous" optical efficiency of the GaN-based materials, which is exceptional in light of the high concentration of threading dislocations (~ 10^8 - 10^{10} cm^{-2}) that currently plague this material system. The localization imposed by the perhaps natural nanostructure of the GaN materials may make the dislocation largely irrelevant to the purely optical (but not to the electrical) behaviour of the material.

PHOTONIC CRYSTALS

A photonic crystal is a periodic structure of materials with differing refractive indices. Therefore, the optical multilayer films widely used as anti-reflection films *etc.* for the surface of television monitors, glasses and other optical components can be considered to be one dimensional photonic crystals. However, generally the term photonic crystal refers to two dimensional and three dimensional structures.

The pitch of the structure is usually chosen such that it is half the wavelength of the light for which the element is designed for. Typically, photonic crystals used in the visible optical region are designed and fabricated so that the period is about 300nm.

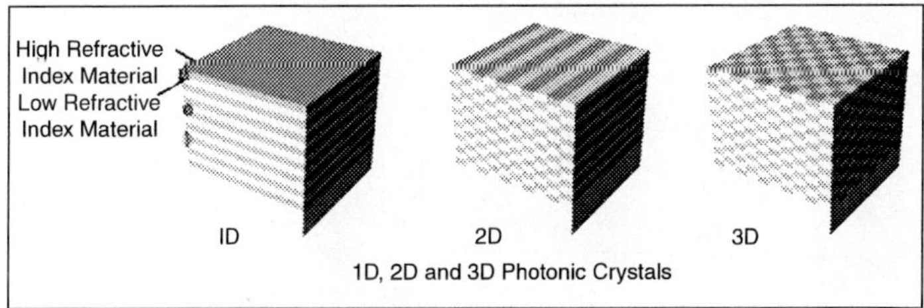

1D, 2D and 3D Photonic Crystals

Like optical multilayer films, photonic crystals are used to control optical transmission, reflection and refraction characteristics. Using 2D and 3D photonic crystals it is possible to control the propagation of light at arbitrary angles of incidence and not only the light normally incident as is the case for conventional optical films.

Further, using photonic crystals, it is possible to achieve optical functionality not possible using conventional optical materials. Examples of such functionality are introduced briefly on the functionality page.

PHOTONIC CRYSTAL STRUCTURES

Two different optical materials (or one optical material and air) periodically arranged as above constitute a photonic crystal. A periodic one dimensional structure can easily be fabricated by sequential deposition of two materials. Though, various structures and fabrication methods have been proposed for photonic crystals of two dimensions and three dimensions, these are considerably more difficult to fabricate.

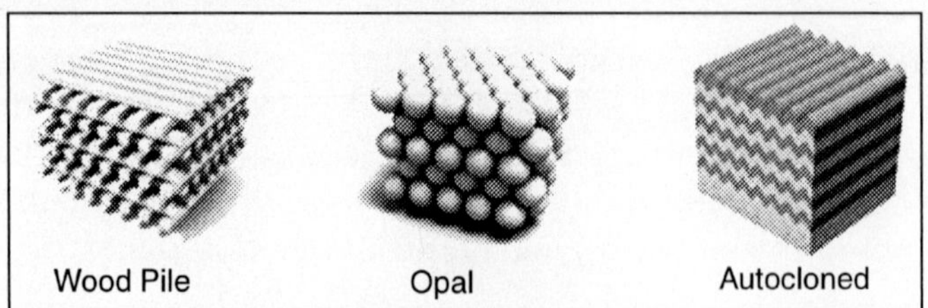

The wood pile photonic crystal was one of the first three dimensional photonic crystals to be proposed. The origin of the name is the manner in which the optical material is piled up to form a support structure similar to that of a well. The fabrication process of piling up the "logs" is complex and the use of semiconductor fabrication processes is necessary. The opal type photonic crystal stacks beads. These techniques generally require a large amount of time, though there are techniques such as self-assembly; where grains of an optical material are placed in a liquid.

This autocloned photonic crystal is composed of a patterned multilayer film, and is the only, multidimensional, photonic crystal that can be fabricated by conventional sputtering deposition, and established mass production technologies.

PHOTONIC CRYSTAL ESSENTIALS

The key to understanding photonic crystals is the dependency of the interaction between light waves and the structure on the period of the structure. The most widely know example of a multilayer optical film is the Bragg reflector.

The Bragg reflector employs the phenomenon of total reflection, which occurs as a result of the intensity of the reflected light increasing at each layer as shown in the figure below.

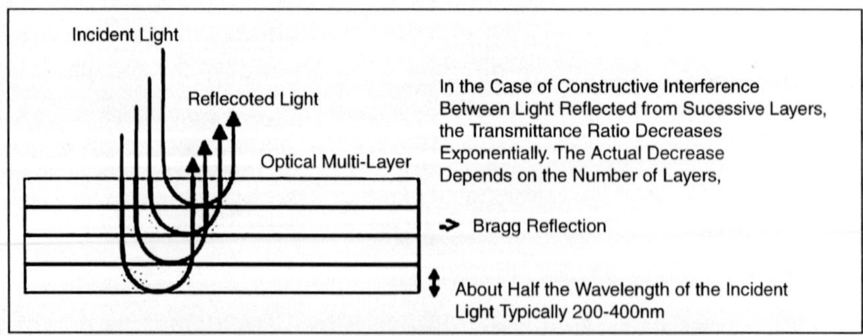

Figure : The Behaviour of Light in an Optical Multi-layer.

Research on the electron transmitted as (an electromagnetic wave) in periodic atomic structures of the atom is established as band theory. A similar model is applicable to the interaction of a photonic crystal and light. Despite the periodicity of photonic crystals being on the order of 100nm, they are referred to as crystals by analogy to molecular structures with a periodicity on the order of angstroms (1 Angstrom = 0.1nm).

The wavelength range over which light is transmitted through the crystal is called the "Photonic Band" and the range over which light is not transmitted the "Photonic Bandgap".

HISTORY OF PHOTONIC CRYSTALS

Although photonic crystals have been studied in one form or another since 1887, the term "photonic crystal" was first used over 100 years later, after Eli Yablonovitch and Sajeev John published two milestone papers on photonic crystals in 1987.

Before 1987, one-dimensional photonic crystals in the form of periodic multi-layers dielectric stacks (such as the Bragg mirror) were studied extensively. Lord Rayleigh started their study in 1887, by showing that such systems have a one-dimensional photonic band-gap, a spectral range of large reflectivity, known as a stop-band.

Today, such structures are used in a diverse range of applications; from reflective coatings to enhancing the efficiency of LEDs to highly reflective mirrors in certain laser cavities. The concept of three-dimensional photonic crystals was then discussed by Ohtaka in 1979, who also developed a formalism for the calculation of the photonic band structure.

However, these ideas did not take off until after the publication of two milestone papers in 1987 by Yablonovitch and John. Both these papers concerned high dimensional periodic optical structures – photonic crystals. Yablonovitch's main motivation was to engineer the photonic density of states, to control the spontaneous emission of materials embedded within the photonic crystal; John's idea was to use photonic crystals to affect the localisation and control of light.

After 1987, the number of research papers concerning photonic crystals began to grow exponentially. However, due to the difficulty of actually fabricating these structures at optical scales, early studies were either theoretical or in the microwave regime, where photonic crystals can be built on the far more readily accessible centimetre scale. By 1991, Yablonovitch had demonstrated the first three-dimensional photonic band-gap in the microwave regime. The structure that Yablonvitch was able to produce involved drilling an array of holes in a transparent material, where the holes of each layer form an inverse diamond structure - today it is known as Yablonovite.

In 1996, Thomas Krauss made the first demonstration of a two-dimensional photonic crystal at optical wavelengths. This opened up the way for photonic crystals to be fabricated in semiconductor materials by borrowing the methods used in the semiconductor industry. Today, such techniques use photonic crystal slabs, which are two dimensional photonic crystals "etched" into slabs of semiconductor; total internal reflection confines light to the slab, and allows photonic crystal effects, such as engineering the photonic dispersion to be used in the slab. Research is underway around the world to use photonic crystal slabs in integrated computer chips, to improve the optical processing of communications both on-chip and between chips.

Although such techniques have yet to mature into commercial applications, two-dimensional photonic crystals have found commercial use in the form of photonic crystal fibres (otherwise known as holey fibres, because of the air holes that run through them). Photonic crystal fibres were first developed by Philip Russell in 1998, and can be designed to possess enhanced properties over (normal) optical fibres.

The study of three-dimensional photonic crystals has proceeded more slowly than their two-dimensional counterparts. This is because of the increased difficulty in fabrication; there was no inheritance of readily applicable techniques from the semiconductor industry for fabricators of three-dimensional photonic crystals to draw on. Attempts have been made, however, to adapt some of the same techniques, and quite advanced examples have been demonstrated, for example in the construction of "woodpile" structures constructed on a planar layer-by-layer basis.

Another strand of research has been to try and construct three-dimensional photonic structures from self-assembly – essentially allowing a mixture of dielectric nano-spheres to settle from solution into three-dimensionally periodic structures possessing photonic band-gaps. The first demonstration of such "inverse opal" structures came in 2000 by researchers at the University of Toronto, Canada. The ever expanding field of biomimetics - the study of structures from nature in order to better understand and use them in design - is also helping researchers in photonic crystals. For example, in 2006 a naturally-occurring photonic crystal was discovered in scales of a Brazilian beetle.

COMPUTING PHOTONIC BAND STRUCTURE

The photonic band gap (PBG) is essentially the gap between the air-line and the dielectric-line in the dispersion relation of the PBG system. To design photonic crystal systems, it is essential to engineer the location and size of the bandgap; this is done by computational modeling using any of the following methods.

- Plane wave expansion method
- Finite element method.
- Finite difference time domain method
- Order-n spectral method
- KKR method
- Bloch wave – MoM method.

Essentially these methods solve for the frequencies (normal models) of the photonic crystal for each value of the propagation direction given by the wave vector, or *vice-versa*. The various lines in the band structure, correspond to the different cases of n, the band index. For an introduction to photonic band structure.

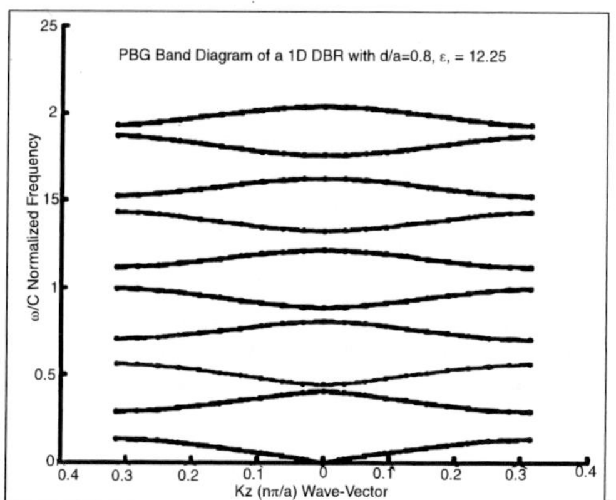

Figure: Band Structure of a 1D Photonic Crystal, DBR air-core Calculated using Plane Wave Expansion Technique with 101 Planewaves, for d/a=0.8, and Dielectric Contrast of 12.250.

The plane wave expansion method, can be used to calculate the band structure using an eigen formulation of the Maxwell's equations, and thus solving for the eigen frequencies for each of the propagation directions, of the wave vectors.

It directly solves for the dispersion diagram. Electric field strength values can also be calculated over the spatial domain of the problem using the eigen vectors of the same problem. For the picture shown to the right, corresponds to the band-structure of a 1D DBR with air-core interleaved with a dielectric material of relative permittivity 12.25, and a lattice period to air-core thickness ratio (d/a) of 0.8, is solved using 101 planewaves over the first irreducible Brillouin zone.

In order to speed up the calculation of the frequency band structure, the Reduced Bloch Mode Expansion (RBME) method can be used. The RBME method applies "on top" of any of the primary expansion methods. For large unit cell models, the RBME method can reduce the time for computing the band structure by up to two orders of magnitude.

OPTICAL SENSORS

Proximity Sensors

Proximity Sensors are designed for use in detecting the presence of an object or motion detection in various industrial, mobile, electronic appliances and retail automations. For example, optical proximity sensors detect the presence or absence of an object like paper in a printer to detect an out-of-paper condition, the mobile phone next to your face which will dim the display and save battery life.

Ambient Light Sensors

Ambient Light Sensors provide precise light detection for a wide range of ambient brightness and are commonly used in LCD Backlight Control in mobile phones, LCD TV/Panel and Notebook applications.

This series of products are part of the Sensor Solution. Ambient Light sensors detect the amount of light available and signal as to whether or not backlighting is necessary and also control the turning on and off of the lighting system.

What is an Optical Sensor?

An optical sensor is a device that converts light rays into electronic signals. Similar to a photoresistor, it measures the physical quantity of light and translates it into a form read by the instrument.

Usually, the optical sensor is part of a larger system integrating a measuring device, a source of light and the sensor itself. This is generally connected to an electrical trigger, which reacts to a change in the signal within the light sensor.

One of the features of an optical sensor is its ability to measure the changes from one or more light beams. This change is most often based around alterations to the intensity of the light.

When a phase change occurs, the light sensor acts as a photoelectric trigger, either increasing or decreasing the electrical output, depending on the type of sensor.

Optical sensors can work either on the single point method or through a distribution of points. With the single point method, a sole phase change is needed to activate the sensor. In terms of the distribution concept, the sensor is reactive along a long series of sensors or single fibre-optic array.

Other features of optical sensors include the distinction of whether it is placed internally or externally in a device.

External transducers register and transmit the necessary quantity of light. These are known as extrinsic sensors. Intrinsic sensors are those that are embedded within a optical fibre or device.

These are generally used to measure smaller changes such as a bend or slight change in direction. The major importance to the proper use of an optical sensor is that it retains certain facets of measured properties. It must always remain sensitive to the property. To the same point, it must be insensitive to any other property. In addition, it cannot influence what measurement is normally being made.

That is, it cannot alter the amount of light impacting the photoelectric property. Optical sensors have a variety of uses. They can be found in everything from computers to motion detectors. For example, when the door to a completely darkened area such as the inside of a copy machine is opened, light impacts the sensor, causing an increase in electrical productivity. This will trigger an electric response and stop the machine for safety. Due to the nature of photoelectric sensors, the registration head of the device must remain clean at all times.

Things like dust and materials can prevent the proper reception of light, limiting the success of the sensor to perform its job. Without the proper level of light, a photosensitive device cannot create or limit adequate amounts of electricity.

REFERENCES

- Bose, G.K. : *A-Z Optics*, Centrum Press, Delhi, 2009.
- Ghatak, Ajoy and K. Thyagarajan: *Fiber Optics and Lasers : The Two Revolutions*, Macmillan Publishers India, Delhi, 2006.
- Ghosh, Krishnapada and Anandamoy Manna: *Textbook of Physical Optics*, Macmillan, Delhi, 2007.
- Hall, J Storrs : *Nanofuture : What's Next for Nanotechnology*, Manas, Delhi, 2006.
- Jha, D.K. : *Contemporary Optics*, Discovery Publishing House, Delhi, 2009.
- Joseph, Subita: *The Illustrated Encyclopaedia of Nano-Photonics*, Dominant Pub., Delhi, 2008.
- Suri, Shalini : *Nano Technology : Basic Science to Emerging Technology*, APH, Delhi, 2006.
- Tuchin, Valery: *Tissue Optics: Light Scattering Methods and Instruments for Medical Diagnosis*, New Age International, Delhi, 2010.

5

Nanoscale Optical Microscopy

Far-field microscopy, scanning confocal optical microscopy in particular, is discussed because the size of the focal spot routinely reaches the diffraction limit. Many of the experimental concepts that are used in confocal microscopy have naturally been transferred to near-field optical microscopy. In a near-field optical microscope a nanoscale optical probe is raster scanned across a surface much similar as in AFM or STM.

We will see that there is a variety of possible experimental realizations for scanning near-field optical microscopy while *e.g.* for AFM and STM a (more or less) unique setup exists. The main difference between AFM/STM and near-field optical microscopy is the fact that in the latter in general an optical near-field has to be created at the sample or at the probe apex in the first place. This is done by external illumination either in the far field or in the near field. A similar situation exists for the detection of a near-field signal. It can either be detected directly in the near-field zone or it may be converted to propagating waves which can then be detected in the far field. The resulting combinations of near-field/far-field illumination and near-field/far-field detection define the possible experimental realizations of nanoscale optical microscopies.

NONLINEAR EXCITATION AND SATURATION

The possibility that a transition in a quantum system could be achieved by simultaneous absorption of two or more photons was first investigated theoretically by Maria Goeppert-Mayer in 1931. The phenomenon could only be demonstrated experimentally in 1961 after the invention of the laser which provided the necessary high photon densities. Today, with the availability of fs-pulsed lasers two- and multi-photon excitation is a standard tool in high-resolution confocal microscopy. Chromophores with transitions in the blue and green can be excited by using infrared light. At the same time, multi-photon microscopy leads to improved and simplified optical sectioning capabilities since excitation only takes place at the regions of highest intensity, *i.e.* in a tight focus, which makes the technique an indispensable tool, not only in biology, for studying the three-dimensional morphology of samples.

NANOSCALE RESOLUTION IN LIGHT MICROSCOPY

In the early 1870s, the German physicist Ernst Karl Abbé formulated a rigorous criterion for being able to resolve two objects in a light microscope. According to his equation, the best resolution achievable with visible light is about 200 nanometers. This theoretical resolution limit of conventional optical imaging methodology was the primary factor motivating the development of recent higher-resolution scanning probe techniques. The interaction of light with an object results in the generation of what is called 'near-field' and 'far-field' light components.

The far-field light propagates through space in an unconfined manner and is the visible light utilized in conventional light microscopy. The near-field (or evanescent) light consists of a nonpropagating field that exists near the surface of an object at distances less than a single wavelength of light. So called near-field microscopy beats light's diffraction limit by moving the source very close to the subject to be imaged. When the first theoretical work on a new technique called "scanning near-field optical microscopy" (SNOM or NSOM) appeared in the 1980's, Abbé's classical diffraction limit was overcome, and resolution even down to single molecule level became feasible. However, light microscopy is still the only way to observe the interior of whole, or even living, cells.

The use of fluorescent dyes makes it possible to selectively obtain images of individual cell components, for example, proteins. Today, the wavelength dogma has been overcome with the development of the stimulated emission depletion (STED) microscope. Now, the German team that developed STED is reporting layer-by-layer light microscopic nanoscale images of cells and without having to prepare thin sections with a technique called *optical 3D far-field microscopy*.

They use a chemical marker for fluorescence nanoscopy that relies on single-molecule photoswitching. "Well thought-out modifications of the fluorescence process have opened a back-door for circumventing Abbé's paradigm" explains Dr. Stefan Hell, head of the nanobiophotonics department at the Max Planck Institute for Biophysical Chemistry in Göttingen, Germany. "In a scanning microscope, improving the resolution is equivalent to decreasing the spatial extent of the fluorescent spot that is scanned through the sample. Hence, we are scrutinizing physical phenomena that bear the potential of cutting down the size of the fluorescence spot. The focal spot is termed 'the point-spread-function'. But non-scanning microscopes have Point-Spread-Functions, too. In fact virtually all of the physical ingredients that we use in a scanning microscope can be employed in a non-scanning CCD camera based system as well, with some modifications and advantages and disadvantages." Hell explains that the states of the fluorescent marker was not just used for generating the signal, but also for breaking the diffraction barrier.

"In fact, all the methods that have successfully outperformed diffraction have so far relied on selected pairs of molecular states – specifically, a 'bright' one to generate the signal and a 'dark' one to ensure that the measured signal stems from a subdiffraction-sized region." "An alternative way of using molecular photoswitching to break the diffraction barrier is to stochastically switch on, read out the fluorescence, and switch off isolated marker molecules such that simultaneously emitting ('on') markers are further apart than the minimal distance resolved by the microscope" says Hell. "In this case, the spatial confinement of the fluorescence is down to the size of a single molecule by definition." Hell's group has now been demonstrating the power of this concept. They report the use of molecules that are not only transferred but can be 'switched' from fluorescent to non-fluorescent and back.

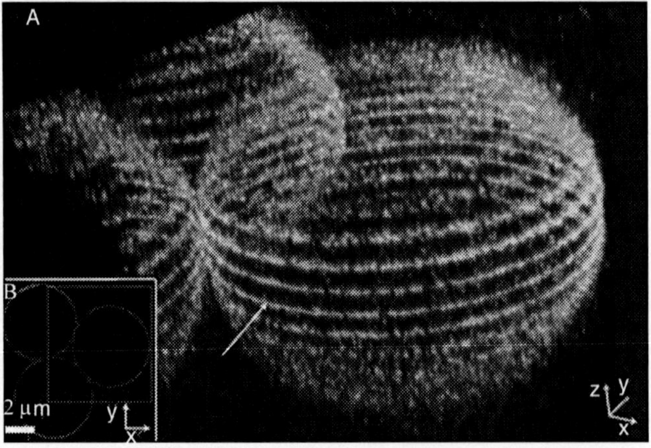

Figure: (A, B) Imaging of 5-mm Silica Beads Surface Stained with 5-NHSS. A) 3D Reconstruction from 17 Slices in the z direction; B) Equatorial Slice (arrow in (A)). For Clarity, a Smaller Area was Selected for the 3D Reconstruction (dotted-line box in (B)). (Reprinted with Permission from Wiley-VCH Verlag) In Contrast to STED, only Separate, Isolated Marker Molecules are Randomly Switched on at the Same Time. Their Fluorescence is Registered, and then they get Switched off again Automatically. In this way, the Simultaneously Fluorescing (Switched on) Markers are Farther Apart from each other than the Minimum Distance that the Microscope can Resolve. This is only Possible using Switchable Molecules that Emit many Photons, one after the other, when Switched on.

If these photons are captured with a camera, the centres of the individual fluorescing dots can be distinguished. After the exposure, the molecule becomes dark again (switches off), allowing further, neighbouring molecules to be photographed. This process is repeated many times, until many dots become a picture. The full distribution can be reconstructed – at a resolution not limited by the wavelength of light.

The researchers have now found a class of substances that fulfill all the requirements of this technique: rhodamine amides. At the core of these molecules lies a system of five rings. In this form, the compound is colourless and does not fluoresce. Irradiation with light induces an isomerization in which one of the rings is opened. This form of the molecule is red and can be excited several times. Most importantly: rhodamine amides can be switched on by either a UV photon or two photons in the red part of the spectrum. This two-photon excitation can be focused onto a thin plane, which allows biological samples to be photographed layer by layer. The individual images can then be reconstructed into a single multilayer image. The resolution reached in the focal plane is far beyond the diffraction barrier (10–30 nm). Hell has also published a comprehensive review of the state of far-field optical nanoscopy.

In it he concludes: "Along with improved fluorescence schemes, nonfluorescent schemes should enable us to take another step in the not-too-distant future: nanoscale 3D imaging at high speed. In any case, the works reviewed herein already broke the barrier of perception of what a lens-based light microscope is able to accomplish. With human perseverance focused on this matter, 3D imaging of live cells with electron microscopy resolution should be possible."

NEAR-FIELD MICROSCOPY OF SHG

Near-field SHG microscope can be exploited in both apertured and apertureless realizations. We will consider these two techniques in turn.

Aperture-Based SHG SNOM

There are three possible experimental configurations for near-field secondharmonic measurements with an apertured probe: (1) the secondharmonic (SH) collection mode when the excitation is achieved by direct illumination of a sample and SH light is collected with a fibre tip, (2) the SH excitation mode when a surface is illuminated with fundamental light passing through a fibre tip and SH light is collected in the far field, and (3) the double-pass mode when SH excitation is achieved through a fibre tip and second-harmonic light is collected in the near-field with the same tip. In addition to subwavelength resolution, SHG SNOM provides additional parameters for local optical characterization such as the polarization of fundamental and SH light. Thus, the described techniques allow one to excite SHG locally at a specifically chosen surface area (excitation mode) or to detect local SHG before the SH contributions from individual surface features are averaged in the far-field zone (collection mode).

Collection mode SHG SNOM is most often used because SH generation requires high-intensity excitation light, which is achieved by a focused laser beam. In most experiments, femtosecond Ti:sapphire lasers are used, which provide high excitation power at a low pulse energy, thus effectively generating SHG but introducing low heating and sample damage. The SH signal collected by a fibre tip is usually filtered by an interference filter and detected by a gated photon counter. Shear-force distance regulation is conventionally used to keep a tip in a near-field proximity and at a constant distance from the surface, thus providing simultaneous topographic imaging.

In most SHG SNOM experiments, uncoated fibre tips are used. The biggest disadvantage of metal-coated fibre tips in nonlinear optical measurements is a strong perturbation introduced by the tip in the local electromagnetic field distribution and SH generation from the metal coating. Uncoated fibre tips introduce a much smaller perturbation since the dielectric constant of a fibre is much less than that of a metal. However, even for fibre (glass) tips, some degree of the electromagnetic field enhancement is associated with a tip apex.

This is crucial in nearfield second-harmonic measurements for which the signal depends on the fourth power of the local excitation field. Due to the field enhancement, the surface area just below the tip dominates the detected SH signal. As a result, theoretically, the resolution of the SHG SNOM can be estimated to be as small as nm for 50 nm radius of a fibre tip apex. In fact, more severe limitations on the resolution of a SHG SNOM are imposed by the small number of SH photons collected by the tip and the related shot noise and thermal drifts due to long signal acquisition times.

The high repetition rate of pulsed Ti:sapphire lasers allows the image acquisition time to be kept in a reasonable range even at a low number of detected SH photons per pulse. The characteristic SH photon count rate in SNOM measurements is, on average, of the order of one SH photon count per 100–300 laser pulses. However, a long signal averaging time, which would help to register more photons, is not desirable due to possible drifts in the experimental apparatus. For that reason, the resolution of the SHG SNOM is in many cases limited by a low detected SH signal rather than the geometry of a probe tip and is determined by a trade-off in the choice of SH signal averaging time.

In the SHG measurements with uncoated fibre tips, the experimentally achievable spatial resolution is usually about 80–100 nm; however, it can be much higher in some specific cases. The resolution and perturbation introduced by a fibre tip depends on the sharpness and size of the taper via the tip-induced field enhancement. It is possible to reduce the influence of the tip by choosing an appropriate tip apex shape. With less sharp tips, the influence of the tip-induced effects will be smaller and SHG will be directly related to the surface. Although the theoretical resolution in this case will be smaller than that for sharper tips due to a smaller electromagnetic field enhancement, the practical resolution will remain about the same (~100 nm) due to a better overall collection efficiency of such tips. Thus, the choice of the tip depends on the task at hand. For example, for studies of SHG from metal surfaces, perturbation introduced by a tip should be as small as possible. On the other hand, for local crystal analysis with SHG, the field enhancement at a probe tip that depends on the excitation light polarization can be advantageous. In the case of polarization measurements with uncoated fibre tips, SNOM resolution down to 40 nm was reported, which is related to the strong field enhancement in the vicinity of high refractive index materials.

Apertureless SHG SNOM

Another type of second-harmonic SNOM makes total use of the effect of the field enhancement at a metallic tip to achieve a local enhancement of nonlinear optical processes in the vicinity of the tip, similar to apertureless scanning near-field optical microscopy (ASNOM) schemes based on two-photon excited fluorescence

and Raman scattering. It provides much stronger light confinement and better spatial resolution than an aperture-based SHG SNOM. This technique, however, poses the problem of image interpretation.

In SHG ASNOM, the excitation light illuminates the tip-surface junction and the scattered SH light is detected in the far field. The signal related to nonlinear optical processes is determined by the superlinear dependencies on the intensity of the driving field (the fourth power of the local fundamental field in the case of SHG). This makes such techniques more sensitive to the field distribution over a surface than ASNOM based on linear optical processes. Not only the magnitude of the signal but also the spatial resolution of ASNOM is related to the field enhancement since it determines the effective surface area most strongly contributing to the measured signal.

As discussed by Kawata, Xu, and Denk, in contrast to the scattering or conventional fluorescence, the light associated with nonlinear processes can be confined (falling faster than with a distance from a tip apex) even without localized surface plasmon excitation. An approach based on SHG is advantageous compared to other ASNOM schemes because there are almost no restrictions on tip and sample materials since all materials exhibit surface SHG. Only one tip or sample needs to posses strong SHG to provide second-harmonic ASNOM imaging.

In different experimental realizations, the technique of apertureless second-harmonic SNOM allows one to probe the local second-harmonic field generated at the surface if the nonlinear susceptibility of the sample is higher than that of the probe tip (a nonlinear background-free analogue of ASNOM based on local scattering). In the opposite case, one can achieve a strongly confined second-harmonic light source at the tip apex with properties dependent on the topographic and dielectric properties of the surface under investigation.

RESOLUTION LIMIT IN LIGHT MICROSCOPY

What is Resolution?

The term 'resolution' has to be discriminated from 'sensitivity', 'sampling' and 'precision'. Subsequently we will try to clarify the meaning of these terms.

One can have a very sensitive light microscopic system which makes it possible to see single virus particles or even single fluorescent molecules. This only means we would have single molecule sensitivity, but the size of such a molecule in the image could, for example, still correspond to 0.5μm in the sample coordinate system, which would be a relatively poor optical resolution.

Another common confusion is the difference between resolution and sampling, which relates to magnification. It is easy to magnify an image of the sample by optical means to any desired degree, however, the process of magnification does not increase the optical resolution; at best it preserves it. When an image of the sample is detected by our eyes or for example a CCD camera, it gets 'sampled' into a discrete set of measured intensity values, one for each detector element (such as a photosensitive cell in the brain). These sampling points correspond to nominal positions in the coordinate system of the object under investigation, without a limit to their density. The magnification has to be adjusted such that the finest level of detail present in the image is still measured (sampled) by at least two such detector elements, but any denser sampling will not yield new information about the sample. Magnification significantly beyond this limit is sometimes called 'empty magnification'. When there is a noise attributable to the readout process of the detector ('read noise'), such empty magnification should be avoided as it deteriorates the image quality.

In many cases there can be very specific questions in biology which are to be answered by microscopic imaging. Such questions could be: 'What is the spatial distance between two specific genetic loci in the nucleus or between two molecules on the cell surface?' To answer these questions, there is no need for a resolution in the order of this distance, but the error of localization (the reciprocal of the localization precision) needs to be below this expected distance. That means that the distance between the estimated and the true centre position of the object has to be below the distance between the two objects. Localization precision can be far higher than the optical resolution (*e.g.* the localization error of single molecules can be smaller than 10nm on a system of 200nm optical resolution), and we can use this even for relative position determination between multiple loci, as long as we are able to discriminate between the target genes (or other small structures) in our image. Such discrimination can be achieved by using multiple colours, differences in fluorescence lifetime, photo-bleaching, the individuality of statistical blinking events and photoactivation. The higher the resolution the better the localization precision, but how much better than the optical resolution depends strongly (with a square-root dependence) on the number of photons collected from each target. An impressive application of the localization idea was to observe the molecular steps of Myosin V. Another interesting application was to follow the kinetics of the receptor-mediated entry of NTF2 and transport of NTF2 and transporting 1, with and without transport substrate, into the nucleus. The method of speckle microscopy uses precise localization for the tracking of molecular clusters to reveal their temporal behaviour. Nanosizing can be used to address the question of the size of biological particles of known shape.

Even when objects are not point-like structures, the position of features like straight edges (*e.g.* tubulin fibres) or planes (*e.g.* plasma membrane) can often be determined very precisely by microscopy methods. All of these methods are based on localization precision, which should not be confused with the term resolution as discussed subsequently. When we talk about high resolution of an optical instrument in this article, we mean the ability to see a structure at a high level of detail. There are various ways to define the term resolution more precisely which are discussed in the two separate boxed sections. Although, the definitions of resolution mentioned in the boxed section 'The PSF' may be useful in some cases, we would like to use a different limit called the Abbe limit within this article, as this limit has a very direct relationship to which light rays are captured by the objective lens of the microscope.

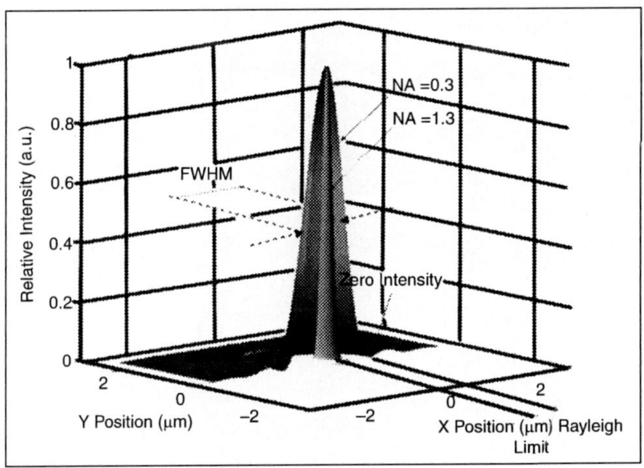

Figure : The Point Spread Function (PSF). Two PSFs are Shown, one for High Numerical Aperture (NA = 1.3 in grey) and one for NA = 0.3 (Slightly Transparent Shades). As can be Seen the Low NA PSF is Wide and has well-defined Positions of Zero Intensity, Leading to the Definition of the Rayleigh Limit.

It is interesting to note that the ability to precisely localize can be turned into a genuine 'resolution' when many closely spaced particles can be discriminated. From the precise localization of these particles a picture can be constructed by painting dots at each localized position. A method termed 'Pointillism'. A very successful way of discriminating closely spaced molecules is using a repetitive cycle of faint photo-activation followed by bleaching of the activated molecules.

In every cycle only very few molecules get activated, so the chance of PSF overlap is small, and the localization thus very precise. In fixed samples it was possible to reconstruct impressive images of cryo-prepared thin section from a COS-7 cell expressing CD63 (lysosomal transmembrane protein) tagged with Kaede and dEosFP-tagged cytochrome-C oxidase import sequence in mitochondria of COS-7 cells down to a resolution of about 20nm. Currently a problem is the rather long imaging time well above an hour for a single slice image, but further development should obviate this.

Detection Abbe Limit

A number of modern microscopy techniques such as confocal, 4Pi and structured illumination microscopy achieve their high resolution by a spatially non-uniform illumination. Non-uniform illumination implies that adjacent positions in the sample can receive different illumination intensity. Subsequently, we will describe in detail how the resolution enhancement is achieved for these methods.

In confocal microscopy the sample is illuminated with a focused beam, which is then raster scanned over it. The emitted fluorescent light is imaged onto a pinhole, whose diameter can be adjusted. Due to the combined effects of illumination and detection, this procedure does indeed enhance the resolution beyond that of a standard widefield fluorescence microscope. If the pinhole is imagined to be very small, the light distribution used for illumination is very similar (except for the usually small difference in wavelength caused by the Stokes shift) to the map of how sensitive the detection system (with a small pinhole) can detect from each position in the sample (called the detection PSF).

For light particles (photons) to make it to the detector a particular sample point needs to be illuminated (*e.g.* 1/10 probability) *and* light emitted from this point needs to be detected (*e.g.* 1/10 detection probability). This chance of illumination and detection (in the above example 1/10 * 1/10 = 1/100) in combination with a linear response of the sample leads to the resulting PSF of the overall system to be the product of the illumination and detection PSFs. Discussed in terms of periodic distributions inside the sample, this confocal system is now in principle able to pick up a 2-fold finer periodicity.

However, in practice this is almost never achieved as the fine details (high spatial frequency) are suppressed so much that they get lost in the noise. To make matters worse, a closed pinhole would not detect any light, which leads to pinhole diameters in the range of the diameter of an Airy disc (where the first theoretical minima occur in the image of a point) to be used in biological experiments.

In this case, even features slightly finer than the standard uniform widefield illumination Abbe limit cannot be resolved as they get lost in the noise. This reduces the in-plane performance of a confocal microscope back to that of a uniform illumination system, however the increased sectioning ability of confocal microscopes is important and lead to their success.

By altering the intensity and phase of the light which gets sent into or detected from the sample under different angles, a technique often referred to as 'apodization' or pupil plane filtering, the high spatial frequencies can be emphasized, getting the practical performance a bit closer to the theoretical limit of twice the widefield resolution with uniform illumination. Nevertheless, to our knowledge, no such system has ever been successful enough to make it into the market.

The I^5M and the 4Pi type C microscope take both the illumination aperture and the detection aperture to their extreme. As the name 4Pi suggests (a full sphere has a solid angle of 4π), the idea is to get as close as possible to illumination (and detection) from all sides of the sample. For practical reasons the best approximation is achieved by combining two objective lenses opposed to another along the Z (optical axis) direction to act as a single lens. To do so, the two optical paths to the sample (through the top and though the bottom objectives) need to be identical in length, both for the illumination and detection wavelengths (4Pi type C microscope). It can be seen that the maximal possible numerical aperture is obtained along the optical axis (Z) to which the two objectives are perpendicular (becoming identical to the refractive index), whereas little changes along the X and Y directions. Thus for illumination the resolution limit is $\lambda_{ex}/(2n)$ and for detection it is $\lambda_{em}/(2n)$, with λ being the respective wavelengths of excitation and emission. This means the finest periodical fluorescent structure which can be detected in a 4Pi (type C with visible excitation) microscope is about $\lambda/(4n)$, which amount to approximately 100nm along the optical axis for glycerol embedding and a wavelength around 500nm.

However, considering the in-plane directions (X and Y) this realization of the 4Pi concept has not increased the numerical aperture (and thus the resolution) as compared to a standard confocal microscope. Noticeably the lateral resolution, albeit in theory improved by a factor of ~2, suffers also from the same problem as standard confocal microscopes, with practically no XY-resolution improvement in comparison to a uniform illumination microscope, detectable for reasonable pinhole sizes. Nevertheless the enormous improvement along the axial direction is worthwhile. Features in only slightly different planes would normally blur together in one image plane, but are now separated nicely in a 4Pi microscope. This leads also to a much cleaner appearance of individual slices of a 4Pi data set. Similar arguments apply to the I^5M method, where incoherent illumination is used.

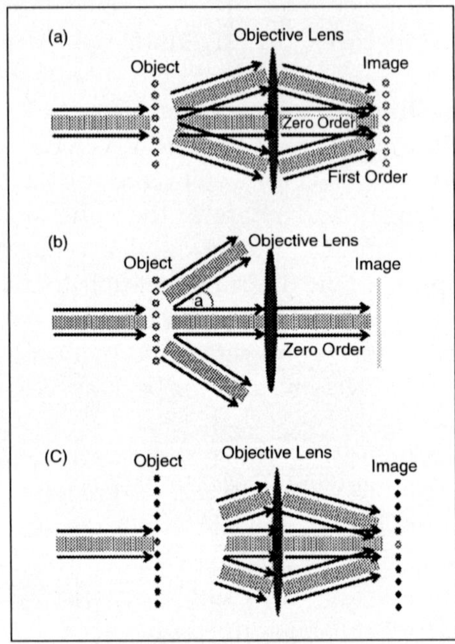

Figure: The Abbe Limit (a) Diffraction of a Parallel Light Beam from a Periodic Grating Structure (only the zero and first diffraction order are shown) (b) Diffracted Light from a Fine Periodical Structure already Fails to Fall within the Angular Acceptance Range of the Objective Lens. The Image can thus only be Formed from the Undiffracted Zero order Yielding a Flat Image of Equal Intensity. (c) For Fluorescence the Emitters have to be Considered Independently and Intensity will be added. Nevertheless, the Finest Resolution (Constituent of the PSF) is still Defined by the Biggest Angle of Emitted Light Rays that were Captured and Refocused.

The methods of structured illumination exploit the fact that the excitation and emission Abbe limit can be added. The highest possible spatial frequency is intentionally introduced during excitation via a fine sinusoidal illumination pattern in the sample volume.

This is achieved by projecting a fine grating into the sample under coherent illumination conditions, which allows the contrast of the excitation structure to be maximal. A high-resolution image can be reconstructed from a series of images under different positions (and orientations) of the excitation pattern.

In comparison to confocal microscopy, structured illumination captures high spatial frequencies of the sample much more efficiently, and thus allows for a true factor of two resolution improvement, albeit at the expense of requiring extensive computational reconstruction.

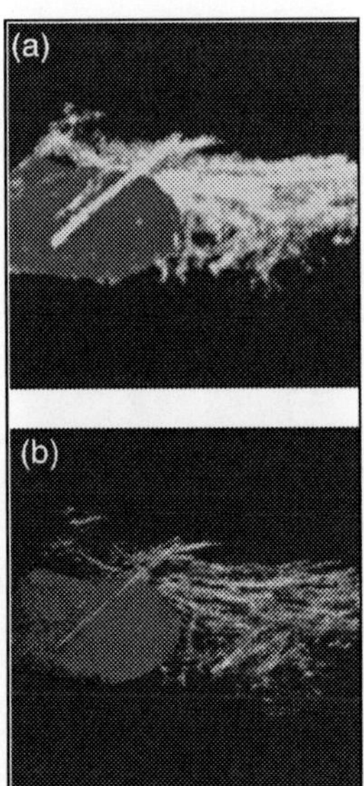

Figure : Confocal and 4Pi Image. Comparison of Volume Rendered Images of the Actin Cytoskeleton and the Nucleus of Muscle Support Tissue Endothelial Cells. The Nucleus was Stained with DAPI and the Actin Cytoskeleton with Alexa488-phalloidin. (a) 2-Photon Confocal View of the Sample (b) 2-photon 4Pi (type A) View (after 3-point Deconvolution) of the Sample. The Improved Resolution and Better Discrimination of Biological Structures is Visible. This Data Set Corresponds to a Width of 23.4 μm, a Height of 2.9 μm and a Depth of 30.0 μm. Sample Prepared by Elisabeth Ehler.

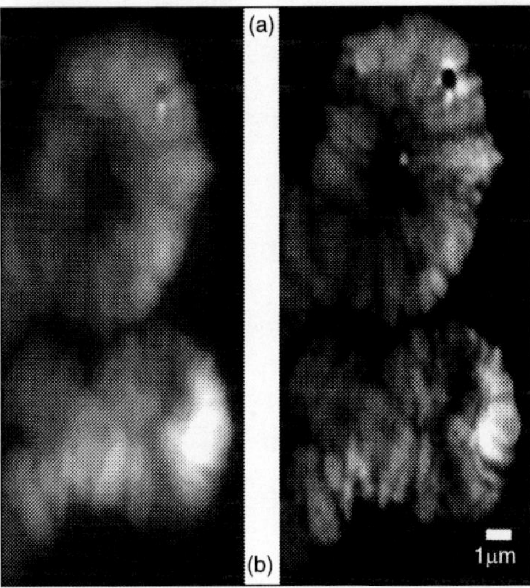

Figure : Linear Structured Illumination of *Drosophila* Polytene Chromosomes Stained with Sytox Green (a) Widefield Image Obtained by Summing all the Individual Images under Structured Illumination. (b) Reconstruction of the Sample from the Structured Illumination Data.

Methods Circumventing the Abbe Limit

The most advanced high-resolution microscopy methods use some form of nonlinearity to avoid the restrictions of the Abbe limit. In the following section, the importance of the non-linearity is stressed and exemplified with two-photon microscopy, nonlinear structured illumination and stimulated emission depletion microscopy.

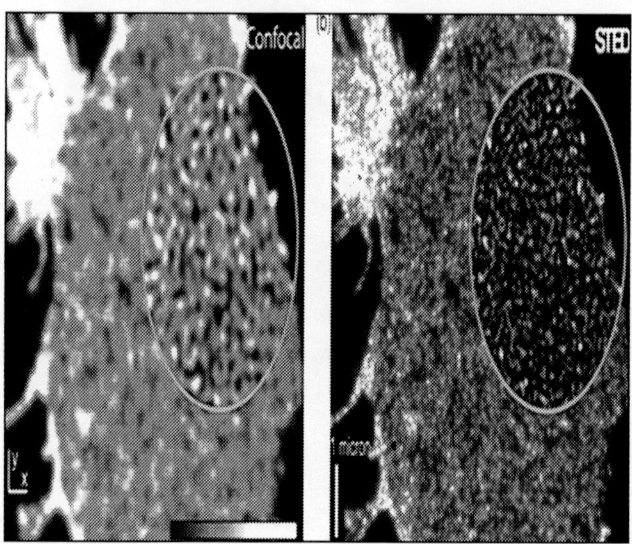

Figure : Stimulated Emission Depletion Image.

As mentioned in the boxed section 'The Abbe Limit', the linear response of the sample led us to define the combined illumination and detection Abbe limit as a fundamental limit in far field microscopy. This limit, however, does not apply if this linear relationship can be violated. Abbe's theory has to be modified if the emitted light can only be described by a nonlinear dependence of emission on the local excitation intensity. As it turns out, the highest exponent present in a polynomial expansion of the nonlinear dependence between excitation and emission defines the resolution limit. Every exponent in such an expansion of the nonlinear dependence of emission on local excitation can be thought of a number of additional multiplications, each adding another excitation Abbe limit to the dependence.

The simplest non-linear dependence is encountered in two-photon microscopy. Due to the probabilistic nature of the two-photon absorption process (both photons need to interact at the same time) the chance for a fluorescent molecule to be in the excited state is proportional to the square of the incident intensity at a specific position in the sample.

Thus also the emitted light intensity is proportional to this square of the excitation intensity. As a result two-photon microscopy indeed circumvents the Abbe limit by a factor of two in the excitation pathway. However, this improvement is bought at the expense of a roughly 2-fold longer excitation wavelength leading to essentially no additional gain, when compared with a standard single photon excitation in a confocal microscope. However, the significant advantage of the two-photon microscopy is the reduced absorption and scattering, allowing to image deeper into tissue and the inherent confocality due to the squaring of the excitation intensity, obviating the need for a pinhole on the emission path. Another advantage of the square dependence is the reduction of sidelobe height, which is particularly useful in 4Pi microscopy.

However, other processes can achieve this nonlinear dependency with the help of visible light, and are also not limited to a maximal exponent of two in the polynomial expansion. The simplest case is saturation of the excited state of fluorophores by intense illumination. Practically this can be realised as a microscopy method by combining the nonlinearity with the concept of structured illumination, or by detecting high harmonics in the time domain under temporally sinusoidally modulated illumination intensity in a confocal system.

Also the spatial non-uniformity of a depletion, bleaching or light-induced inactivation process (dynamic saturation optical microscopy,) can be used, which is again a nonlinear saturation effect. All of these methods are still in a developmental stage and some of them (nonlinear structured illumination, dynamic saturation optical microscopy) require a substantial amount of image processing, since the high-resolution information contained in the measured data has to be computationally extracted.

The most successful method so far is stimulated emission depletion (STED) microscopy. The method itself is slightly more complex, but has the advantage of obtaining a directly scanned image without the need of image processing. The idea is to deactivate fluorophores surrounding the centre of the scanning spot by stimulated emission by the help of a second beam (the doughnut-shaped STED beam), with a wavelength being within the emission spectrum of the fluorophore. If this was only done with a relatively dim STED beam this method would add another linear term, leading only to a moderate resolution improvement.

However, driving this STED beam into saturation establishes the required nonlinearity of high order. In the centre of the spot to detect this STED beam is expected to have zero intensity, leaving the fluorophores unchanged in their excitation state. If a little bit of STED beam is present at a location only slightly away from its centre (zero intensity) position, its long enough interaction with the fluorophore pushes the latter into the ground-state. This will force emission of light at the wavelength of the STED beam, which can be specifically blocked by appropriate filters allowing most of the spontaneous fluorescence emission to pass. Also a temporal discrimination is possible. It has recently been demonstrated that STED can be applied to imaging of endogenous markers such as green fluorescent protein (GFP). By allowing the triplet states to relax between successive excitations, the fluorescence yield in STED microscopy could be dramatically improved.

A practical rule of thumb can be established as alteration to Abbe's equation, with the aim to estimate a resolution criterion considering the limited signal to noise ratio available. However, such a rule of thumb should not be taken to be a new Abbe limit, as the Abbe limit sets a fixed limit to linear resolution methods, whereas the suggestion in is an approximation to the width of the directly obtained PSF, which may or may not be a good indicator of the smallest periodical structure that can be discriminated for a given noise level in the system. In principle all of the saturation-based nonlinear methods have infinite resolution, but the high-resolution information becomes obscured by the noise.

A Comment on Near Field Methods and Pendry's Lense

Obviously the distribution of emitting fluorophores in the sample can have any structure, and is not constrained by the Abbe limit. The sample is just as it is with no limit to its periodical constituents, but the light intensity is known to fall off by an inverse square law (in the scalar approximation) with the distance to each emitting molecule. However, such a light distribution within the sample would also not be limited by the Abbe limit. The discrepancy to the light detection discussed so far arises from the fact that in the aforementioned theory, we only considered light waves that are able to propagate (so called far field waves).

However, Maxwell's equations in a homogeneous medium allow also for a different kind of oscillating electro-magnetic fields called evanescent (or near field) waves. These solutions of the Maxwell equations in a homogeneous medium decay exponentially in a direction of space. Even worse; the finer the (diffracting) sample structure causing these evanescent waves is, the quicker their decay over distance. In the optical regime these evanescent waves become usually undetectable after about one micrometer.

Scanning near-field optical microscopy (SNOM) methods place a probe (a tip, a scatterer of particles that can fluoresce or resonantly be excited) close enough to the sample structure to measure these evanescent fields. From the aforementioned it becomes clear that near-field methods are constrained to imaging surfaces.

An additional problem is that at these small distances to the sample the interaction of the tip with the electromagnetic field can change the local field and the interpretation of the images may become difficult. Like with the described nonlinear methods, the Abbe limit does not apply to the near field. Practically SNOM achieves a resolution (expressed as half width of the direct PSF) of down to about 30nm in a watery environment. An interesting perspective is the measurement of tip-enhanced Raman scattering.

Another particular example of a near-field effect is evanescent wave microscopy. Here an exponentially decaying intensity distribution can be generated close to a surface by illumination above the critical angle of total internal reflection (TIRF) of a planar interface between an optically dense medium (*e.g.* glass, $n = 1.518$) and a less dense medium (*e.g.* water, $n = 1.33$). In contrast to SNOM, this evanescent wave just extends along Z and the resolution improvement gained is also just oriented along Z. However, the main advantage of TIRF is a relatively sharp discrimination between (usually membrane and vesicular) sample structures close to the optical interface (depending on the illumination angle closer than about 200nm) and structures further away from this interface. For cells which are adherently grown to this interface, this method is interesting as vesicles in the interior of the living cell can be suppressed, and only structures tethered to or fused with the plasma membrane are present in the image.

Recently the idea of the 'perfect lens' has arisen, which is also based on using non-propagating waves. Pendry realised that a medium of negative refractive index (*e.g.* a thin slab of silver) would lead to exponentially growing waves within this medium. This effect can, in theory, make up for the exponentially decreasing evanescent waves outside the medium such that they theoretically have similar levels at the focus position like in the sample space. This effect has recently been demonstrated experimentally for light at small distances, but it still remains to be shown how much the non-ideal media used for such lenses can re-amplify the 'lost' evanescent waves and to which maximal distance such an ideal lens could work.

DEPARTMENT OF ELECTRONIC AND ELECTRICAL ENGINEERING

The technology of THz near-field imaging experienced significant advances since the first demonstration in the late 1990's: novel near-field probes have been developed and various applications have been demonstrated. THz near-field microscopy is now making an impact on the development of THz technology. High spatial resolution and the possibility of time-domain analysis make this method invaluable for research on THz devices. The near-field method helps understand the underpinning device physics by visualising THz waves with a sub-wavelength spatial resolution, greatly surpassing the diffraction limit. It allows measurements of device characteristics that are difficult or impossible to assess with far-field methods.

We recently applied the near-field microscopy technique for characterisation of THz waveguides. The multimode nature of most THz waveguides makes the time-domain spectroscopy analysis difficult in application to waveguides due to mode interference. The near-field microscopy solves the multimode problem by providing space-time maps of the propagating wave in which the waveguide modes can be distinguished. The visualization of the propagating THz waves showed for the first time the mode structure in the dielectric-lined hollow metallic waveguides. It also showed the impact of the spatial mode profile on the loss and dispersion characteristics. We also applied the near-field method for imaging of THz surface plasmon waves. This type of waves is in the core of many interesting phenomena, such as the extraordinary transmission of sub-wavelength hole arrays and the negative value of the refractive index in metamaterials. The surface waves provide a useful possibility to guide and confine THz waves near metallic surfaces without radiation. The non-radiative nature prevents observation of the surface waves by far-field methods. They however can be detected locally by the near-field probe and their role in operation of THz devices can be studied with the near-field microscope.

NEAR-FIELD MICROSCOPY SYSTEM

Most THz near-field probes that have been developed recently can be used for device research. Each probe type has specific advantages: electro-optic probes provide a remarkable possibility to detect individual components of the electric field; scattering needle probes allow mapping of the THz field with an unprecedented spatial resolution. In our studies, we use an integrated sub-wavelength aperture probes, which provide excellent sensitivity and a wavelength-independent spatial resolution better than 10 microns. This probe has a well-understood frequency response, covering a broad band of frequencies from about 0.2 THz to 3 THz. The near-field probe consists of a GaAs-based photoconductive antenna detector and a small aperture in a metallic screen. A short distance between the aperture and the photoconductive antenna allows detecting evanescent waves that couple through the aperture and therefore improve the detector sensitivity. For near-field imaging and spectroscopy studies, device samples are placed on a sample platform and illuminated by a THz pulses delivered through a cylindrical waveguide.

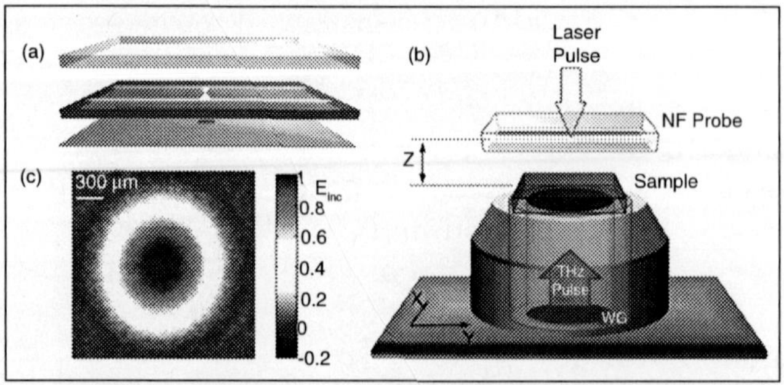

The near-field microscopy system: (a) schematic diagram of the integrated sub-wavelength aperture THz near-field probe; (b) typical arrangement of the near-field probe and the waveguide platform with a sample; (c) distribution of the THz illumination field delivered by a THz waveguide.

Mode Imaging and Dispersion Analysis in The Waveguides

The spatial profile of the dominant mode in a waveguide can give intuitive understanding of the waveguide loss and dispersion characteristics. Experimental studies of the mode profile require the near-field imaging of the waveguide output. In most THz waveguides, the problem of mode characterization however becomes difficult due to mode interference.

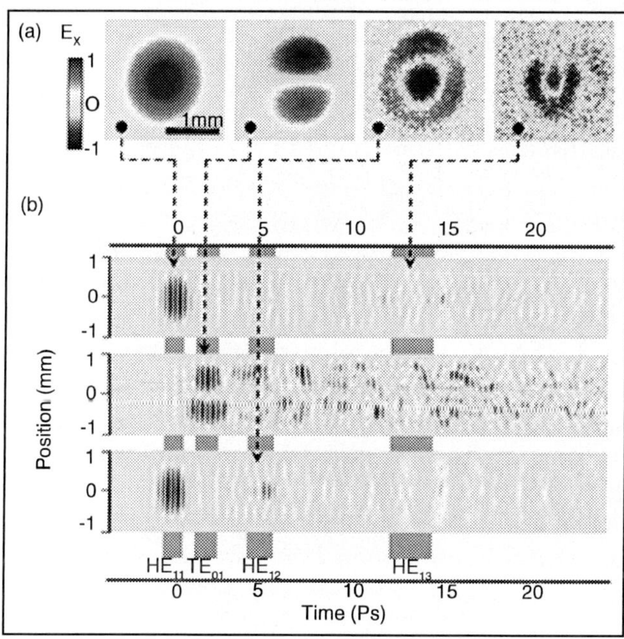

Figure: Mode Mapping in THz Wavesguides: (a) Time-resolved Near-field Spatial Maps Showing the Normal Modes of the Dielectric-lined Cylindrical Metallic Waveguide; (b) Space-time Maps at the Output of a 133.5 mm Long Waveguide for Three Different Coupling Configurations.

The time-resolved capability of pulsed THz spectroscopy systems combined with near-field imaging solves the multimode problem by providing space-time maps of the propagating wave at the waveguide output. The time-resolved imaging allows distinguishing the waveguide normal modes and mapping their spatial profiles because the modes separate in time as they travel through the waveguide. This approach avoids the problem of mode interference occurring in the frequency domain. The spatial resolution of the probe also allows measurements of dispersion and losses for individual modes by positioning the probe in a location where the mode of interest has high amplitude whereas other modes have minimal amplitudes.

The near-field spectroscopy system provided first experimental measurements of dispersion for main modes in the dielectric-lined hollow cylindrical metallic waveguide. In this waveguide, the mode profile explains very low loss (1 dB/m) and relatively low dispersion (6 ps/THz/m).

Surface Plasmon Wave Imaging

It was recently discovered that surface plasmon waves can be detected with the integrated aperture probe. This result is not obvious because the detector antenna in the probe is orthogonal to the polarization of the field. Nevertheless, the surface nature of plasmon waves allows them to couple through the probe aperture and to induce a current in the photoconductive antenna detector.

This probe was employed for mapping a previously undetected phenomenon, a surface wave formed on a metallic surface by a tightly focused THz beam.

The near-field probe can be also applied for studies of surface plasmon resonances in THz devices, such as a THz antenna. The formation of a standing resonant wave in a bow-tie antenna illuminated by a THz pulse. Transient distributions of the field can be detected by the near-field probe as well. Such images provide an insight into the physics of the antenna response.

Our experimental investigations allowed us to deduce the response characteristics of the integrated sub-wavelength aperture probe to the surface plasmon wave (polarized perpendicular to the aperture place): the probe detects a spatial derivative of the surface plasmon field, rather than the field itself. The understanding of the coupling mechanisms now provides a complete framework for analysis and interpretation of images in near-field microscopy.

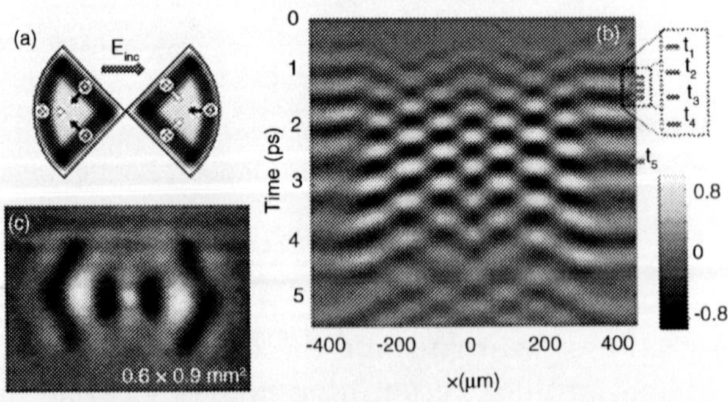

Figure: The Surface Plasmon Waves formed on a Bow-tie Antenna: (a) Schematic Diagram of the Bow-tie Antenna showing the Surface Waves Generated by the incident THz Pulse; (b) a Space-time Map Measured along the Antenna Axis (y = 0); (c) THz Image of a Resonant Surface Wave formed on the Antenna about 2 ps after the Arrival of the THz Pulse.

Further Applications of Near-field Microscopy

Outline a potential impact of THz near-field microscopy on THz device research. In these applications, the underlying physics of device operation is uncovered through the detection of non-radiative components of the THz electric field and through local THz spectroscopy. The combination of high-resolution imaging and the time-resolved measurements allows addressing difficult problems in device characterization.

Presently, many THz near-field approaches provide sufficient spatial resolution for studies of THz devices. The interaction between a near-field probe and an object however requires further investigations in order to determine which near-field probe is more appropriate for a given application. The interaction can be complex, particularly for near-field probes that contain metallic components. For the integrated sub-wavelength aperture probe, however the coupling mechanism yields a simple analytical relationship between the field near the object and the detected field. The understanding of the coupling mechanism allows further applications of this near-field probe for THz device research.

FAR-FIELD PHOTOMETRIC MEASUREMENTS

The concept of "photometric distance" in light source measurements is very important, yet frequently poorly understood. The light output from luminaires is measured using an illuminance (lux) meter and the luminous intensity (in candelas) calculated using the inverse squared rule. By definition, such measurements must be performed in the photometric far-field. Think of it this way: in the far-field, a light source behaves as a point source of light and the illuminance varies with the square of the distance between source and receiver. In other words, the beam from the source is fully formed and the luminous intensity is a constant.

As you move closer to the light source, you transition from the photometric far-field to near-field. In the near-field, the size of the light source starts to have an effect and the measured illuminance (and intensity) also starts to depend upon the size of the detector. In the near-field, the inverse squared rule does not apply and intensity is not a constant. If you were to make an illuminance measurement in the near-field, the calculated luminous intensity would be erroneous.

All of which means that you have to ensure that when measuring the intensity of a luminaire for the purpose of generating standard photometric data, you must be sure to place the photometer in the far-field. Formulae exist that allow you to calculate the photometric distance for a given light source, but as a rule of thumb, you will generally be in the far-field if you place the photometer at a distance from the source some 5-10 times the maximum dimension of the source. Care should be taken with this 10x rule though – some light sources (particularly LED-based solid state lighting) feature complex reflectors and/or lenses that cause the photometric distance to be much larger than expected. The best approach is to establish that you are working in the far-field by making illuminance measurements at different distances and confirming that they follow the inverse squared rule.

NEAR-FIELD PHOTOMETRIC MEASUREMENTS

In recent years, a new type of light meter has been developed called an imaging photometer. Similar to a digital camera, an imaging photometer is based upon a high resolution, wide dynamic range CCD sensor that comprises a two dimensional array of several million detector pixels that can be used to record the spatial luminance or illuminance from a light source. Compared to an ordinary digital still camera, an imaging photometer employs a system of filters which precisely matches the photopic or colourimetric sensitivity of the human eye. An imaging photometer is also calibrated such that the digital image contains absolute photometric and chromaticity values for all points on the device under test.

In a near-field goniophotometric system, an imaging photometer replaces the illuminance meter which greatly reduces the space required for the dark room. The imaging photometer can be placed at any distance from the light source under test, typically at a working distance of between 0.5 and 3m. For a given size of luminaire, the actual working distance of the imaging photometer can be chosen based upon the lens fitted to the photometer. A lens is chosen that causes the object to fill as much of the photometer's field of view as possible which ensures good image resolution. The overall size of the dark room required for a near-field system is typically no more than 4 x 5m. This in itself means that goniophotometric testing of luminaires becomes a practical and economically viable proposition for the majority of lighting manufacturers.

Figure: Prometric Imaging Photometers

A near-field measurement of a light source or luminaire comprises a sequence of high resolution digital images of the device under test recorded at each angle of orientation. For a goniometer step size of 5° in inclination and 15° in azimuth, the resultant near-field data file comprises about 400 images. Each image contains the luminance (in cd/m2) from each spot on the light source. This way, we know how much light the luminaire is emitting per unit area and how the amount of light varies as a function of angle. This sequence of images is saved as a data file which describes the spatial and angular light output of the light source.

It is interesting to note that a near-field goniophotometric measurement can be used to calculate illuminance at any distance from a light source, including those distances that are not in the far-field for that luminaire. Using near-field luminance data, it is possible to model the illuminance from a source using various optical design software programmes. In this way, the effect of reflections from ceilings and walls (which are often in the near-field relative to the luminaire) can be taken into account. This improves the predictive accuracy of the lighting design model compared to that which would be obtained from standard lighting design software in which only far-field illuminance data can be used.

REFERENCES

- Ahmad, Lone: *Nanoscience: Nanotechnologies and Nanophysics,* Jammu Kashmir: South Kashmir Press, 2006.
- Awasthi, K.P. : *Advanced Magnetic Nanostructures,* Cyber Tech Publications, Delhi, 2008.
- Cerofolini, G.F.: *Nanoscale Devices,* Springer-Verlag, Berlin, 2009.
- Geckeler, K.E. : *Biomedical Applications of Nanostructured Materials,* Macmillan Publishers India, Delhi, 2011.
- Jenkins, Francis A. and Harvey E. White: *Fundamentals of Optics,* Tata McGraw Hill, Delhi, 2011.
- Lukas Novotny and Bert Hecht: *Principles of Nano-Optics,* Cambridge University Press, Cambridge, 2006.
- Ranganathan, S.: *Handbook of Nanoparticles,* Chennai: T.T. Ranganathan Clinical research Foundation, 2008.

6

Theoretical Methods in Nano Optics

NANOOPTICS : THE DETERMINATION OF ELECTROMAGNETIC FIELD

A major problem in Nanooptics is the determination of electromagnetic field distributions near nanoscale structures and the associated radiation properties. A solid theoretical understanding of field distributions holds promise for new, optimized designs of near-field optical devices, in particular by exploitation of field-enhancement effects and favourable detection schemes. Calculations of field distributions are also necessary for image reconstruction purposes. Fields near nanoscale structures have often to be reconstructed from experimentally accessible farfield data. However, most commonly the inverse scattering problem cannot be solved in a unique way and calculations of field-distributions are needed to provide prior knowledge about source and scattering objects and to restrict the set of possible solutions.

Analytical solutions of Maxwell's equations provide a good theoretical understanding but can be obtained for simple problems only. Other problems have to be strongly simplified. On the other hand, a pure numerical analysis allows to handle complex problems by discretization of space and time but computational requirements (usually given by cpu time and memory) limit the size of the problem and the accuracy of results is often unknown. The advantage of pure numerical methods, such as the finite-difference time-domain (FDTD) method or the finite-element (FE) method, is the ease of implementation.

We will not review these pure numerical techniques since they are well documented in the literature. Instead we will review two commonly used semi-analytical methods in nanooptics: the multiple-multipole method (MMP) and the volume integral method. The latter exists in different implementations such as the coupled-dipole method, the dipole-dipole approximation, or the method of moments. Both MMP and the volume integral method are semiclassical methods since they render an analytical expansion for the electromagnetic field by numerical means.

FAST MULTIPOLE METHOD

The fast multipole method (FMM) is a mathematical technique that was developed to speed up the calculation of long-ranged forces in the n-body problem. It does this by expanding the system Green's function using a multipole expansion, which allows one to group sources that lie close together and treat them as if they are a single source.

The FMM has also been applied in accelerating the iterative solver in the method of moments (MOM) as applied to computational electromagnetics problems. The FMM was first introduced in this manner by Greengard and Rokhlin and is based on the multipole expansion of the vector Helmholtz equation. By treating the interactions between far-away basis functions using the FMM, the corresponding matrix elements do not need to be explicitly stored, resulting in a significant reduction in required memory. If the FMM is then applied in a hierarchical manner, it can improve the complexity of matrix-vector products in an iterative solver from $O(N^2)$ to $O(N)$. This has expanded the area of applicability of the MOM to far greater problems than were previously possible.

The FMM, introduced by Rokhlin and Greengard, has been acclaimed as one of the top-ten algorithms of the 20th century. The FMM algorithm dramatically reduces the complexity of matrix-vector multiplication involving a certain type of dense matrix which can arise out of many physical systems.

The FMM has also been applied for efficiently treating the Coulomb interaction in Hartree–Fock and density functional theory calculations in quantum chemistry.

OPTICAL MICROCAVITY-MIRROR

The theory of image charges offers a useful method for solving the electric and magnetic field distributions of either a charge or a magnet near the flat surface of a perfect electric conductor (PEC). In particular, the field distribution for a charge is obtained by replacing the conductor by a fictitious charge placed at the mirror image of the real charge, but with opposite sign. For a magnet near a PEC surface, the mirror magnet has a magnetization vector that is just the specular reflection of the magnetization vector produced by the real source. We therefore conclude that the force between a charge and a PEC surface is always attractive, whereas the force between a magnet and a PEC surface is always repulsive. This simple analysis finds application to electrostatic adsorption and magnetic levitation.

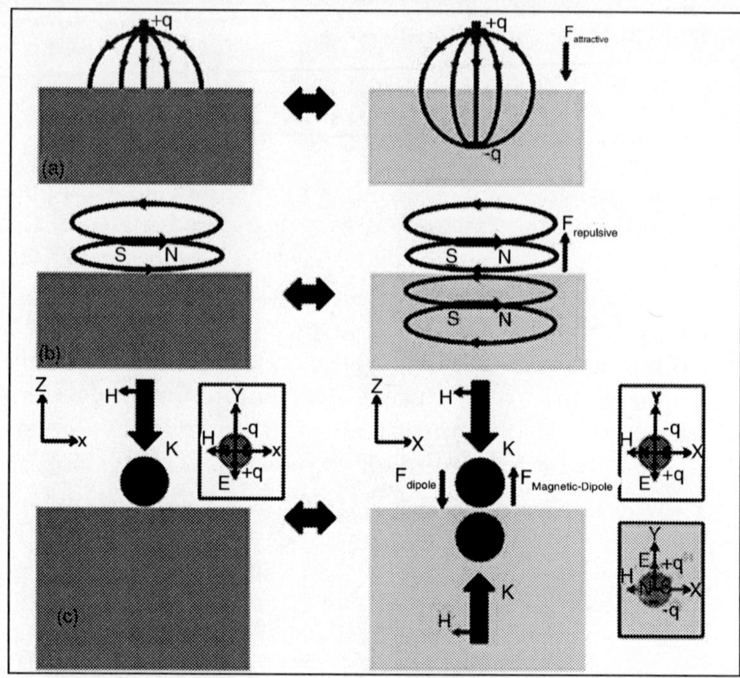

Figure: Schematic View of the Mirror Image Method.

Here, we show that the method of charge and magnet images can be extended for understanding the effect of flat metallic surfaces on the modes of neighbouring dielectric microcavities, and theoretically show the application of these concepts to a new form of optical levitation. We follow a procedure similar to the simulation of electromagnetic (EM) modes in resonant antenna near earth ground. Specifically, we use silicon colloids as high-refractive-index nanometre-sized spherical microcavities because they display well-defined, sharp Mie resonances with huge scattering cross-sections. The microcavity modes can be envisaged as electric and magnetic oscillating multipoles. The presence of a flat metallic surface induces mirror EM multipole images. In particular, the electric-dipole (ED) mode parallel to the surface is the easiest to understand: in an electrostatic picture, the metallic surface induces an imaginary ED with antiparallel dipole moment, and consequently, the microcavity undergoes attraction towards the conductor. In contrast, a parallel magnetic dipole (MD) induces a mirror parallel magnetic dipole, which results in a repulsive photonic force. Generally speaking, multipolar optical modes in microcavities near good conductors induce attractive or repulsive forces depending on the type of resonance that is involved. The net photonic force acting on the microcavity under illumination by a laser impinging perpendicularly to the metal surface emerges from the balance between attractive and repulsive forces acting on it. The photonic force is stronger for modes producing larger scattering cross-section (*e.g.*, modes in cavities with high refractive index such as our silicon microspheres). However, magnetic (TM) and electric (TE) resonances usually are close in wavelength, and consequently, attractive and repulsive forces tend to cancel out, except for low-order modes, which tend to be well separated. More precisely, magnetic-like photonic forces are dominant for low-order TM modes with strong scattering and well apart from TE modes. TM modes also exhibit large magnetic response. Here we demonstrate a dominant magnetic repulsive force over the electrical attractive force in silicon nanospheres when the external light is tuned near the b11 resonance. This repulsive photonic force is larger than other competing forces such as van der Waals (vdW) attraction and Brownian motion.

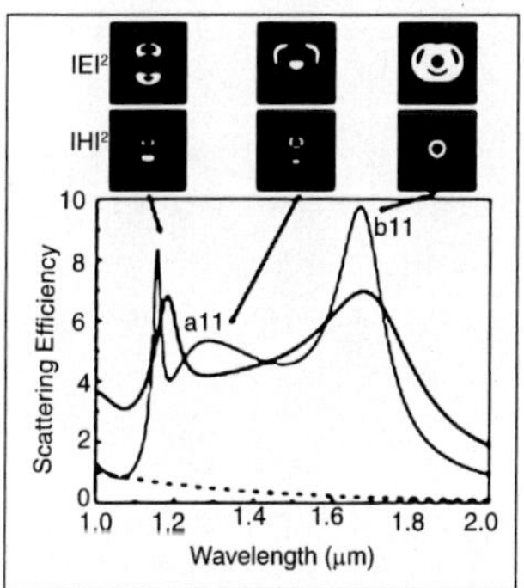

Figure: Scattering Efficiency of a Single Silicon Sphere.

Magnetism at optical frequencies has recently been achieved through elaborate metamaterials designs, and even direct evidence of the resulting magnetic field has been reported using an artificial magnetic atom attached to a probing tip. In this context, our silicon nanospheres constitute a new class of magnetic atoms, capable of amplifying the magnetic component of the EM waves at optical frequencies, which in turns produces repulsive forces near a metal surface.

Photonic forces are extensively used in optical tweezers for manipulating microparticles. Recently, plasmons have been realised to produce intense gradient forces capable of trapping small nanoparticles. In these studies, low-refractive-index dielectrics are used, producing relatively weak scattering. This is in contrast to our high-index nanospheres, from which we derive completely new phenomena under external illumination for wavelengths in resonance with low-order magnetic and electric optical modes near a metallic film.

In this work, we show that EM field distribution of the low-order modes of a spherical optical microcavity near a metallic surface under external laser irradiation is identical to the EM field of a particle dimer in which each particle is illuminated with counter-propagating laser beams, as deduced from the mirror image method. We also report strong repulsive photonic forces four orders of magnitude larger than the particle weight (F = mg = $1.2*10^{-3}$ pN) over a broad range of laser wavelengths. Importantly, the optical force acting on a silicon nanosphere can be tuned from repulsive to attractive and *vice-versa* by selecting either the light wavelength or the gap between the sphere and the metal.

NUMERICAL METHODS

We calculate both EM fields and optical forces through two methods; a) by rigorously solving Maxwell's equations using a highly convergent multiple-scattering method based upon a multipolar expansion of the fields near the sphere and an expansion in plane waves elsewhere, including reflection at the planar metal surface (MESME). This is similar to previous approaches based on the multiple multipole method and specific implementations for aggregates of up to thousands of spherical particles; b) by finite difference time domain simulation (FDTD) calculation using the public software (Lumerical FDTD Solutions). The results from both methods are in full agreement with each other. The dielectric constant of Au is taken from Ref, while we assume a frequency independent refractive index of 3.5 and 1.6 for silicon and polystyrene (PS), respectively. The radius of the silicon sphere is set to 230 nm, so that the fundamental optical modes appear in the near-infrared (NIR) region. The optical force is obtained upon integration of the Maxwell stress tensor, which results in an analytical formula in terms of the multipolar field components near the sphere. A plane EM wave at normal incidence with 10 mW/m^2 is presumed. However, due to the small size of the microcavity only a tiny power fraction (1.6 mW) is directly impinging on the sphere.

Results and Discussion

We start by inspecting the EM field distribution in the microcavity placed near a PEC surface, as shown schematically in the insets. Because of the high refractive index of silicon, a sphere as small as 0.46 m is capable of supporting well-defined low-order Mie resonance modes. The lowest-order Mie mode resonances correspond to transverse magnetic (TM) and transverse electric (TE) modes. We have plotted the EM field distribution for two wavelengths (1434 nm and 1744 nm) at which TM and TE are dominant. The upper plots, show the distribution of Ex, Ez and Hy fields at 1434 nm, where the TM mode b11 is dominant. In particular, Ex shows a typical dipole feature, which demonstrates that the incident light induces an electric dipole resonance. Interestingly, Ez shows strong enhancement in the gap region, which illustrates the strong coupling existing between the sphere and the metal surface. The Hy component indicates that the incident light also induces a very strong magnetic dipole resonance in the silicon sphere. Both modes contribute to the scattering cross-section but compete with each other to produce a net photonic force. The lower plots show the EM field at a wavelength of 1744 nm, for which the TE mode becomes dominant (TE dipole, TED). The EM field distribution at the same wavelengths, but for the case of a dimer consisting of two identical microcavities located symmetrically with respect the (now absent) metal interface and illuminated by counter-propagating laser beams, phase-shifted with respect to each other, as suggested by direct application of the mirror image method. The EM fields in both microcavities at 1434 nm (upper plots) and 1744 nm (lower plots) have a plane of symmetry at the planar surface, while the fields in the upper semi-space are identical. Comparison between the full calculation and the image method for different modes and parameters thus results in identical field distributions. Therefore, the image method can be extended to electrodynamics when microcavities near a perfect conductor are considered.

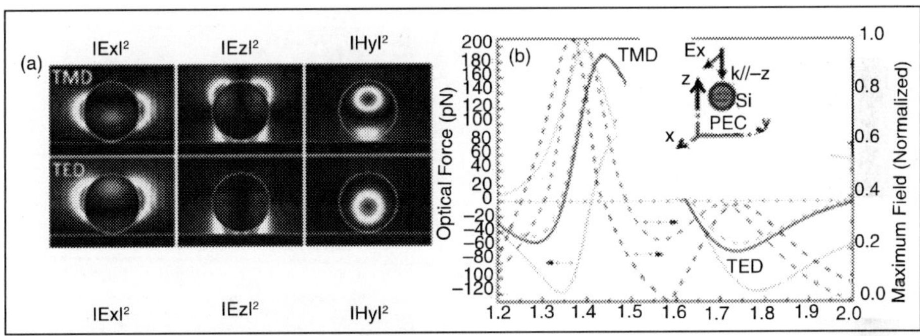

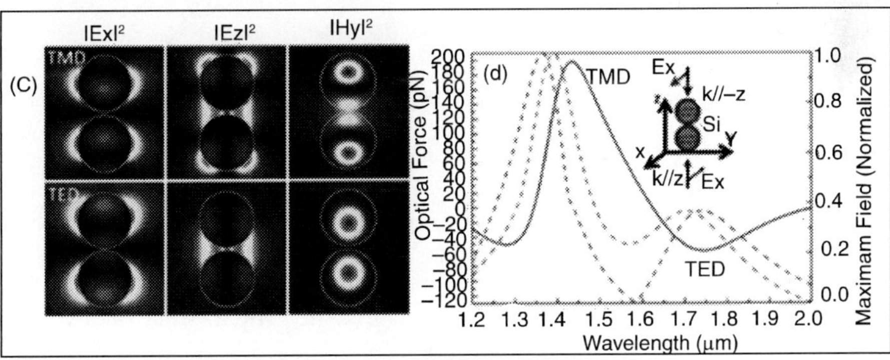

Figure: Optical Force on a Silicon Nanosphere near a PEC Surface.

Now, we examine the resulting optical forces. The black solid curve (FDTD results) and the grey dash curve, the optical force acting on a silicon sphere near a PEC surface (gap equal to 10 nm) as a function of incident wavelength. Positive values correspond to repulsive forces. A strong repulsive optical force opposite to the laser wave vector direction emerges in the region between 1400 nm and 1600 nm. The photonic force acting on two spheres irradiated by counter-propagating EM beams as the mirror image method states. The beams have a phase difference between them. The photonic force acting on the upper sphere is identical to the one calculate. It clearly proves that the strong photonic force of silicon sphere on PEC originates from the strong photonic interaction between silicon sphere and its mirror image. Red and blue dashed curves, the maximum fields Hy (responsible for the repulsive force) and Ez (responsible for the attractive force), indicating that the strong repulsive force dominated by the magnetic resonance. It is very difficult separating the electric and magnetic dipole contributions because strong magnetic resonances induce strong displacement E fields. However, in order to deeper understand the optical process we have separated both the electric and magnetic field component of Maxwell tensor, and we have obtained similar results.

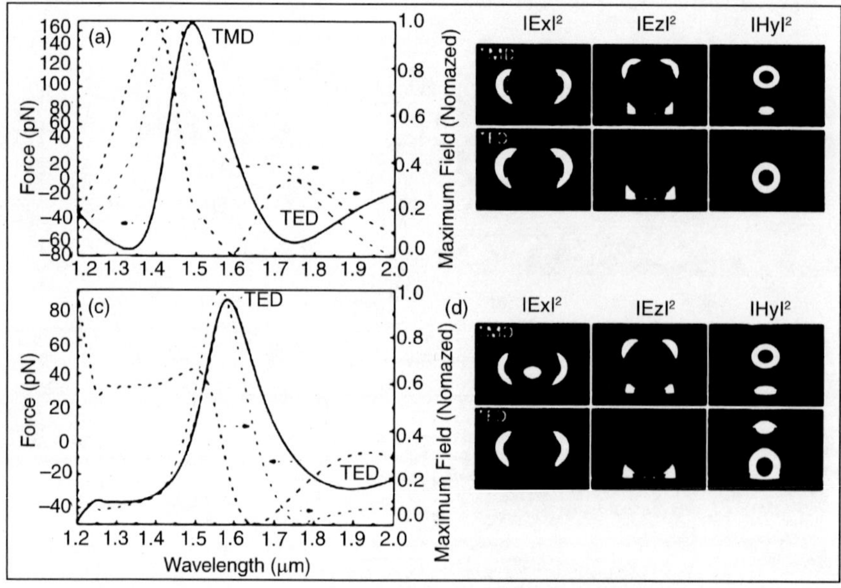

Figure: Optical Force on a Silicon Nanosphere near a Real Metal

A real gold mirror with the microcavity suspended in either vacuum or water. The black solid curve (FDTD results) and the grey dash curve (MESME results) the optical force on the silicon sphere as a function of incident wavelength. Clearly, a strong repulsive optical force appears in the wavelength between 1408 nm and 1624 nm. At 1490 nm, the repulsive force is about five orders of magnitude larger than the weight of the silicon sphere itself. Because most optical tweezers work in liquid phase, a more realistic scenario, in which the metal is gold and the spheres are suspended in water. The results are similar to vacuum or air. The repulsive force maximum appears slightly red shifted because of the high refractive index of water. However, the photonic force with water is about half the value of that appearing with vacuum or air. From the maxima of the Ez and Hy fields, it is again clear that this strong repulsive force dominated by the magnetic resonance. For a low-refractive-index cavity (*e.g.*, a PS sphere) of the same size as the silicon microspheres, Mie modes are much less well defined, and as expected, only very tiny photonic forces are observed (not shown).

Finally, we estimate the degree of detectability of the forces under consideration calculating them as a function of the gap between the sphere and the mirror. The photonic repulsion as a function of gap distance. We use two different wavelength, 1490 nm, and 1580 nm, for vacuum and water medium, respectively.

The vdW force acting on the sphere on surface can be approximated as, $F = Ar/(6s^2)$, where A is the Hamaker constant, r is the sphere radius, and s is the distance between the sphere and the surface. The overall Hamaker constant is approximated as $A \approx (A_{11}$——√—A_{33}——√$)(A_{22}$——√—A_{33}——√), where A_{11}, A_{22} and A_{33} are the individual Hamaker constants for gold, silicon and the vacuum or water medium, respectively. The vdW force is significantly weaker than the optical force (up to 2 orders of magnitude). Down to very small separations ~10 nm, the vdW force is more than twice smaller than the optical repulsive force for a laser intensity of 10 mW/5m^2. Even at perfectly reachable lower intensity ~1 mW/5m^2, photonic repulsion is still stronger than vdW in water case. Additionally, Brownian force is completely negligible at room temperature (10 fN). We deduce that the particle is subject to either repulsive or attractive forces depending on the gap distance. Therefore, particles of 460 nm near a gold mirror and illuminated with an external laser of the appropriate wavelength should undergo oscillatory motion around an equilibrium position of zero total force. With the particle suspended in water, friction force should dampen these oscillations and lead to stable levitation of the particle.

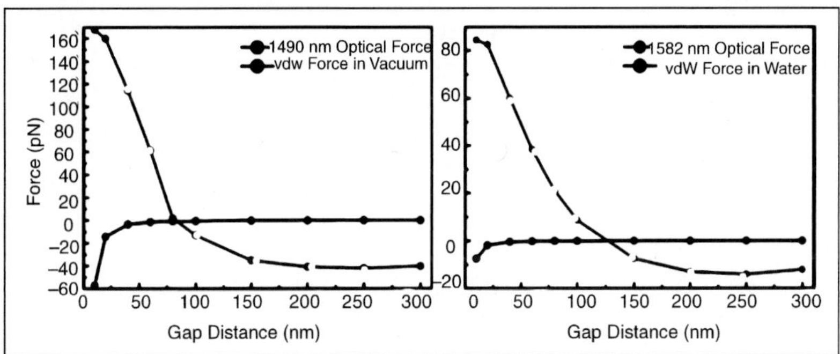

Figure: Optical Force in Different Media as a Function of Sphere-metal Separation and Sphere Dynamic Motion

In conclusion, we have shown that the mirror image method can be extended to deal with the optical modes of a silicon microcavity excited with a laser beam and placed near a good conductor mirror, provided a proper treatment of the image is made by exciting the image cavity with the specularly reflected external light beam. The excited optical modes of the cavity strongly interact with its image beneath the metal surface, which results in attractive or repulsive forces, depending on the symmetry of the cavity modes. The sign of the force is repulsive (attractive) when the resonant mode is TM dominant (TE dominant).

Remarkably, the optical force is detectable because it is much larger than the vdW force or the Brownian force. Our results have important implications for the optical manipulation of nanoparticles, which can then be used as probes to perform subwavelength microscopy. The combination of two lasers tuned to repulsive and attractive modes of the cavity, respectively, should allow exerting full control over the position of the particle by changing the ratio between the intensities of the two lasers. Additionally, the particles can be decorated with biomarkers capable of attaching to specific biomolecules. This suggests a new way of performing molecule-specific biosensing, with the advantage of a high degree of spatial control over the position of the probe. In a separate direction, a trapped particle can be regarded as a switch which interacts with probing light of a colour differing from the trapping laser, thus resulting in the control of light by light. This type of switch or modulator can reach high speeds above MHz for particles trapped in air.

VOLUME INTEGRAL ANALYSIS OF NANOPARTICLES

Surface plasmons can be excited when nanoparticles are illuminated at an appropriate optical frequency. They are a collective oscillation of free electrons along the surface of metallic structure. A strong enhancement of near-field amplitude and a large scattering cross section for a narrow wavelength band can be observed at the surface plasmon resonance. For a simple structure, such as a circular cylinder or a sphere, the resonance can be calculated analytically. However, for a finite structure with a complicated geometry, a solution of the governing Maxwell's equations is needed to obtain its optical properties at resonance.

In addition to the geometry of the structure, the computation of surface plasmon resonance of metals such as silver and gold is difficult due to their large refractive index at optical wavelengths. Such material properties produce a very short effective wavelength inside the metals and strong field discontinuities at the boundary. Various numerical methods have been proposed to characterize the surface plasmons of metallic structures. These numerical methods can be generally categorized into time domain method and frequency domain method. Time domain method such as finite-difference time-domain requires a dispersive model to incorporate the frequency-dependent material properties. The accuracy of the method is highly dependent on the approximation of the dispersive model in comparison to the actual material properties over the simulation bandwidth. Besides, the finite-difference time-domain method usually uses strucutred mesh and this produces staircase error at curved surfaces.

Various frequency domain methods, such as multiple multipole method, discrete dipole approximation, boundary element method and volume integral equation (VIE) method, have been developed to study the surface plasmon problem of piecewise homogeneous nanoparticles. In contrast to the time domain method, these frequency domain methods do not require any dispersive model for the frequency-dependent material properties. They permit direct use of experimental permittivity data in the computation and thus provide the flexibility to solve scattering problems of material with no appropriate dispersive model. In addition, some frequency domain methods allow the use of unstructured mesh that is able to approximate curved surfaces more accurately.

The VIE approach presented in this paper offers the flexibility to model arbitrarily shaped geometry as well as an easy approximation of inhomogeneous material properties. Although the boundary element method generates less unknown, the VIE is more attractive for electromagnetic problem involving complicated material properties since a single VIE is sufficient to handle the complex problem. In this section, we use the VIE to characterize the surface plasmon resonance behaviours of arbitrarily shaped nanoparticles. The geometry of the nanoparticles is discretized by an unstructured mesh and the material properties of the nanoparticles are obtained directly from experimental data. Different from previous work, we have adopted divergence-conformingvector basis functions in the discretization of the resultant integral equation. By using divergence-conforming vector basis functions, the continuity of fields is ensured and no artificial charges are generated during the simulation. These are important features for solving VIE accurately, especially for nanoparticles with complicated geometry and material properties. Several numerical examples will be presented to demonstrate the applicability of the proposed method for the analysis of surface plasmon resonance of complicated nanoparticles.

Our study is organized as follows. We give a brief description on the VIE formulation and the divergence-conforming vector basis functions. We demonstrate the accuracy and efficiency of our computational scheme, follow by the results of different shaped nanoparticles such as hollow and star-shaped cylinders. The effects of illumination directions and surrounding media to surface plasmon resonance.

FORMULATION

Consider an arbitrarily shaped two dimensional (2D) scatterer which is a structure with a translation symmetry along the *z*-axis and consists of inhomogeneous dielectric material properties. The object is embedded in an isotropic homogeneous background medium with permeability $\mu_b = \mu_0$ and permittivity ε_b. The scatterer is illuminated by an incident transverse-electric TEz wave (electric field in *xy*-plane), which is excited by impressed sources in the background medium. The scatterer is assumed non-magnetic (μ_r =1) and has permittivity $\varepsilon(r)$ at location *r*.

By invoking the volume equivalence principle, the dielectric material can be removed and replaced by equivalent volume electric current densities J_V. The equivalent current densities are radiating in the unbounded background medium. By using the equivalent current densities, the total electric field *E* in the general dielectric material region can be written as the sum of the incident electric field and the scattered electric field due to J_V. Thus the VIE can be expressed as

$$Einc = D\varepsilon + jkb\eta b \int VJVG + 1k2b\nabla\nabla \cdot (JVG)dV'$$

where D=εE is the electric flux density, $\eta b = \mu b / \varepsilon b \text{———} \surd$ is the intrinsic impedance of the background medium, $kb = \omega\mu b\varepsilon b \text{——} \surd$ is the wavenumber of the background medium and $G = 14jH(2)0(kb\,|\,r - r'\,|)$ is the two dimensional scalar Green's function in the background medium. It is also noted that the equivalent current densities are related to the electric flux density through

$$JV(r) = j\omega\kappa e(r)D(r)$$

where κ_e is the contrast ratios of the permittivity and is defined as

$$\kappa e(r) = \varepsilon(r) - \varepsilon b\varepsilon(r).$$

The resultant VIE can be solved numerically by using method of moments. The 2D cross section of the structure is discretized by using triangular elements that permit flexible modeling of arbitrarily shaped cross section. For a highly curved surface, finer triangular elements are needed to approximate the curvature.

It is also possible to use parametric triangular elements that conform to the curved surface.

The dielectric property in each individual triangular element is assumed constant for inhomogeneous materials. When solving integral equations using the method of moments, the equivalent current density is normally used as the unknown quantity to be determined. However, for solving VIE, the equivalent current density is not directly used as the unknown quantity. Instead, the electric flux density *D* is used as the continuity of the normal components of *D* can be ensured by using proper basis functions. For triangular elements, the suitable basis functions are the triangular rooftop basis functions.

The length of the nth edge, $A^{\pm}$ represents the area of the triangles $T^{\pm}$, and ρ^+ and ρ^- are vectors pointing away and toward the free vertex respectively. An example of triangular rooftop basis function.

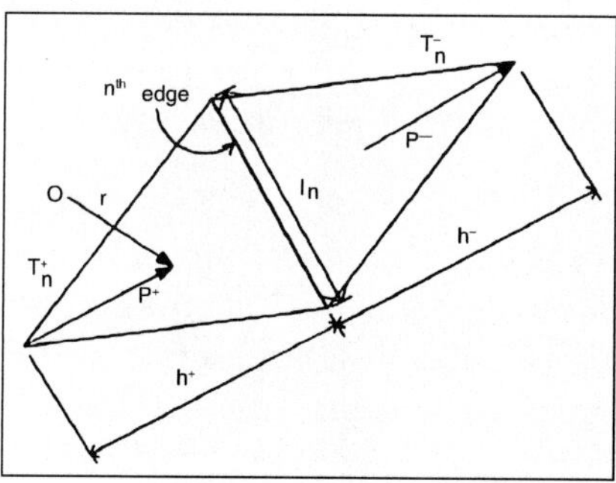

Figure: A Triangular Rooftop Basis Function.

On every element, only a maximum of three basis functions, corresponding to the three edges, will exist. The features of triangular rooftop basis functions, such as divergence conforming and continuity of the normal component of flux density to the interior face, make them suitable to be implemented in the volume integral equation. The electric flux density is then expanded using the triangular rooftop basis functions f_n

$$D(r) = \Sigma n = 1NDnfn(r)$$

where D_n is the coefficients to be determined. For the nth edge located at the exterior boundary of the object, an auxiliary triangle is introduced in the exterior region where the free vertex of the auxiliary triangle coincides with the centre of the nth edge. J_V can be expressed as;

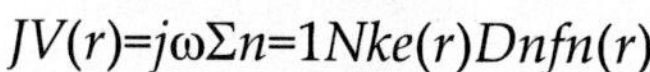

$$JV(r)=j\omega\Sigma n=1Nke(r)Dnfn(r)$$

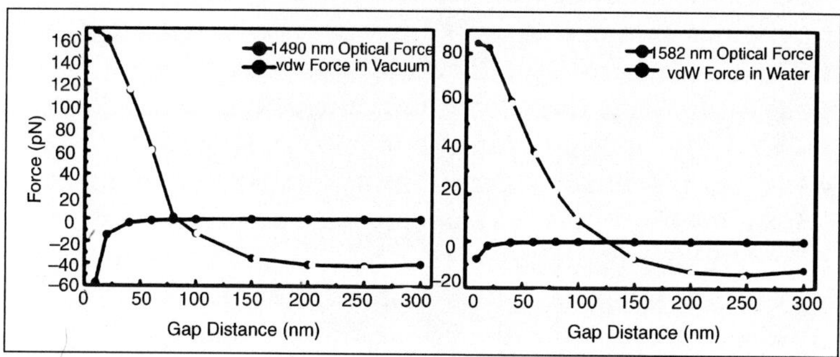

Figure: SCS as a Function of Wavelength for a Silver Circular Cylinder with Radius = 50 nm. The Plasmon Resonance at λ=347 nm is well Reproduced by our Method.

Subsequently, we substitute (7) into (1), and test (1) with f_m. This results in a $N \times N$ linear system which can be written in a matrix form as

$$Z\,I = V,$$

where matrix Z represents the interaction between equivalent sources and column vectors V and I are respectively the excitation vector and unknown coefficients vector to be determined. The elements of matrix Z can be computed using

$$Zmn = \int Tm\mathbf{f}m \cdot fn\varepsilon dV - \omega kb\eta b \int Tm\mathbf{f}m \cdot \int Tn\kappa e\mathbf{f}nG + 1k2b\nabla\nabla \cdot (\kappa e\mathbf{f}nG)dV'dV,$$

and the elements of vector V can be computed using

$$Vm = \int Tmfm \cdot EincdV.$$

Once all the elements are obtained, the matrix equation can be solved either by a direct method like Gaussian Elimination or by an iterative matrix solver like Conjugate Gradient. For a large matrix problem, the solution can be accelerated using fast algorithms such as fast multipole method and adaptive integral method.

Numerical Results

Four numerical examples are presented to demonstrate the capability of the proposed method for analysing the surface plasmon resonance properties of nanoparticles. All the frequency-dependent material properties used in the computation are directly obtained and interpolated from experimental data. In all examples, the light incident angle is the angle between the incident light and the horizontal axis. The first example is a silver circular cylinder with a radius of 50 nm. The scattering cross section of the circular cylinder is computed. The results obtained by using Mie series are also plotted for comparison and a very good agreement with the results obtained by using our method is observed. The plasmon resonance at 347 nm has been well reproduced by using our method.

In the second example we consider the surface plasmon resonance of elliptical, rectangular and square cylinders which have the same cross section area as the circular cylinder described in example 1. The eccentricity of the ellipse is 0.8 and the height-to-width ratio of the rectangle is 0.6. The scattering cross sections of these cylinders with light incident from angle 180° and 90° are computed. The scattering cross sections of the cylinders obtained using boundary element method are also computed and shown in square symbols for comparison. A very good agreement has been observed between the results computed by VIE and boundary element method. When the light incident angle varies from 180° to 90°, the resonance for the elliptical cylinder is redshifted from 343 nm to 375 nm. We also found that the full width at half maximum (FWHM) of the resonance for the elliptical cylinder has been expanded from 36 nm to 156 nm when the incident angle varies from 180° to 90°. The rectangular cylinder also exhibits same optical behaviours as elliptical cylinder, *i.e.* redshift of the resonance from 373 nm to 378 nm and increment of the FWHM from 28 nm to 153 nm, when light incident angle varies from 180° to 90°. The near-field distributions of these cylinders at their respective resonant frequencies are plotted for light incident angle 180° and 90°, respectively. For both incident angles, strong fields can be observed at the surface of the nanoparticles, in particular at the corners of the rectangular and square cylinders.

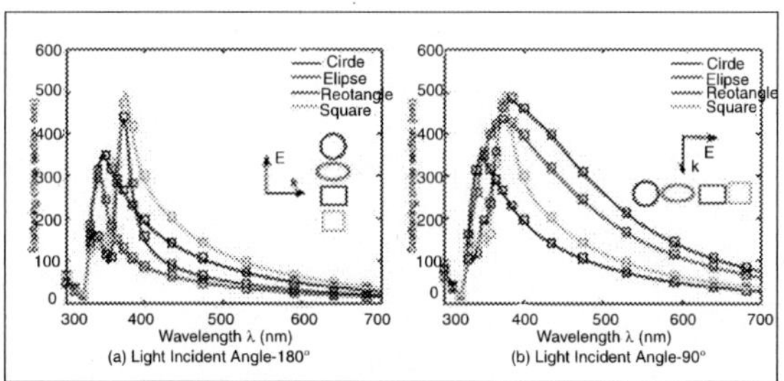

Figure: SCS as a Function of Wavelength for Different Shaped Cylinders Computed by VIE (solid lines) and Boundary Element Method (square symbols).

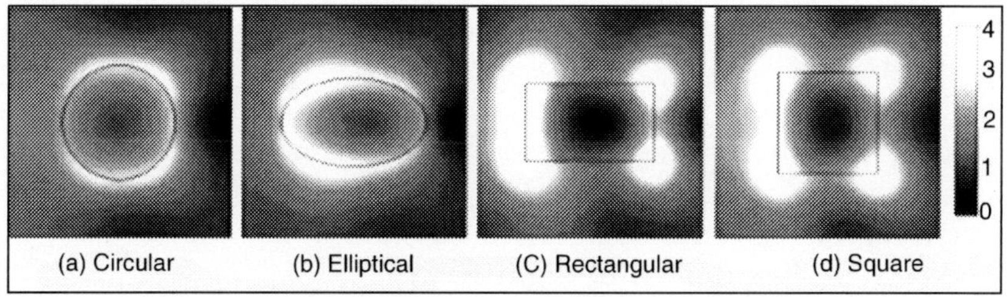

Figure: The Normalized Near-field Distribution of Nanoparticles at their Corresponding Resonant Wavelengths with Light Incident from Incident Angle 180°.

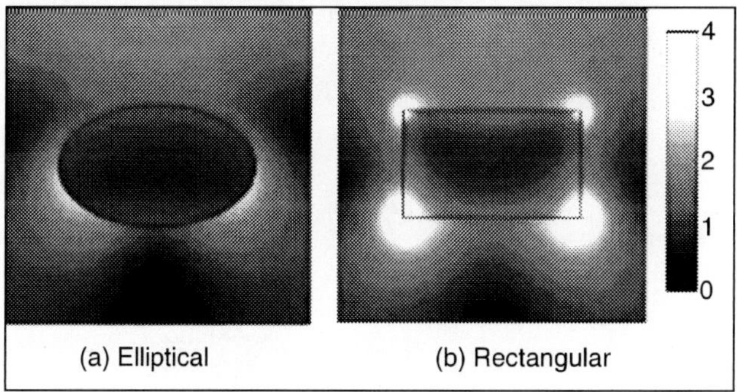

Figure: The Normalized Near-field Distribution of Nanoparticles at their Corresponding Resonant Wavelengths with Light Incident from Incident Angle 90°.

The third example we consider is a hollow cylinder. The inner and outer radii of the cylinder are 75 nm and 50 nm, respectively. The circular void is shifted 10 nm to the right of the centre. The hollow cylinder is illuminated by light propagating along the x-axis. The scattering cross sections of the hollow cylinder with background and void filled with different materials, *i.e.* air, water and glycerol. We also computed and plotted the scattering cross section of a solid cylinder by filling the void with silver. Generally, the solid cylinder produces a higher magnitude of scattering cross section than hollow cylinders in all three different background materials. For the solid cylinder immersed in different materials, we only recorded a single resonance in the computation and observed a redshift of the resonance when the refractive index of the background medium increases. However for the hollow cylinder, we obtained multiple resonances in the computation and observed a redshift of the resonances when the cylinder is placed in a higher refractive index background.

The multiple resonances are probably due to the resonances of the void and the coupling between the void and the core of the cylinder. The near-field distributions of the solid and hollow cylinders immersed in different background materials are computed at their respective principal resonances, respectively. In the near-field plots of the hollow cylinders, we found that a strong field enhancement can be obtained inside the void for all three different background materials but the field strength is weakened when higher refractive index background materials are used.

The last example we considered is five- and six-pointed star-shaped cylinders. The length *d* of these two star-shaped cylinders are set equal to 100 nm and the corners of the stars are filleted with radius *r*. Two values of *r*, *i.e.* 1 nm and 5 nm have been considered in this example.

The stars are illuminated by light propagating along -*y*-axis and their scattering cross sections. The surface plasmon resonances of the five- and six-pointed stars with fillet radius $r = 1$ nm are 435 nm and 400 nm, respectively. As for the stars with fillet radius $r = 5$ nm, the resonances of the five- and six-pointed stars are blueshifted to 393 nm and 383 nm, respectively.

The near-field distributions of the five- and six-pointed stars are computed at their respective resonant frequencies. In the near-field plots for the stars with $r = 1$ nm, we observed a very strong field near the corners transverse to the illumination direction. However, a more evenly distributed field surrounding the corners are observed in the near-field plots for the stars with $r = 5$ nm.

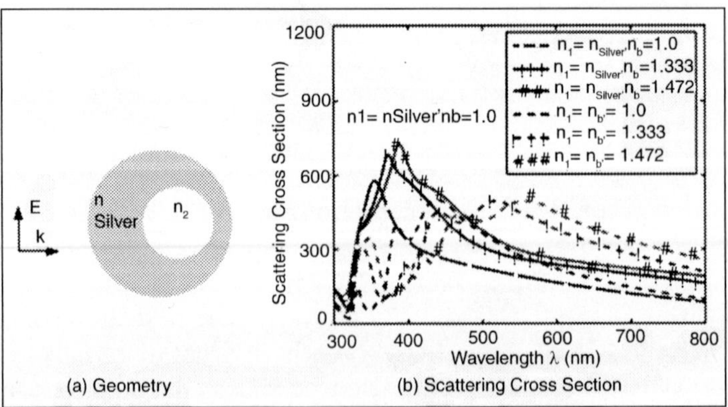

Figure: Geometry and Scattering Cross Section as a Function of Wavelength for a Hollow (solid) Silver Sylinder. The Cylinder is Considered Solid when $n_1 = n_{silver}$. The Solid and Hollow Cylinders are Immersed in Different Background Materials.

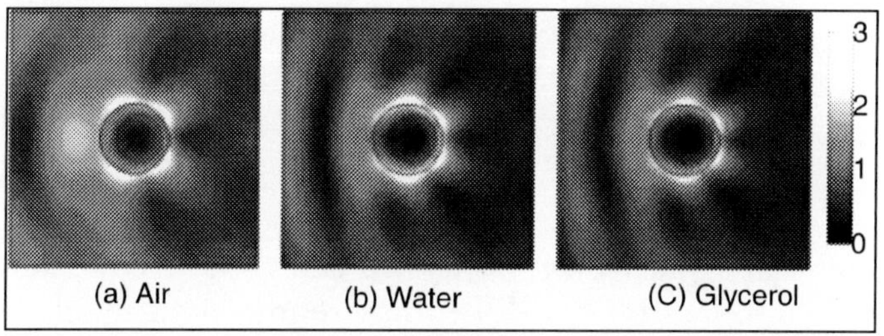

Figure: The Normalized Near-field Distribution of a Solid Silver Cylinder ($n_1 = n_{silver}$) Immersed in Different Background Materials.

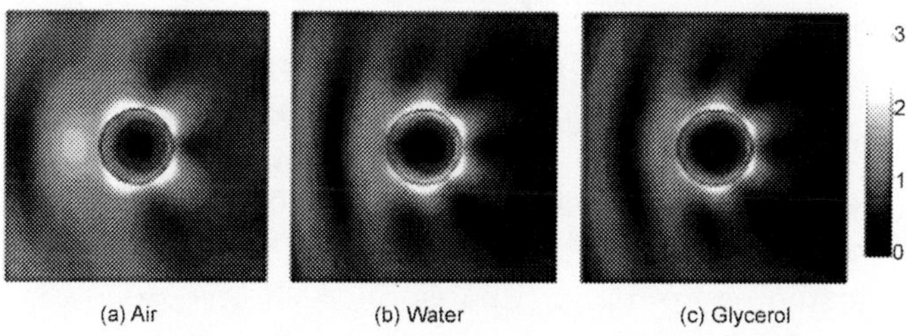

Figure: The Normalized Near-field Distribution of a Hollow Silver Cylinder ($n_1 \neq n_{silver}$) Immersed in Different Background Materials.

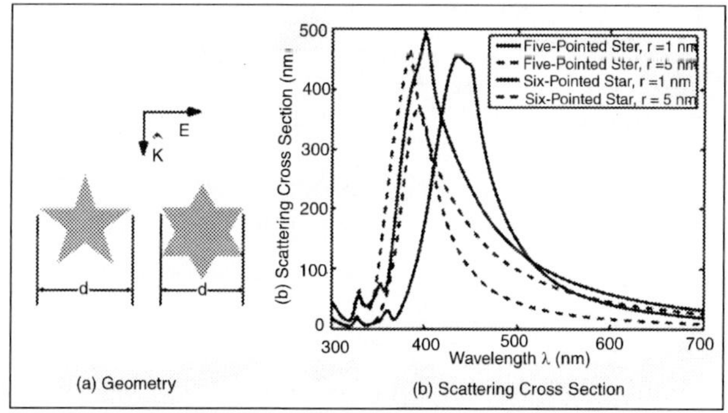

Figure: Geometry and Scattering Cross Section as a Function of Wavelength for Silver Five-pointed and Six-pointed Stars. The Corners of the Stars are Filleted with Different Radii *r*.

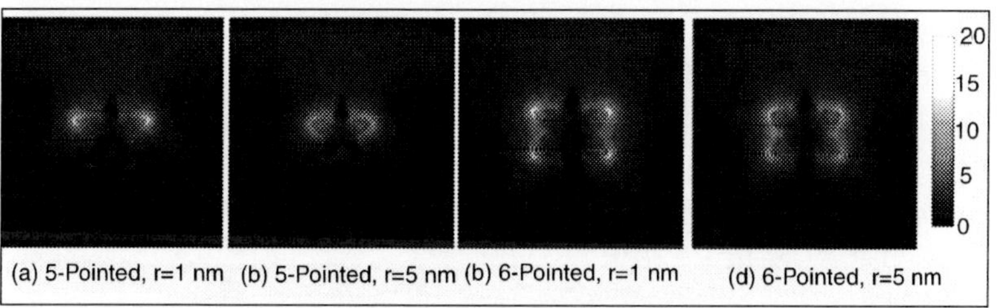

Figure: The Normalized Near-field Distribution of Five- and Six-pointed Stars with Different fillet Radii *r*.

INTEGRATING NANOOPTICS

Nanooptics support multiple approaches of integration — monolithic and hybrid —into optical components, allowing for a wide diversity of applications.

While there are many clever ways to design and build networks and systems, the fundamental breakthrough required to meet dramatically reduced cost points is the introduction of viable photonic ICs. Although there are physical differences between photons and electrons that make exact comparison an area for debate, the evolution of electronic ICs does provide a guide for the direction required for photonics. Paralleling the evolution of electrical circuits, the development of photonic ICs promises to increase functionality and density and significantly reduce costs when compared to optical components assembled from discrete devices. The result will open more general systems and service applications for optical networks.

There are three core criteria any integrating technology must meet:

- The technology must be capable of creating a broad range of optical functions out of a single fabric or process.
- The means must exist for it to be readily manufactured at low cost in high volume.
- The capability must exist to aggregate individual optical functions into more complex arrangements within the technology and with other optical technologies.

The third criteria may be the most challenging. While investigation of many technologies is underway, recent work on nanooptics shows that it meets these challenges, including the second, with surprising breadth, suggesting clear first steps to viable photonic ICs.

NANOOPTIC DEVICES AND MANUFACTURING

Standalone nanooptic devices typically consist of nano-scale structures, <1 mm thick, with individual feature sizes controlled to dimensions of <10 nm, on an optically transparent substrate. As incident light passes through the nano-structures, a portion of the light is transmitted and a portion is reflected.

Changing the arrangement and material of the nano-structures modifies the resulting optical function as well as the degree to which it is applied to the transmitted and reflected components of light. Demonstrated functionality created from nano-structures includes polarizers, polarization-beam splitters and combiners, waveplates, fixed and tunable filters, microlenses, anti-reflection coatings, and photodetectors.

In simpler applications, including polarization, phase management, and filtering, the nano-structure is a diffraction grating with critical dimensions an order of magnitude smaller that the wavelength of incident light — on the order of a few hundreds to tens of nanometers. Since modifying the arrangement and materials will tailor their optical performance, nanooptic structures can be considered a core set of photonic building blocks, addressing the first challenge.

Nanooptic devices are manufactured on a wafer scale in a semiconductor-like manufacturing process that results in optical chips in sizes from 1x1 mm and 0.2 mm thick up to tens of millimetres to a side and 1 mm thick. The first step in realizing a nanooptic device is to design the nano-pattern that yields the desired optical properties. That can be done through a combination of simulation and prototyping. Once a design is finalized, a mold is prepared that has the complement of the desired pattern inscribed in it. Various lithography techniques, including holographic lithography and e-beam lithography, can be used to create the mold.

In production, the mold is contacted with a prepared substrate — consisting of a pliable polymer resist layer, a layer of the target material for the nano-structure, and an optically transparent substrate — and the nano-structure pattern is transferred to a pliable polymer resist layer. This mode of pattern transfer is used to avoid the errors that result due to reflection, diffraction, and scattering effects associated with using higher-energy lithography methods at such fine dimensions. Since the manufacturing method is the same for any nanooptic produced — fundamentally only the mold and the settings for each processing step are varied — the same manufacturing line is used for the full range of devices. By combining well-understood nano-structure pattern transfer methods and semiconductor processing methods, nanooptics meet the second challenge.

Monolithic Integration

Because of their physics and the manufacturing process used to create them, nanooptic structures can be combined in multiple ways, both with themselves (monolithic integration) and with other materials and processes (hybrid integration) to yield integrated photonic devices.

When processing a single beam of light (or multiple beams identically), the straightforward approach to monolithic integration is to sequence optical functions. In the nanooptic paradigm, that is achieved by stacking nano-structures on a single substrate using multiple passes through the manufacturing process to create a summation of optical functions. By appropriately choosing the optical functions and their sequence, useful optical functions can be created.

A simple application is to combine waveplate and polarizer layers to create a circular polarizer. More complex configurations are possible, incorporating spacer layers to achieve walkoff in multiple beam components and reflective layers or waveguide layers to redirect both reflected and transmitted beams. The benefit of layered nanooptic devices is the ability to combine adjacent optical functions into much smaller, denser devices that can either encompass the full function of an optical component or allow the optical component to be significantly reduced in size.

A second type of monolithic integration requires the processing of multiple parallel beams of light discretely in a single device. That's accomplished by using an array of differing optical functions on a single horizontal layer. A master mold consisting of an array of nanooptic functions is created by using multiple masking and lithography steps. Since this complex sequence is captured in the mold and does not need to be repeated in regular manufacture, it lends itself to ready replication. That is distinctly different from technologies whereby optical materials are created from thin films, crystals, or self-assembling structures, since those methods produce uniform layers.

In a broad range of optical components, the incident optical beams are split into two beams, which are differentially processed, passed through waveplates to compensate for introduced phase differences, then recombined. For multiple parallel beams, that creates complexity in design and assembly because of the working size and alignment of individual waveplates. A simple application for nanooptics can arrange waveplate functions on a single chip in a desired (say, checkerboard) pattern. The resulting benefit is a compact, pre-aligned array of optical functions that can be assembled much more easily, resulting in increased density and reduced manufacturing costs.

Hybrid Integration

The breadth of integration opportunities is vastly expanded by viewing nanooptics not merely as a set of devices, but rather as a process that can be integrated into a more complex manufacturing sequence. Generally, nanooptics can be treated as an optical process, akin to thin films, that can be applied to other materials. While the materials used in the nanooptic layer do need to be matched with the materials used in the substrate — taking into account optical interactions, thermal expansion coefficients, and adhesion — the fact that nanooptics derive their optical properties from flexibility in material and structure lessens the challenge this factor might normally pose. The resulting hybrid integration enables nanooptic structures to be combined with other materials to simplify assembly, improve overall optical functionality, and create new optical combinations. A simple application is to apply a polarization nano-structure directly to a garnet substrate, creating a semi-isolator.

The physics of nanooptics allows the optical functionality of nanooptic devices to be modified, or tuned, in real time by utilizing an adjacent dynamic layer. By dynamic layer, we mean a material whose optical properties can be changed by the application of an electric or magnetic field such as lithium niobate, liquid crystal, or micro-electromechanical systems (MEMS). In a simple sense, changing the optical properties of the substrate on which the nanooptic structure rests has a synergistic interaction with the electromagnetic fields in the nano-structure layer.

As an example, a nanooptic filter consists of a nano-structure on top of a waveguide layer, both on top of a substrate. Modifying the optical properties of the waveguide layer, or the effective birefringence of the entire structure by incorporating a dynamic layer, results in a tunable filter. The benefit to the optical-component designer is the ability to create dense, high-performance tunable optical components.

Component Architectures

The modes of integration introduced here have several implications for component architectures. First, because nanooptic devices often have unique optical properties and a small form factor, they can be assembled into unusually compact optical devices. From the simplest perspective, a discrete nanooptic component — whether single-function or a monolithically integrated device — can be assembled by itself into a fibre pigtailed optical component. A simple example is a packaged polarization beam splitter/combiner (PBS/C) that is extremely compact. The device takes advantage of the nanooptics' 180 degrees of effective beam separation. Introducing nanooptic devices with more complex functionality — stacked or arrayed, monolithic or hybrid — can continue to use this component architecture.

The second and even more interesting implication is combining multiple modes of integration to create very complex optical devices. As an example, direct layering of nanooptic PBS and waveplate arrays can be used to create the core functionality in a highly compact polarimeter, providing real time readings of the polarization and phase of incident light. More sophisticated arrangements are also possible, for example, integrating optical materials (via hybrid integration) and dynamic materials into a layered stack.

The breadth of integration approaches — stacking, arrays, hybrid integration, and dynamic synergy — alone and in combination and the new possibilities in component architecture satisfy the third challenge for a core technology for photonic ICs. The various modes of integration enabled by nano-optic-component physics and manufacturing processes enable the creation of a broad range of optical components, spanning both passive and dynamic functions. In more complex optical components, rethinking optical-component architectures to take advantage of the strengths of nanooptics offers a path to reduce system costs and open new optical-networking opportunities.

Optical Integration

Subwavelength optical elements provide a new set of building blocks for integrating optical functions. Two approaches to integration are possible: hybrid and monolithic. Hybrid integration takes dissimilar technologies and combines them into optical components; monolithic integration self-combines nanooptic elements into more complex, layered optical components.

Discrete nanooptic devices add breadth to the toolkit of the hybrid integrated optical circuit designer, creating multiple advantages. Individual nanooptic devices often provide density, optical power handling, optical coupling, and compatibility advantages over the bulk optic counterparts. Appropriate selection of materials provides uniform performance over a broad range of optical power and environmental requirements. In addition, material selection or direct integration with other technologies can simplify cross-technology interfaces. By providing a broad range of fundamental building blocks for optical component and module design, nanooptic components enable both new functions and new combinations of functions.

Nanooptic components also offer advantages for manufacturing hybrid integrated optical components. The wide acceptance angle of many of these devices simplifies the alignment process, reducing manufacturing time and cost. Additionally, the robustness of nanooptic devices—with proper selection of materials they can be subjected to temperature ranges of –200 to 400 °C—makes them tolerant of a broad range of manufacturing operations.

POLARIZABILITY AND PLASMON RESONANCES

The optical properties of small structures have fascinated scientists for a long time. In 1857, Michael Faraday studied the interaction of light with colloidal metals and in 1871 Lord Rayleigh presented an analytical analysis of light scattering from small spherical particles. A full analytical analysis of spherical particles can be made using the Lorenz-Mie scattering theory. Light scattering of a normal incidence plane wave by a small cylinder was addressed by Lord Rayleigh in 1918 and extended to oblique incidence by J. R. Wait in 1955. Since then the problem of light scattering from small particles and cylinders has been revisited many times and is today well understood.

Due to the recent progress in the development of nanotechnologies, which has enabled fabrication of structures on the scale of a few nanometers, the interest in the optical properties of such structures is today enormous. In particular, metal nanostructures are intensively studied because they allow for resonant excitation of localized surface plasmons, *i.e.* collective excitations of the free conduction electrons that are resonantly coupled to the electromagnetic field. Both metal nanowires and-particles have several interesting applications as optical antennas, within photovoltaics, and many others. Optimal design and fabrication of metal nanostructures for applications rely on accurate numerical simulations and theoretical predictions. Full numerical solution of the Maxwell equations for metal nanostructures is a demanding and time-consuming computational task. The huge field-gradients near *e.g.* metal corners require a very fine meshing in order to obtain reliable results. Thus, for these cases, accurate analytical models that predict the optical properties are very important, both for applications and as benchmarks for numerical methods.

In the present work, we analytically calculate the polarizability and the plasmon resonances of a geometry, which we name "the partly buried double half-cylinder". The geometry is illustrated schematically. It consists of a double half-cylinder (a pair of conjoined half-cylinders) with optical properties described by the dielectric constants ε_1 and ε_2, respectively. The double half-cylinder is placed in two semi-infinite half-spaces with optical properties given by the dielectric constants ε_3 and ε_4. The simpler case of a half-cylinder or a double half-cylinder in a *homogenous* surrounding has, to some extent, been analysed before. Also, a double half-sphere consisting of two joint hemispheres with different dielectric constants has been studied. However, for this three dimensional problem no closed-form analytical solution was found. In Ref., two conjoined half-cylinders were analysed using Kelvin inversion and Mellin transforms yielding a solution for the potential, however, only for a special resonant case where the two dielectric constants are of similar magnitude and opposite sign. Using the bipolar coordinate system, cosine- and sine-transformations, a more general analysis of the double half-cylinder has recently been presented by Pitkonen. To some extent, our analysis follows the lines of Pitkonen, in that we use bipolar coordinates, as well as cosine- and sine-transformations. However, we consider a much more general geometry. Our approach *e.g.* enables analysis of a nanowire with the shape of a half-cylinder supported by a substrate. This is important because nanowires are often supported by a substrate and usually have cross-sectional shapes approximating half-circles. Compared to Pitkonen we also obtain much more compact expressions for the polarizabilities, *i.e.* we avoid using poly-logarithms as after reduction, only the natural logarithm is needed. Finally, we also analyse the plasmon resonances of the more general geometry in detail.

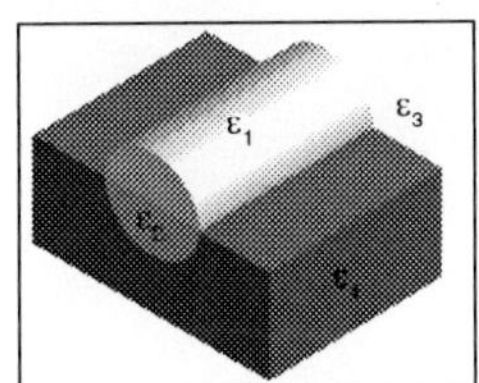

Figure: Geometry of the Partly Buried Double Half-cylinder.

THEORY

We start our analysis by assuming that the dielectric constants are linear, isotropic, homogenous, and frequency dependent. To simplify the notation, the frequency dependence of dielectric constant $\varepsilon_i = \varepsilon_i(\omega)$ is implicitly assumed. We also assume that the cross section of the double half-cylinder is small compared to the wavelength, *i.e.* we take an electrostatic approach. In a static theory, ∇ E(r) = 0 and the electrostatic field can be expressed by means of the electrostatic potential $\mathbf{E}(\mathbf{r}) = -\nabla\phi(\mathbf{r})$. In each domain, the electrostatic potential must fulfil Laplace's equation $\nabla^2\phi(\mathbf{r}) = 0\forall\mathbf{r}$, with the boundary conditions $\phi_i = \phi_j$ and $\varepsilon_i n\hat{}\cdot\nabla\phi_i = \varepsilon_j n\hat{}\cdot\nabla\phi_j$ on *S*, where the subscripts *i* and *j* refer to the different domains (1,2,3, and 4) and *S* to the boundaries.

The first step of our analysis is to switch to bipolar coordinates *u* and *v*, which are connected to rectangular coordinates via x = sinh$u$/(cosh u – cos$v$) and y = sin$v$/(cosh$u$ – cos$v$). The domains of *u* and *v* for the different regions. First we consider the case where the incident field is polarized along the *y* axis. Such a field will induce a vertically oriented dipole moment in the double half-cylinder. Thus, we look for solutions that are even functions of *x* and, hence, *u*.

In each of the four regions, the scattered part of the potential is expanded as

$$\phi_i = (u, v) = \int_0^\infty \overline{\phi_i}(\lambda, v)\cos(\lambda u)d\lambda,$$

with $\overline{\phi_i}(\lambda, v) = c_i(\lambda u)\cosh(\lambda v) + s_i(\lambda)\sinh(\lambda v)$ and the inverse transfor-mation given as

$$\overline{\phi_i} = (\lambda, v) = \frac{2}{\pi}\int_0^\infty \phi_i(u, v)\cos(\lambda u)du.$$

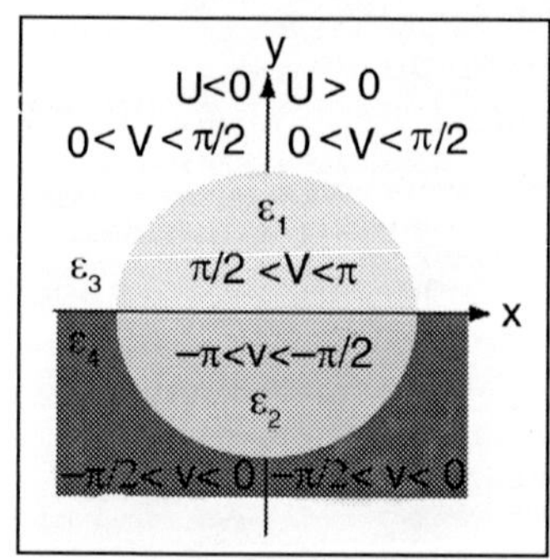

Figure: Cross Section of the Partly Buried Double Half-cylinder Geometry.

The vertical polarizability α_v of the partly buried double half-cylinder can be found as

$$\alpha v = 4\pi\int_0^\infty \lambda s_3(\lambda)d\lambda.$$

The normalized polarizability. To convert into standard units α_v should be multiplied by the radius of the cylinder squared. By solving the equation system formed from the boundary conditions of the potential and its normal derivative, $s_3(\lambda)$ can be calculated, and by performing we find the vertical polarizability as

$$\alpha_v = \frac{2\pi}{\varepsilon_1\varepsilon_2(\varepsilon_3+\varepsilon_4)+(\varepsilon_1+\varepsilon_2)\varepsilon_3\varepsilon_4} \times \left[\varepsilon_1\varepsilon_2(\varepsilon_3+\varepsilon_4)-(\varepsilon_1+\varepsilon_2)\varepsilon_3\varepsilon_4 - \frac{2(\varepsilon_1+\varepsilon_2)\varepsilon_3\varepsilon_4}{\pi^2}\ln^2\left(\frac{A+B}{A-B+2i\sqrt{AB}}\right)\right],$$

where $A = (\varepsilon_1+\varepsilon_2+\varepsilon_3+\varepsilon_4)\left[\varepsilon_1\varepsilon_2(\varepsilon_3+\varepsilon_4) + (\varepsilon_1+\varepsilon_2)\varepsilon_3\varepsilon_4\right]$ and $B = (\varepsilon_1\varepsilon_4-\varepsilon_2\varepsilon_3)^2$. It should be noted that the result presented by Pitkonen for the simpler case of $\varepsilon_3 = \varepsilon_4$ is consistent. This can be demonstrated using identities for polylogarithms to rewrite Pitkonen's result in terms of the natural logarithm. The resonance condition can be identified as

$$\varepsilon_1\varepsilon_2(\varepsilon_3+\varepsilon_4) + (\varepsilon_1+\varepsilon_2)\varepsilon_3\varepsilon_4 = 0.$$

For ordinary dielectric surroundings $\varepsilon_3, \varepsilon_4 > 0$, the resonance condition can only be fulfilled if the dielectric constant of the partly buried double half-cylinder is negative *e.g.* if the cylinder is made of a free electron-like metal as silver or gold. In this case, the resonances are dipole surface plasmon modes that arise due to the interaction of the free conduction electrons in the metal cylinder with the time-dependent incident field. The plasmon resonance conditions for a vertically polarized incident field for some special geometries. Note that the well known result $\varepsilon = -\varepsilon_h$ for a homogenous cylinder is obtained from our general result.

Table : Plasmon Resonance Conditions for Vertically induced Dipole Moments

Description	*Dielectric constants*	*Resonance condition*
cylinder	$\varepsilon_1 = \varepsilon_2 = \varepsilon$ and $\varepsilon_3 = \varepsilon_4 = \varepsilon_h$	$\varepsilon = -\varepsilon_h$
partly buried cylinder	$\varepsilon_1 = \varepsilon_2 = \varepsilon$ and $\varepsilon_4 = \varepsilon_h$	$\varepsilon = -2\varepsilon_3\varepsilon_h/(\varepsilon_3 + \varepsilon_h)$
half-cylinder	$\varepsilon_1 = \varepsilon$ and $\varepsilon_2 = \varepsilon_3 = \varepsilon_4 = \varepsilon_h$	$\varepsilon = -\varepsilon_h/3$
half-cylinder on substrate	$\varepsilon_1 = \varepsilon$ and $\varepsilon_2 = \varepsilon_4 = \varepsilon_h$	$\varepsilon = -\varepsilon_h\varepsilon_3/(2\varepsilon_3 + \varepsilon_h)$

For a horizontally polarized incident field *i.e.* $E_0(r) = \hat{x}$ we have

$$\phi_0 = (u, v) = -\frac{\sinh u}{\cosh u - \cos v}.$$

Such a field will induce a horizontally oriented dipole moment in the cylinder and we therefore look for solutions that are odd functions of x and, hence, u. In each of the four regions, we expand the scattered potential as

$$\phi_0 = (u, v) = -\int_0^\infty \overline{\phi_i}(\lambda, v)\sin(\lambda u)\,d\lambda.$$

Similarly to the vertically polarized case, the horizontal polarizability may be found as

$$\alpha_h = 4\pi\int_0^\infty \lambda c_3(\lambda)\,d\lambda.$$

From the equations formed from the boundary conditions of the potential and its normal derivative, $c_3(\lambda)$ can be calculated. By performing the integral, we find the horizontal polarizability as

$$\alpha_h = \frac{2\pi}{\varepsilon_1+\varepsilon_2+\varepsilon_3+\varepsilon_4}\left[\varepsilon_1+\varepsilon_2-\varepsilon_3+\varepsilon_4+\frac{2(\varepsilon_1+\varepsilon_2)\varepsilon_3\varepsilon_4}{\pi^2}\ln^2\left(\frac{A+B}{A-B+2i\sqrt{AB}}\right)\right].$$

It should be noted that there exists a simple symmetry between α_h and α_v. By replacing ε_i with $1/\varepsilon_i$ and changing the sign, α_v transforms into α_h and *vice-versa*. For a horizontally polarized induced dipole moment the plasmon resonance condition simply states that the sum of all the dielectric constants must be zero

$$\varepsilon_1 + \varepsilon_2 + \varepsilon_3 + \varepsilon_4 = 0.$$

Special cases for simpler geometries. The principal results of this work.

Table : Plasmon Resonance Conditions for Horizontally Induced Dipole Moments.

Description	*Dielectric constants*	*Resonance condition*
cylinder	$\varepsilon_1 = \varepsilon_2 = \varepsilon$ and $\varepsilon_3 = \varepsilon_4 = \varepsilon_h$	$\varepsilon = -\varepsilon_h$
partly buried cylinder	$\varepsilon_1 = \varepsilon_2 = \varepsilon$ and $\varepsilon_4 = \varepsilon_h$	$\varepsilon = -(\varepsilon_3 + \varepsilon_h)/2$
half-cylinder	$\varepsilon_1 = \varepsilon$ and $\varepsilon_2 = \varepsilon_3 = \varepsilon_4 = \varepsilon_h$	$\varepsilon = -3\varepsilon_h$
half-cylinder on substrate	$\varepsilon_1 = \varepsilon$ and $\varepsilon_2 = \varepsilon_4 = \varepsilon_h$	$\varepsilon = -(2\varepsilon_h + \varepsilon_3)$

Results

We illustrate the polarizability for three important geometries. In order to easily identify the plasmon resonances, we first calculate the polarizability as a function of the real part ε_r of the dielectric constant of the cylinder (or half-cylinder) with a fixed small imaginary part as $\varepsilon = \varepsilon_r + 0.01i$. As we are interested in the plasmon response of the system we consider negative ε_r. First we consider the geometry of a half-cylinder in a homogenous surrounding with $\varepsilon_h = 1$.

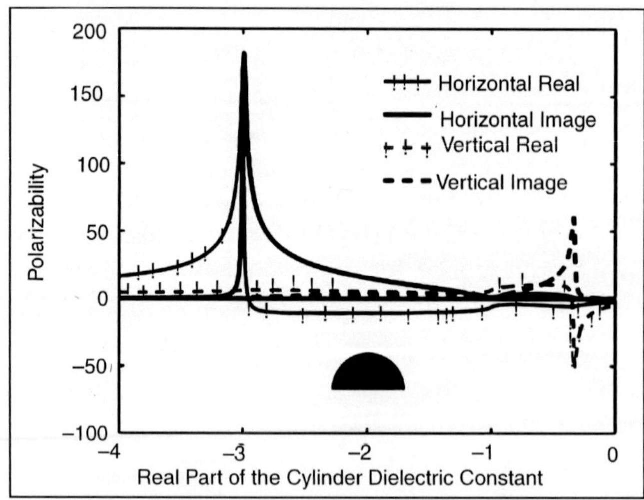

Figure: Polarizability as a Function of ε_r of a Half-cylinder with $\varepsilon_1 = \varepsilon_r + 0.01\, i$ in a Surrounding Medium Described by $\varepsilon_2 = \varepsilon_3 = \varepsilon_4 = 1$.

The result shows that both the real and the imaginary part of the vertical polarizability display a resonant behaviour at $\varepsilon_r = -\varepsilon_h/3 = -1/3$. Such a resonance is commonly referred to as a dipole surface plasmon resonance. Notice how the imaginary part of the polarizability is always positive as it should be for ordinary lossy materials. This is contrary to the numerical results presented by Pitkonen, which display strange negative imaginary parts of the polarizability. Note also how the real part of the polarizability changes sign across the plasmon resonance. For horizontally induced dipole moments the surface plasmon resonance of the metallic half-cylinder is located at $\varepsilon_r = -3\varepsilon_h = -3$. This plasmon resonance is clearly visible in both real and imaginary parts of the polarizability.

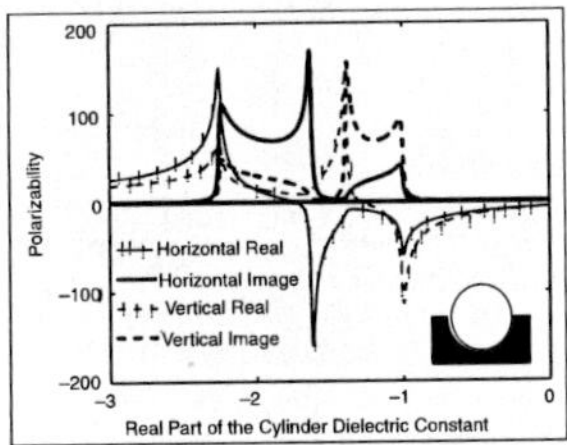

Figure: Polarizability as a Function of ε_r of a Half-cylinder Laying on a Substrate. The di-electric Constants of the Half-cylinder, the Substrate, and the Superstrate are $\varepsilon_1 = \varepsilon_r + 0.01i$, $\varepsilon_2 = \varepsilon_4 = 1.52$, and $\varepsilon_3 = 1$, Respectively.

The second geometry that we consider is a half-cylinder supported by a substrate. Such a configuration is of particular interest in relation to experimental work. Firstly, because nanowires are usually supported by substrates and, secondly, because nanowires often have cross sectional shapes similar to half-circles. Here, we consider the case of an air ($\varepsilon_3 = 1$) superstrate and quartz ($\varepsilon_2 = \varepsilon_4 = \varepsilon_h = 1.5^2$) substrate. For vertical polarization the configuration has a surface plasmon resonance at $\varepsilon_r = -\varepsilon_h\varepsilon_3/(2\varepsilon_3 + \varepsilon_h) \approx -0.53$. This resonance is clearly seen in both real and imaginary parts of the polarizability. For the horizontal case the surface plasmon resonance is located at $\varepsilon = -(2\varepsilon_h + \varepsilon_3) = -5.5$. Note that when compared to the half-cylinder in air, the plasmon resonances shift significantly, in particular, for the case of a horizontally polarized driving field. This is an important finding because it underlines the fact that the substrate plays a crucial role when the spectral location of the plasmon resonances of metallic nanostructures are established.

The third geometry that we consider is a full cylinder ($\varepsilon_1 = \varepsilon_2 = \varepsilon$) partly buried in a quartz substrate ($\varepsilon_4 = 1.52$ and $\varepsilon_3 = 1$). The polarizability is present. For the case of vertical polarization this configuration has a dipole plasmon resonance at $\varepsilon = -2\varepsilon_3\varepsilon_4/(\varepsilon_3 + \varepsilon_4) \approx -1.38$. For horizontal polarization the plasmon resonance is located at $\varepsilon = -(\varepsilon_3 + \varepsilon_4)/2 = -1.625$. For this geometry the polarizability is more complicated. Three peaks in both the real and the imaginary part of the polarizability can be identified from the figure, resulting in a large polarizability over a wide range of ε_r.

Next, we have calculated the polarizability as a function of wavelength for the same three geometries in the case of silver in air or on quartz. The data for the dielectric constant of silver is taken from the experiments of

Johnson and Christy. For the half-cylinder the result is present. The horizontal polarizability displays a clear resonance around 370 nm.

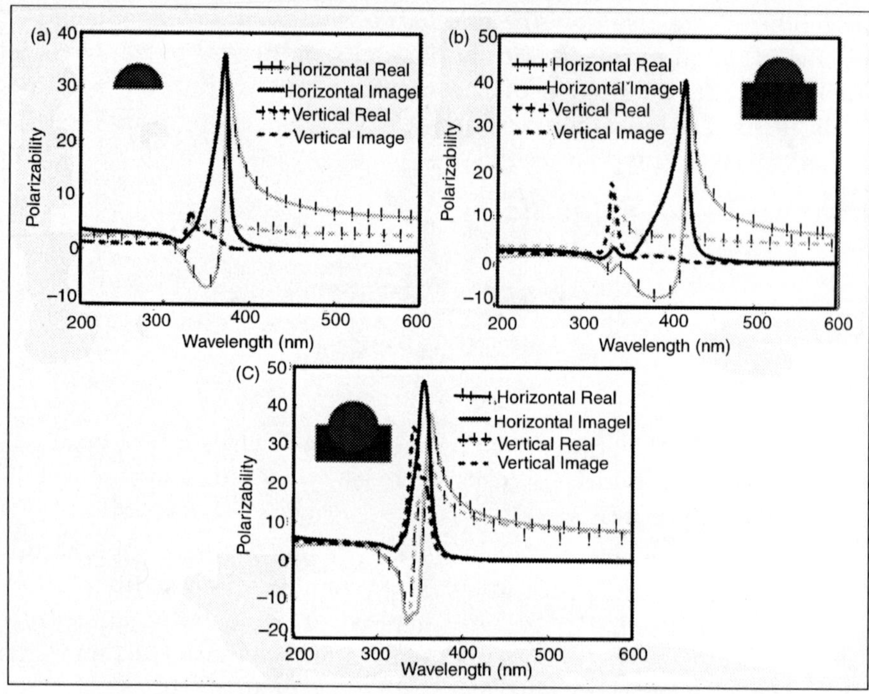

Figure: Polarizability as a Function of Wavelength for (a) a Silver Half-cylinder, (b) a Silver Half-cylinder on a Quartz Substrate, and (c) a Silver Cylinder Partly buried in a Quartz Substrate.

This corresponds nicely to the wavelength where the real part of the dielectric constant of silver equals –3. The vertical response shows a resonance at approximately 330 nm, which is where the real part of the dielectric constant of silver is –1/3.

For a half-cylinder supported by a quartz ($\varepsilon_h = 1.5^2$) substrate the polarizability is present. Note how the resonance in the horizontal polarizability has red-shifted approximately 50 nm due to the presence of the quartz substrate.

The resonance wavelength is now close to 420 nm, which is where the real part of the dielectric constant of silver is approximately –5.5. When compared to the half-cylinder in homogenous space, the vertical resonance of the half-cylinder on quartz does not red-shift significantly. The explanation for the larger red-shift seen in the horizontal polarizability is that a larger part of the field is within the substrate for a horizontal dipole moment than for a vertical one. For the partly buried cylinder the polarizabilities as a function of wavelength. The result shows that the resonances in the vertical and the horizontal polarizabilities are similar to each other. This might be expected because for both polarizations approximately half of the plasmon field will be within the substrate.

RESONANT PLASMONIC NANOPARTICLE SYSTEMS

Plasmon resonant metal nanoparticles have been widely employed as building blocks of diverse nano-optical applications, such as surface enhanced Raman spectroscopies, optical nanocircuits, photovoltaic devices and nanoparticle-based transmission lines. Recent advances in nanofabrication techniques have made it easy to construct complex nanostructures such as dimers and one or two dimensional arrays using plasmonic nanoparticles as components. When particles are close, coupling between nearby particles affects the local plasmon resonance behaviour of single particles and subsequently the overall optical property of the complex structures they are involved in. The generalized Mie theory provides a rigorous framework for the description of interacting nanoparticle systems whose complexity prohibits an intuitive physical understanding. Other theoretical approaches have been proposed for more simplistic descriptions.

In particular, the plasmon hybridization method has successfully explained interactions between plasmon modes in a coupled cluster of particles such as dimers by using an analogy with quantum systems. The Maxwell-Garnet (MG) effective medium theory provides a simple model for calculating the macroscopic optical properties of materials with a dilute inclusion of spherical nanoparticles.

It takes into account the dipolar interaction between particles through the averaged Lorentz local field. Unlike plasmon hybridization methods, the MG theory is not valid near the plasmonic resonances of metal particles. Together with other restrictions on the MG theory, such as the small particle size and the low concentration, this off-resonant restriction discourages the use of the MG effective medium approach for resonant plasmonic particle systems.

In this part, we resolve this problem by showing that even highly concentrated, resonant plasmonic particles can be correctly described by an effective medium theory by using a more elaborate treatment of the dipolar interaction between nearby particles. We utilize the notion of dressed polarizability and present a simple yet reasonably accurate description of the dipolar particle interaction as well as the macroscopic optical properties. A monolayer of nanoparticles is taken as an exemplary system and the dressed polarizability is evaluated explicitly. This results in an effective permittivity of the monolayer modifying the MG theory and gives absorbance spectrums that agree nicely with the more rigorous Finite Difference Time Domain (FDTD) results.

THE DRESSED POLARIZABILITY

In the quasi-static limit, scattered light by a single spherical nanoparticle is described by the electric field of a point dipole located at the centre of a sphere of radius *a*. The dipole moment p of the point dipole, which is the volume integral of polarization P, is induced by the incident electric field E0 such that

$$p = \varepsilon - 1\varepsilon + 2a_3E0 \equiv \alpha 0E0,$$

where α0 is the polarizability of the sphere and ε is the relative permittivity.

The polarization P is proportional to the internal electric field inside the sphere, which is constant in the quasi-static limit. For a fine sized particle, dynamical effects come into play through changing the local field and subsequently the polarizability α0 with a perturbative correctional factor. In addition to the dynamical effect, nearby particles also change the constant internal field, the intensity profile and electric field vector map calculated using the Finite Difference Time Domain (FDTD) method.

The vector map clearly shows that the internal electric field is significantly enhanced by nearby particles while its spatial distribution remains approximately constant. This enhances the particle dipole moment p. The 'dressed polarizability' is defined as the ratio between the enhanced dipole moment and the incident field. Two different situations of dressed polarizability where nearby particles either enhance or quench the internal field depending on the location.

In order to evaluate the dressed polarizability, we first note that the dipole moments of interacting particles are not solely induced by the incident field E0, but by the local field E*loc* at the particle location which is a superposition of the field E0 and the fields generated by dipole moments of other particles. For simplicity, we assume that the particles are of spherical shape and form a lattice so that all particles have the same dipole moment p.

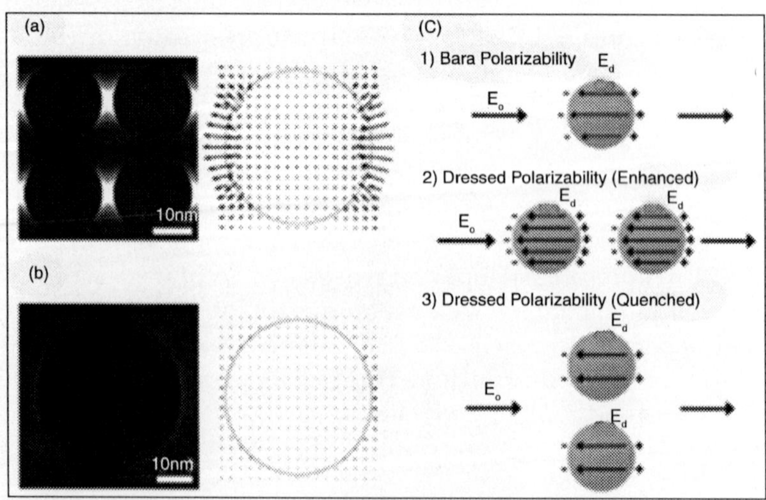

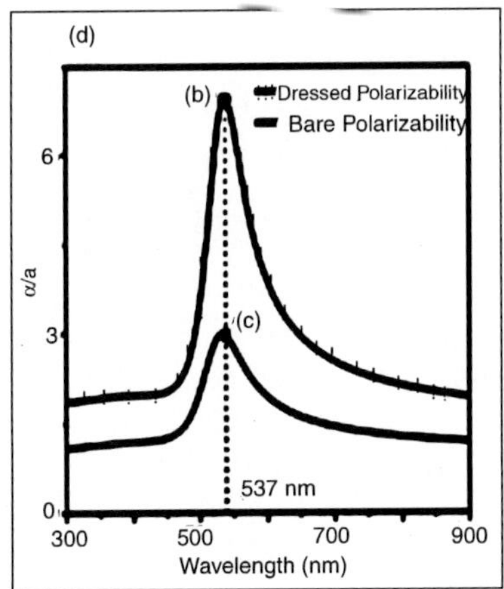

Figure: (a) The Field Intensity Profile (left) and The Vector Field Plot (right) of a Monolayer with a Lattice Constant of 23 nm at 537 nm Resonant Wavelength, and (b) the Field Intensity Profile (left) and the Vector Field Plot (right) of a Single Non-interacting Particle at the Same Wavelength (c) Schematic of Plasmonic Interaction between Metallic Nanoparticles which Enhance or Quench Polarization of Each Particle, and (d) the bare Polarizability (blue line) and the Dressed Polarizability Normalized by the Cubic of the Particle Radius, $a3$.

Then the local field Eloc is given by

$$\mathrm{E}loc = \mathrm{E0} + G\mathrm{p},$$

where G is the coupling coefficient determined by the geometrical configuration of particles. The dipole moment p is induced by the local field Eloc so that

$$\mathrm{p} = \alpha 0\mathrm{E}loc = \alpha 0(\mathrm{E0} + G\mathrm{p}) = (1 - \alpha 0G) - 1\alpha 0\mathrm{E0}.$$

From this, we obtain the dressed polarizability and the dressed internal field Eint, d,

$$\alpha\alpha d = (1 - \alpha 0\mathrm{G}) - 1\alpha 0 \equiv \gamma\alpha 0,$$

$$\mathrm{Eint}, d = (1{-}\alpha 0a3)\gamma\mathrm{E0},$$

where $\gamma = (1 - \alpha 0G){-}1$ is the dressing factor and we assume that the dressed internal field is constant.

Derivation of the Coupling Coefficient G

In the following, we focus on a monolayer configuration as a test system. Identical particles of radius a and relative permittivity ε form a two dimensional rectangular array with the lattice constants dy and dz, which are smaller than the wavelength λ of the incident light. For this monolayer, the coupling coefficient G can be found by a direct evaluation of the retarded dipolar field resulting from the neighbouring particles. Here, we evaluate the coupling coefficient G define by considering only nearest neighbouring spheres. The dipole moment $p = (4\pi/3)a_3P$ is the volume integral of polarization P which is assumed to be a constant vector oriented along the z-direction inside all spheres. Since the local field E*loc* is the superposition of the incident field E0 oriented along the z-direction and the depolarization field generated by nearby induced dipoles and also the self-generated dynamical field, we can obtain the *G*p- term by evaluating these depolarization fields explicitly. We calculate the depolarization field at the origin *O*. generated by a dielectric sphere of radius *a* centreed at O_2 which is located at r0 = (*r*0, θ0, ϕ0). As components perpendicular to P cancel on integration due to symmetry, we need to integrate only the parallel component of the depolarization field generated by a dipole moment *d*p(r)=P*dV*. Expanding the retarded dipolar field up to the second order in *k*, the parallel component is given by

$$dEd,\|= [1r3(3\cos 2\theta - 1) + k22r(\cos 2\theta + 1)]PdV,$$

where (*r*, θ, ϕ) specifies the location of the volume element.

In order to derive the depolarization field of a monolayer, we consider only neighbouring spheres and perform the integration separately for the enhanced and quenched arrangements.

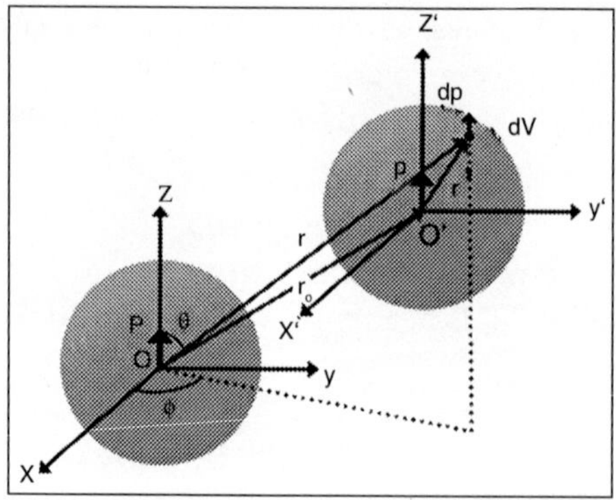

Figure: Original Spherical Coordinate System (*r*, θ, ϕ) centreed at *O* and the Spherical Coordinate System (*r*′,θ′,ϕ′) centreed at O_2 shifted from *O* by r0 = (*r*0, θ0, ϕ0).

ENHANCED ARRANGEMENT

Consider a monolayer of spheres forming a rectangular lattice in the yz-plane with lattice constants *dy* and *dz*. As the neighbouring spheres aligned along the polarization direction enhances depolarization. Specifically, spheres located at r0 = (*r*0 = *dz*,θ0 = 0) and r0 = (*r*0 = *dz*, θ0 = π) enhance the parallel component of the depolarization field while perpendicular components cancel on integration due to the symmetric configuration.

EFFECTIVE PERMITTIVITY VIA DRESSED POLARIZABILITY

Consider a monolayer of spheres forming a rectangular lattice in the xy-plane with lattice constants *dx* and *dy*. Using the dressed internal field and the coupling coefficient, we define the effective permittivity as the ratio between the averaged E and D fields.

Our modification provides a successful effective medium theory for resonant plasmonic nanoparticle systems. To demonstrate this, we calculated the absorbance spectrum of gold nanoparticle monolayers using the Fresnel equation for a slab with effective permittivity εeff and compared our result with the rigorous FDTD results. A contour map of calculated results for the absorbance spectrum of gold nanoparticle in square lattice monolayers. While keeping the particle size constant (*a*=10 nm), we varied the lattice constants $dx = dy = d$ from 20 nm (the touching limit) to 30 nm.

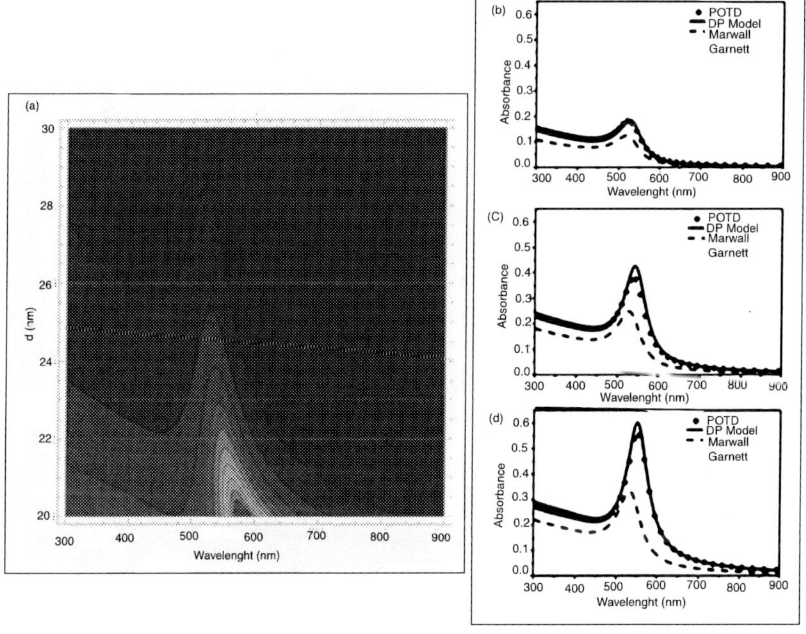

Figure: (a) Contour Map of Absorbance Spectrum of Gold Nanoparticle Monolayers, and the Corresponding Absorbance Spectrum at a Lattice Constant of (b) 26 nm, (c) 23 nm and (d) 22 nm Corresponding to the Filling Factor of f = 0.24, 0.34, 0.39 Respectively. The Absorbance Spectra show Increased Absorbance of the Resonance for Monolayers with Shorter Lattice Constants.

The permittivity of gold is modelled through the Drude-critical point model by fitting the experimental data of Johnson and Christry. The contour map shows that plasmonic resonance become stronger and the peak is red-shifted as the gap decreases. The specifically compare absorbance spectra with the predictions of the MG theory and FDTD calculations. There is an excellent agreement between our model and the FDTD result while the MG theory fails to predict the absorbance spectra near resonance. The peak absorbance depict, the margin of deviation of 2%, 14% and 9% respectively which result from the neglect of higher order pole excitations and interactions with other neighbouring particles.

The agreement ceases to exist near the touching limit due to the quantum tunneling which comes into play below sub nanometer the gap size and also at the touching limit where the layer becomes conductive. The MG theory not only underestimates the magnitude it also fails to produce the red-shift behaviour of the resonance peaks while our model shows the agreement. We were also able to confirm that dipolar interaction between nearby particles does indeed control the resonant behaviour of dense plasmonic nanoparticle systems.

Equipped with the ability to predict resonance behaviours, our effective medium model provides a wonderful opportunity for tailoring optical response in dense plasmonic systems. In the central resonant region, both the real and imaginary part of the refractive indices become significantly enhanced. Near resonance, the real and imaginary part of the refractive indices are sensitively influenced by the lattice constant *d*. The maximum values of *neff* and *keff* show a 3.5-fold and 12-fold increase, respectively, with corresponding peak shifts from 545 nm to 596 nm and from 514 nm to 564 nm when the lattice constant *d* decreases by only 10 nm. In contrast, in the off-resonant long wavelength region the imaginary part becomes nearly zero, this implies that nanoparticle systems are transparent in this region despite of their extremely high density. This is also shown in the absorbance spectrum plotted. In the off-resonant short wavelength region, the imaginary part becomes relatively large this is clearly.

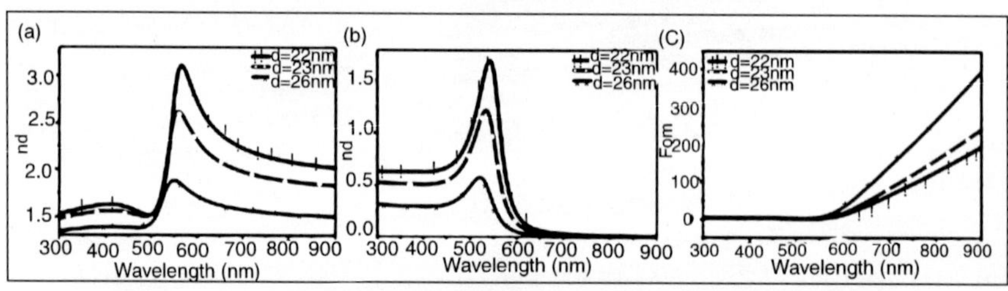

Figure: Effective Indices: (a) Real Part of Effective Refractive Index *neff*, (b) Imaginary Part of Effective Refractive Index *keff* and (c) Figure of Merit (FOM) which is Defined by *FOM=neffkeff*.

If particles form a non-square type rectangular lattice ($dx \neq dy$), the effective medium possesses anisotropic permittivity ($\varepsilon effx \neq \varepsilon effy$). This is in sharp contrast to MG theory which is used to explain the properties of dilute, small, randomly placed spheres utilizing only the overall particle filling factor *f*. Compares absorbance spectrum for three different cases of rectangular layers: enhanced ($dx < dy$), isotropic ($dx = dy$) and quenched ($dx > dy$) layer. The filling factor is fixed to a constant value, $f = 0.24$, as such the MG theory predicts the same spectrum for all three cases. We also note that the isotropic and quenched layers agree quite well with the FDTD results while the enhanced layer case shows deviation in the off-resonant short wave length region.

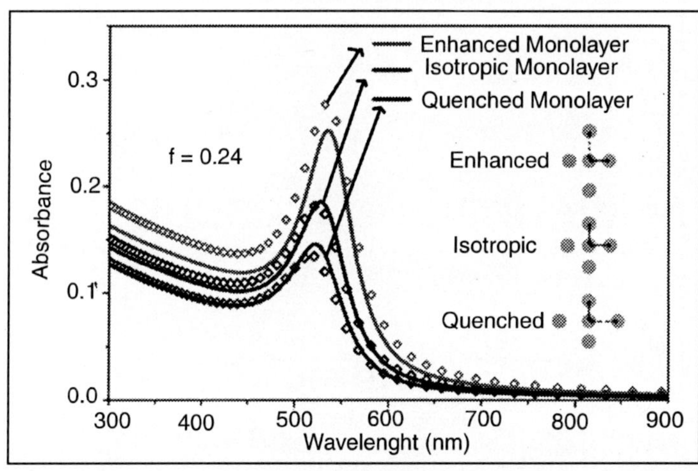

Figure: Absorbance Spectrum for (a) the Enhanced Monolayer with d_x = 23.6 nm d_y = 28.6 nm, (b) the Isotropic Monolayer $d_x = d_y$ = 26 nm, and (c) the Quenched Monolayer d_x = 28.6 nm, d_y = 23.6 nm.

This is due to the neglect of higher-order excitations including the quadrupolar interaction, which become dominant factors between close particles of relatively large size as pointed out in earlier literatures.

REFERENCES

- Ipsita Roy: *Bio-Nano-Geo Sciences : The Future Challenge*, Ane Books, Delhi, 2009.
- Mahan, G.D. : *Many-Particle Physics*, Plenum Press, New York, 1990.
- Singh, Seema: *Optics and Atomic Physics*, Campus Books, Delhi, 2009.
- Srivastava, N.K. : *Electromagnetic Theory and Optics*, ABD Pub., Delhi, 2007.
- Sujata Agrawa: *Small Wonder : The Making of the Nano*, Westland, Delhi, 2010.
- Varadan, Vijay K. : *Microsensors, MEMS and Smart Devices*, Wiley, NY, 2001.

7

Light Scattering and Nanoplasmonics

LIGHT SCATTERING VACUUM

In the real world, we rarely see things in a vacuum, where nothing exists between an object and its observer. In real-time rendering, the effect of participating media on light transport is often subject to low-complexity homogeneous assumptions. This is due to the intractable nature of the radiative transport equation, accounting for emission, absorption, and scattering, in a complex interactive animated environment. In this chapter, we consider the effect of volumetric shadows in the atmosphere on light scattering, and we show how this effect can be computed in real time with a GPU pixel shader post-process applied to one or more image light sources.

CREPUSCULAR RAYS

Under the right conditions, when a space contains a sufficiently dense mixture of light scattering media such as gas molecules and aerosols, light occluding objects will cast volumes of shadow and appear to create rays of light radiating from the light source. These phenomena are variously known as crepuscular rays, sunbeams, sunbursts, star flare, god rays, or light shafts. In sunlight, such volumes are effectively parallel but appear to spread out from the sun in perspective.

Rendering crepuscular rays was first tackled in non-real-time rendering using a modified shadow volume algorithm and shortly after that, an approach was developed for multiple light sources. This topic was revisited in real-time rendering, using a slice-based volume-rendering technique and more recently applied using hardware shadow maps. However, slice-based volume-rendering methods can exhibit sampling artifacts, demand high fill rate, and require extra scene setup. While a shadow-map method increases efficiency, here it also has the slice-based detractors and requires further video memory resources and rendering synchronization. Another real-time method, based on the work of Radomir Mech, uses polygonal volumes, in which overlapping volumes are accumulated using frame-buffer blending with depth peeling. A similar method removes the need for depth peeling using accumulated volume thickness. In our approach, we apply a per-pixel post-processing operation that requires no preprocessing or other scene setup, and which allows for detailed light shafts in animating scenes of arbitrary complexity.

In previous work, a GPU shader for light scattering in homogeneous media is implemented. We extend this with a post-processing step to account for volumetric shadows. The basic manifestation of this post-process can be traced to an image-processing operation, *radial blur,* which appears in many CG demo productions. Although such demos used software rasterization to apply a post-processing effect, we use hardware-accelerated shader post-processing to permit more sophisticated sampling based on an analytic model of daylight.

TYPES OF SCATTERING

- Rayleigh scattering is the elastic scattering of light by molecules and particulate matter much smaller than the wavelength of the incident light. It occurs when light penetrates gaseous, liquid, or solid phases of matter. Rayleigh scattering intensity has a very strong dependence on the size of the particles (it is proportional the sixth power of their diameter). It is inversely proportional to the fourth power of the wavelength of light, which means that the shorter wavelength in visible white light (violet and blue) are scattered stronger than the longer wavelengths toward the red end of the visible spectrum. This type of scattering is therefore responsible for the blue colour of the sky during the day and the orange colours during sunrise and sunset. Rayleigh scattering is the main cause of signal loss in optical fibres.
- Mie scattering is a broad class of scattering of light by spherical particles of any diameter. The scattering intensity is generally not strongly dependent on the wavelength, but is sensitive to the particle size. Mie scattering coincides with Rayleigh scattering in the special case where the diameter of the particles is much smaller than the wavelength of the light; in this limit, however, the shape of the particles no longer matters. Mie scattering intensity for large particles is proportional to the square of the particle diameter.
- Tyndall scattering is similar to Mie scattering without the restriction to spherical geometry of the particles. It is particularly applicable to colloidal mixtures and suspensions.
- Brillouin scattering occurs from the interaction of photons with acoustic phonons in solids, which are vibrational quanta of lattice vibrations, or with elastic waves in liquids. The scattering is inelastic, meaning it is shifted in energy from the Rayleigh line

frequency by an amount that corresponds to the energy of the elastic wave or phonon, and it occurs on the higher and lower energy side of the Rayleigh line, which may be associated with the creation and annihilation of a phonon. The light wave is considered to be scattered by the density maximum or amplitude of the acoustic phonon, in the same manner that X-rays are scattered by the crystal planes in a solid. In solids, the role of the crystal planes in this process is analogous to the planes of the sound waves or density fluctuations. Brillouin scattering measurements require the use of a high-contrast Fabry–Pérot interferometer to resolve the Brillouin lines from the elastic scattering, because the energy shifts are very small ($< 100\ cm^{-1}$) and very weak in intensity. Brillouin scattering measurements yield the sound velocities in a material, which may be used to calculate the elastic constants of the sample.

- Raman scattering is another form of inelastic light scattering, but instead of scattering from acoustic phonons, as in Brillouin scattering, the light interacts with optical phonons, which are predominantly intra-molecular vibrations and rotations with energies larger than acoustic phonons. Raman scattering may therefore be used to determine chemical composition and molecular structure. Since most Raman lines are stronger than Brillouin lines, and have higher energies, standard spectrometers using scanning monochromators may be used to measure them. Raman spectrometers are standard equipment in many chemical laboratories.

Static and Dynamic Scattering

A common dichotomy in light scattering terminology is static light scattering versus dynamic light scattering. The chief distinction is whether the scattering is observed to be changing over time (dynamic) or constant over the observation (static). This terminology is especially commonly encountered in the field of polymer chemistry, though it can obviously be applied to a broad range of situations.

Critical Phenomena

Density fluctuations are responsible for the phenomenon of critical opalescence, which arises in the region of a continuous, or second-order, phase transition. The phenomenon is most commonly demonstrated in binary fluid mixtures, such as methanol and cyclohexane. As the critical point is approached the sizes of the gas and liquid region begin to fluctuate over increasingly large length scales. As the length scale of the density fluctuations approaches the wavelength of light, the light is scattered and causes the normally transparent fluid to appear cloudy.

VOLUMETRIC LIGHT SCATTERING

To calculate the illumination at each pixel, we must account for scattering from the light source to that pixel and whether or not the scattering media is occluded. In the case of sunlight, we begin with our analytic model of daylight scattering.

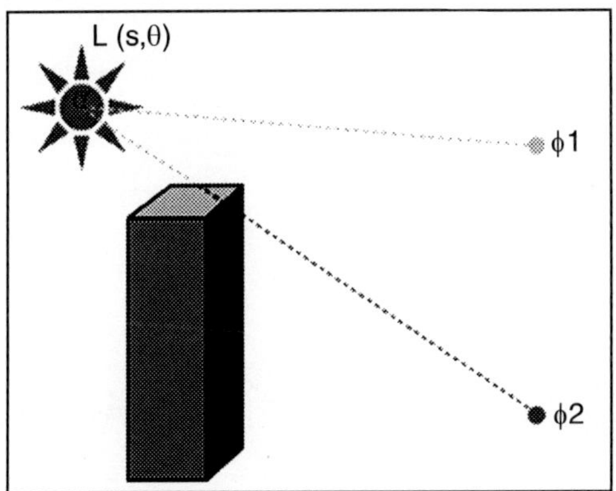

Figure: Ray Casting in Screen Space

We may reduce the bandwidth requirements by downsampling the source image. With filtering, this reduces sampling artifacts and consequently introduces a local scattering contribution by neighbourhood sampling due to the filter kernel. In the demo, a basic bilinear filter is sufficient.

SCREEN-SPACE OCCLUSION METHODS

As stated, sampling in screen space is not a pure occlusion sampling. Undesirable streaks may occur due to surface texture variations. Fortunately, we can use the following measures to deal with these undesirable effects.

The Occlusion Stencil Method

On earlier graphics hardware the same results can be achieved without a pre-pass by using a stencil buffer or alpha buffer. The primary emissive elements of the image (such as the sky) are rendered as normal while simultaneously setting a stencil bit. Then the occluding scene objects are rendered with no stencil bit. When it comes to applying the post-process, only those samples with the stencil bit set contribute to the additive blend.

The Occlusion Pre-Pass Method

If we render the occluding objects black and untextured into the source frame buffer, image processing to generate light rays is performed on this image.

Then the occluding scene objects are rendered with regular shading, and the post-processing result is additively blended into the scene. This approach goes hand in hand with the common technique of rendering an unshaded depth pre-pass to limit the depth complexity of fully shaded pixels.

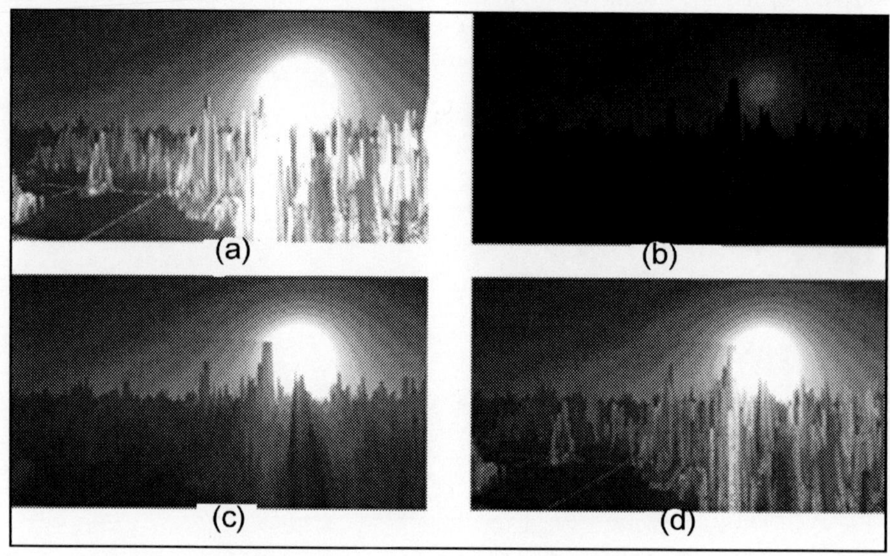

Figure: The Effect of the Occlusion Pre-Pass

The Occlusion Contrast Method

Equally though, this problem may be managed by reducing texture contrast through the texture's content, fog, aerial perspective, or light adaption as the intensity of the effect increases when facing the light source. Anything that reduces the illumination frequency and contrast of the occluding objects diminishes streaking artifacts.

CAVEATS

Figure: Dealing with One Limitation

Although compelling results can be achieved, this method is not without limitations. Light shafts from background objects can appear in front of foreground objects, when dealing with relatively near light sources. In a full radiative transfer solution, the foreground object would correctly obscure the background shaft. One reason this is less noticeable than expected is that it can be perceived as the manifestation of a camera lens effect, in which light scattering occurs in a layer in front of the scene. This artifact can also be reduced in the presence of high-frequency textured objects.

As occluding objects cross the image's boundary, the shafts will flicker, because they are beyond the range of visible samples. This artifact may be reduced by rendering an extended region around the screen to increase the range of addressable samples. Finally, when facing perpendicular to the source, the light's screen-space location can tend toward infinity and therefore lead to large separation between samples. This can be diminished by clamping the screen-space location to an appropriate guard-band region. Alternatively, the effect can be faded toward the perpendicular and is further decreased when using an occlusion method.

THE DEMO

The demo on the book's DVD uses Shader Model 3.0 to apply the post-process, because the number of texture samples needed exceeds the limits of Shader Model 2.0. However, the effect has been implemented almost as efficiently with earlier graphics hardware by using additive frame-buffer blending over multiple passes with a stencil occlusion method.

Figure: Crepuscular Rays with Multiple Additive Frame-Buffer Passes on Fixed-Function Hardware

Extensions

Sampling may occur at a lower resolution to reduce texture bandwidth requirements. A further enhancement is to vary the sample pattern with stochastic sampling, thus reducing regular pattern artifacts where sampling density is reduced.

The method performs the post-process in a single-pass execution of a shader. In one multipass approach, the pixel shader summation may be performed where the results of concentric rectangular bands emanating from the light source may be accumulated into successive outer bands, where $L(s, \theta, \phi) = L_{i-1}(s, \theta, \phi) + L_i(s, \theta, \phi)$. While this may not be most suited to current hardware design, this approach is the minimal required sampling and computation limit.

Creating a balance between light shaft intensity and avoiding oversaturation requires adjustments of the attenuation coefficients. An analytic formula that performs light adaption to a consistent image tone balance may yield an automatic method for obtaining a consistently perceived image. For example, perhaps we can evaluate a combination of average, minimum, and maximum illumination across the image and then apply a corrective colour ramp, thus avoiding excessive image bloom or gloominess.

NANO PLASMONICS

The main distinguishing element of modern nano optics over conventional optics is the strong confinement of plasmon fields at the surface of noble metal nanostructures. Noble metal surfaces and particles with sub-wavelength dimensions exhibit optical eigenmodes called surface plasmon polaritons which can be resonantly excited. Once excited, it is possible to strongly enhance the optical near-field and to resonantly enhance the scattering as well as absorption of light.

The scientific exploration of plasmonic properties is appealing since it promises to overcome limitations imposed by classical optics, such as *e.g.* the Abbe limit of diffraction.

Thus metal-dielectric surfaces and nanoparticles allow tailoring the flux of light in volumes significantly smaller than the wavelength and are therefore very attractive objects to study in nano optics.

To understand, monitor and finally control the plasmonic properties of nanosized objects, the information of the optical nearfield has to be collected by spectroscopic means with high spatial resolution.

However, in the direction normal to the surface, the electric field strength drops off exponentially in most cases and is not transferred into the far-field. Nowadays this challenge is overcome by a variety of different (scanning) nearfield optical microscopy (SNOM) techniques. SNOM exploits the properties of evanescent waves by placing a tip very close to the specimen surface. This allows for the surface inspection with high spatial resolution. With this technique, the resolution of the image is limited by the size of the SNOM tip and not by the wavelength of the illuminating light. In particular, lateral resolution of 20 nm and vertical resolution of 1 nm have been demonstrated.

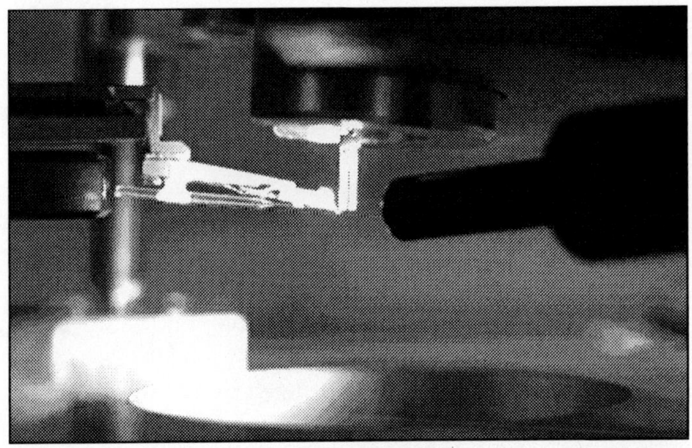

Figure: Two SNOM Tips are Combined into one Instrument to Excite and Detect Light at Plasmonic Nanostructures.

In the group, we develop and operate various nearfield microscopy setups to address different purposes, such as plasmonic waves propagating on noble metal surfaces, locally excited plasmon-polariton eigenmodes in metal-dielectric nanomaterials and metamaterials, or chemically synthesized nanostructures such as semiconductor nanowires. As one particular example, our group investigates the generation of a specific class of non-diffracting surface waves, so called plasmonic Airy beams, propagating on the surface of a gold film. Besides operating commercially available aperture and apertureless (scattering) SNOM setups, our group has put major efforts into the development of new SNOM techniques using two SNOM tips simultaneously on the same sample. Here, one tip is fed by an input signal exciting the probe at one local spot of the sample while the other tip collects the information on the field distribution from a different spot, disclosing the nearfield Green's function. Furthermore we operate heterodyne-based SNOM setups for simultaneous access to amplitude and phase-resolved optical nearfields. With all these techniques, we aim to provide a versatile toolbox that allows for the precise characterization and manipulation of plasmon excitation schemes on arbitrarily complex and heterogeneously nanostructured surfaces.

NANO-PLASMONIC OPTICS

Surface plasmon resonances (SPR) and SERS are two phenomena that intrinsically go together. SPR are excited on the surfaces of high roughness that provide sufficient coupling between the excitation light and the plasmon modes. The particular surfaces of our interest are small aggregates of spherical metallic (gold and silver) nano-particles (NPs) of 10-100 nm in diameter. Owing to their SPR, such NPs have very high absorption in the visible region. When illuminated with laser light, the incident electromagnetic field is strongly enhanced in the near vicinity of the NPs, sometimes also described by the field of the dipole moment induced in the metal.

Control of light properties on the nano-scale via the control of the NP aggregates geometry (*e.g.* dimers and trimers) is one of our research subjects. Recently, managing of the light polarization was demonstrated in a lab, utilizing a molecule adsorbed in the gap between the NPs (hot spots) to detect the light. Another ongoing project is the construction of a nano-lens: an adjustable nano-device, made of metallic NPs, which allows a systematic study of the properties of a molecule located in a hot-spot, as well as the construction of the precise NP geometry required to manipulate various light properties.

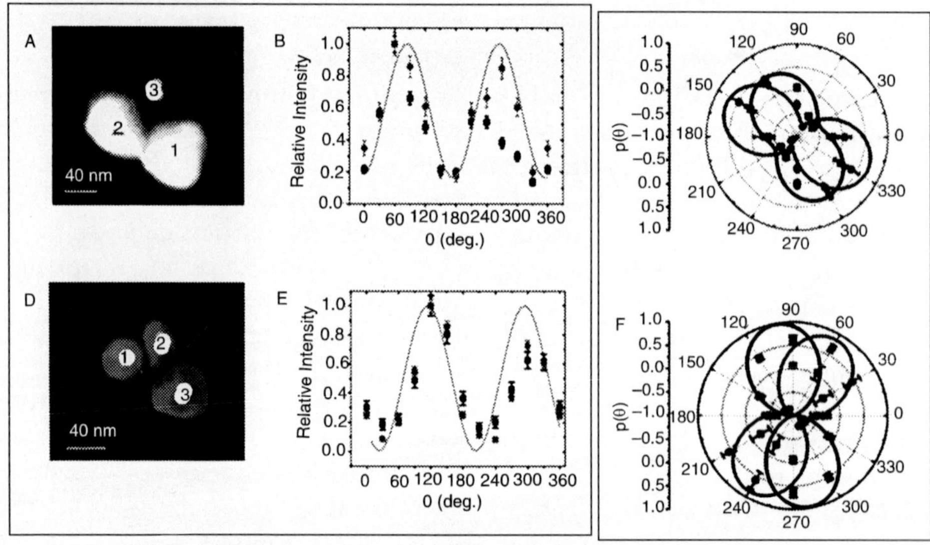

Figure: Polarization Response of a Nanoparticle Trimer :

(A) SEM image of a trimer. A red arrow indicates the position of the molecule that leads to the best agreement between experiment and calculation.

(B) Normalized RS intensity at 555 nm (black squares) and 583 nm (red circles) as a function of the angle of rotation of the incident polarization. The intensities at both wavelengths show approximately the same profile, but the maximal intensity is observed at 75°, which does not match any pair of nanoparticles in the trimer. The green line is the result of a calculation assuming that the molecule is situated at the junction with a gap of 1 nm marked with red arrow in the SEM image, and the corresponding geometrical parameters from the SEM image R_1 = 44 nm, R_2 = 35 nm, and R_3 = 28 nm as the input.

(C) Depolarization ratio (〉) measured at 555 nm (black squares) and 583 nm (red circles). Depolarization profiles are wavelength dependent in this case, and are aligned differently than the intensity profiles. The black and red lines show the result of calculations at the two wavelengths, assuming that the molecule is situated at the junction marked with the red arrow in SEM image.

(D) SEM image of a second trimer. A red arrow indicates the position of the molecule that leads to the best agreement between experiment and calculation.

(E) Normalized RS intensity at 555 nm (black squares) and 583 nm (red circles) as a function of the angle of rotation of the incident polarization. As in B, the intensity profile does not peak along the direction of the axis connecting particles 2 and 3.

(F) Depolarization ratio (〉) measured at 555 nm (black squares) and 583 nm (red circles). As in C, depolarization profiles are wavelength-dependent, and are aligned differently than the intensity profiles. The black and red lines show the result of calculations at the two wavelengths, assuming that the molecule is situated at the junction marked with red arrow in the SEM image.

CHARGE TRANSFER INTERACTION

Some of the details of the SERS mechanism are still not fully understood even after ~30 years of research. These involve charge transfer (CT) processes from the metallic surface to the adsorbed molecule and *vice-versa*. The existence of CT is suggested by experimental observations that cannot be predicted by the electromagnetic mechanism of SERS alone. The CT states originate from mixing between LUMO (HOMO) molecular orbitals and the conduction band of the metal. Recently, fluctuations of the metal work-function (which can modulate the CT process) were associated with SERS signal fluctuations and sampled by control of the surface diffusion rates in various environments. Study of the CT interaction on the single-molecule level, allowed by SERS, is one subject of the research conducted in our lab.

Our optical set-up includes laser-illuminated (CW and ps/fs) inverted microscope operated in the epi-illumination mode and coupled to a spectrograph equipped with back-illuminated CCD camera. Spherical nano-particles are deposited from a colloidal solution on the ITO slide, which acts also as a working electrode of an electrochemical cell used to tune the Fermi level of the metal into the resonance with the molecular states.

COMPLEX METAL OPTICS

Some metal optics makers focus just on the plating and polishing operations, which may be a workable strategy for simple optics. Complex metal optics, however, have specifications that are difficult or impossible to meet without a vertically integrated manufacturing operation.

True vertical integration means that a single supplier performs all the crucial processes of metal optics production, including:

- Machining
- Plating
- Polishing
- Diamond turning
- Coating Assembly
- Testing, metrology and environmental validation.

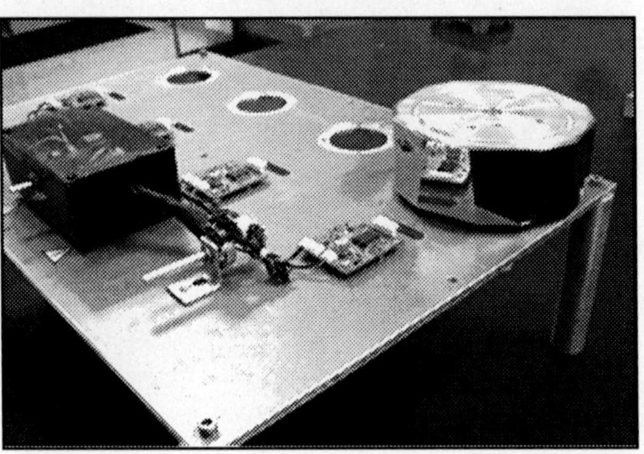

Each of these manufacturing steps has the potential to introduce optical or mechanical defects. And when these optical manufacturing steps are carried out by different suppliers, the defects can be difficult, time consuming and costly to find and fix for two simple reasons:

- Logistical risk. Moving parts back and forth between multiple suppliers adds time and cost to the production process. And all the additional handling during transit increases risk of damaged parts.
- Requirement misinterpretation. Even when governed by stringent specifications, different suppliers can have slightly different interpretations of the long list of requirements that typically define a complex optical component. Many of these requirements, by their nature, can be somewhat subjective. This problem of interpretation has practical implications in the form of inconsistent workmanship and poor optical performance.

SEAMLESS MANUFACTURING SYSTEM

By combining all the optical manufacturing steps into a seamless system, vertically integrated optics suppliers avoid the problems caused by a fragmented supply chain. This integrated approach minimizes threats of supply chain disruption, while optimizing cost, quality, and manufacturing cycle time.

Keep in mind that integrated assembly matters every bit as much as the optical production steps. Sloppy assembly practices can leave even the best-made metal optic susceptible to distortion from CTE mismatches or other sources of mechanical stress. Testing, too, shouldn't be overlooked as a valuable part of an integrated optics manufacturing operation. Aside from saving time and money compared to contract testing, in-house in-process testing provides the immediate feedback needed to fine-tune sequential optical manufacturing steps.

To take one example of the value of a vertically integrated manufacturing approach, we produce optical polygon motor assemblies and balance them in-house. Feedback from the balancing operation has led to improvements to the uniformity of material removed by diamond turning. The improved uniformity, in turn, simplifies the balancing process.

Had the assembly and balancing process taken place downstream of our integrated operation, our customers would not have been in a position to provide the ongoing feedback that led to our diamond turning process improvements. And our customers may have assumed the need for complex balancing procedures was inherent in the optical manufacturing processes.

Risk Mitigation

Vertical integration does more than ensure that all optical and mechanical requirements are satisfied and validated. Vertically integrated optics suppliers remove risk from the supply chain–by eliminating the possibility of non-performing subcontractors.

The technical and supply chain advantages of vertical integration carry much weight with the military, but civilian metal optics users can benefit too. Vertical integration makes sense whenever metal optics need to be at their best.

MIRRORS AND METAL OPTICS

Mirrors and metal optics are a vital part of many optical systems and Nalux has the equipment and expertise needed to create these optics to meet our customer's specifications. Mirrors and metal optics are almost always reflective in nature and are often chosen for one of two main reasons:

- There are physical stresses on the optical system or it is being used in an environment where it can be expected to receive some harsh treatment.
- There are thermal considerations, such as a high power laser being used, that may cause a transmissive optic such as a glass to either heat up to unacceptable temperatures or else the input energy can cause the optic to degrade so mirrors and metal optics are the answer.

As a strict definition, mirrors are an object with at least one highly polished surface. This polished surface usually has a metallic coating applied and then a protective coating is often applied over the metal coating in order to protect the metal coating from being scratched off. The metal coating is often applied by either a process known as sputter coating or by vacuum deposition coating. Nalux has both coating processes available for making mirrors.

Many inexpensive mirrors can be made from plastic as a core material. One side of the plastic mold has a very high polish applied so that this very smooth polished surface is imparted to the moulded plastic parts. After moulding, the plastic mirrors are sent to coating where the metallic and protective coatings are applied with the end result being high quality and inexpensive mirrors.

Another benefit of plastic mirrors is that multiple optical functions can be incorporated into the mirrors as opposed to just reflecting light. For example, by moulding a curvature into mirrors you not only reflect the light but you can also focus it. Additionally, if you superimpose a diffraction grating over the surfaces of curved mirrors, you can reflect the light, focus it, and even create diffraction and interference patterns. All of these things have their uses further making mirrors and metal optics as indispensible parts of optical systems.

SURFACE PLASMON POLARITONS

Surface plasmon polaritons (SPPs), are infrared or visible frequency electromagnetic waves trapped at or guided along metal-dielectric interfaces. These are shorter in wavelength than the incident light (photons). Hence, SPPs can provide a significant reduction in effective wavelength and a corresponding significant increase in spatial confinement and local field intensity.

Collective charge oscillations at the boundary between an insulating dielectric medium (such as air or glass) and a metal (such as gold, silver or copper) are able to sustain the propagation of infrared or visible frequency electromagnetic waves known as surface plasmon-polaritons (SPP). SPPs are guided along metal-dielectric interfaces much in the same way light can be guided by an optical fibre, with the unique characteristic of subwavelength-scale confinement perpendicular to the interface.

Surface plasmons (not SPPs), occur as light induced packets of electrical charges collectively oscillate at the surfaces of metals at optical frequencies. Under specific conditions, the light that radiates the object (incident light) couples with the surface plasmons to create self-sustaining, propagating electromagnetic waves known as surface plasmon polaritons (SPPs). Once launched, the SPPs ripple along the metal-dielectric interface and do not stray from this narrow path. Compared with the incident light that triggered the transformation, the SPPs can be much shorter in wavelength.

In other words, when SPs couple with a photon, the resulting hybridised excitation is called a surface plasmon polariton (SPP). This SPP can propagate along the surface of a metal until energy is lost either via absorption in the metal or radiation into free-space. Application of SPPs enables subwavelength optics in microscopy and lithography beyond the diffraction limit. It also enables the first steady-state micro-mechanical measurement of a fundamental property of light itself: the momentum of a photon in a dielectric medium. Other applications are photonic data storage, light generation, and bio-photonics.

EXPERIMENTAL APPROACHES

Nanoplasmonics

Nanofabricated systems that exploit SPPs demonstrate potential for designing and controlling the propagation of light in matter. In particular, SPPs can be used to channel light efficiently into nanometer scale volumes, leading to direct modification of resonate frequency dispersion properties (substantially shrinking the wavelength of light and the speed of light pulses for example), as well as field enhancements suitable for enabling strong interactions with nonlinear materials.

The resulting enhanced sensitivity of light to external parameters (for example, an applied electric field or the dielectric constant of an adsorbed molecular layer) shows great promise for applications in sensing and switching.

Current research is focused on the design, fabrication, and experimental characterization of novel components for measurement and communications based on nanoscale plasmonic effects. These devices include ultra-compact plasmonic interferometers for applications such as biosensing, optical positioning and optical switching, as well as the individual building blocks (plasmon source, waveguide and detector) needed to integrate a high-bandwidth, infrared-frequency plasmonic communications link on a silicon chip.

In addition to building functional devices based on SPPs, it appears feasible to exploit the dispersion characteristics of SPPs traveling in confined metallo-dielectric spaces to create photonic materials with artificially tailored bulk optical characteristics, otherwise known as *metamaterials.*

SCALAR WAVE EQUATION IN ONE SPACE DIMENSION

FROM HOOKE'S LAW

The wave equation in the one dimensional case can be derived from Hooke's law in the following way: Imagine an array of little weights of mass m interconnected with massless springs of length h. The springs have a spring constant of k:

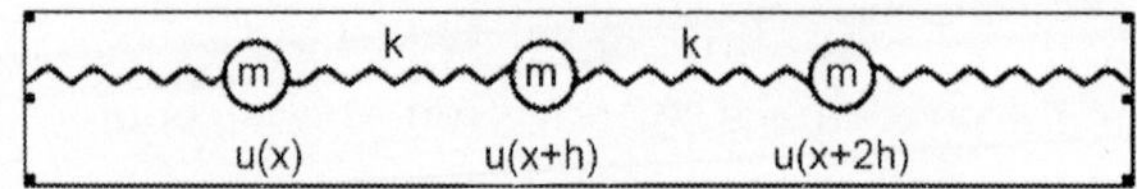

Here *u(x)* measures the distance from the equilibrium of the mass situated at *x*. The forces exerted on the mass *m* at the location $x + h$ are:

$$F_{Newton} = m \cdot a(t) = m \cdot \frac{\partial^2}{\partial t^2} u(x+h,t)$$

$$F_{Hooke} = F_{x+2h} + F_x = k\left[u(x+2h,t) - u(x+h,t)\right] + k[u(x,t) - u(x+h,t)]$$

The equation of motion for the weight at the location *x+h* is given by equating these two forces:

$$m\frac{\partial^2 u(x+h,t)}{\partial t^2} = k[u(x+2h,t) - u(x+h,t) - u(x+h,t) + u(x,t)]$$

where the time-dependence of $u(x)$ has been made explicit.

If the array of weights consists of *N* weights spaced evenly over the length $L = N\,h$ of total mass $M = N\,m$, and the total spring constant of the array $K = k/N$ we can write the above equation as:

$$\frac{\partial^2 u(x+h,t)}{\partial t^2} = \frac{KL^2}{M}\frac{[u(x+2h,t) - 2u(x+h,t) + u(x,t)]}{h^2}$$

Taking the limit $N \to \infty, h \to 0$ (and assuming smoothness) one gets:

$$\frac{\partial^2 u(x,t)}{\partial t^2} = \frac{KL^2}{M}\frac{\partial^2 u(x,t)}{\partial x^2}$$

$(KL^2)/M$ is the square of the propagation speed in this particular case.

General Solution

The one dimensional wave equation is unusual for a partial differential equation in that a relatively simple general solution may be found. Defining new variables:

$$\xi = x - ct \quad ; \quad \eta = x + ct$$

changes the wave equation into

$$\frac{\partial^2 u}{\partial \xi \partial \eta} = 0$$

which leads to the general solution

$$u(\xi, \eta) = F(\xi) + G(\eta) \quad \Rightarrow \quad u(x,t) = F(x - ct) + G(x + ct)$$

In other words, solutions of the 1D wave equation are sums of a right traveling function *F* and a left traveling function *G*. "Traveling" means that the shape of these individual arbitrary functions with respect to *x* stays constant, however the functions are translated left and right with time at the speed C. This was derived by Jean le Rond d'Alembert.

Another way to arrive at this result is to note that the wave equation may be "factored":

$$\left[\frac{\partial}{\partial t}-c\frac{\partial}{\partial x}\right]\left[\frac{\partial}{\partial t}+c\frac{\partial}{\partial x}\right]u=0$$

$$\Downarrow$$

$$\frac{\partial u}{\partial t}-c\frac{\partial u}{\partial x}=0 \quad \text{and} \quad \frac{\partial u}{\partial t}+c\frac{\partial u}{\partial x}=0$$

These last two equations are advection equations, one left traveling and one right, both with constant speed *c*.

For an initial value problem, the arbitrary functions *F* and *G* can be determined to satisfy initial conditions:

$$u(x,0)=f(x)$$

$$u_t(x,0)=g(x)$$

The result is d'Alembert's formula:

$$u(x,t)=\frac{f(x-ct)+f(x+ct)}{2}+\frac{1}{2c}\int_{x-ct}^{x+ct}g(s)ds$$

In the classical sense if $f(x)\in C^k$ and $g(x)\in C^{k-1}$ then $u(t,x)\in C^k$. However, the waveforms *F* and *G* may also be generalized functions, such as the delta-function. In that case, the solution may be interpreted as an impulse that travels to the right or the left.

The basic wave equation is a linear differential equation and so it will adhere to the superposition principle. This means that the net displacement caused by two or more waves is the sum of the displacements which would have been caused by each wave individually. In addition, the behaviour of a wave can be analysed by breaking up the wave into components, *e.g.* the Fourier transform breaks up a wave into sinusoidal components.

SCALAR WAVE EQUATION IN THREE SPACE DIMENSIONS

The solution of the initial-value problem for the wave equation in three space dimensions can be obtained from the solution for a spherical wave. This result can then be used to obtain the solution in two space dimensions.

SPHERICAL WAVES

The wave equation is unchanged under rotations of the spatial coordinates, because the Laplacian operator is invariant under rotation, and therefore one may expect to find solutions that depend only on the radial distance from a given point. Such solutions must satisfy

$$u_{tt}-c^2\left(u_{rr}+\frac{2}{r}u_r\right)=0.$$

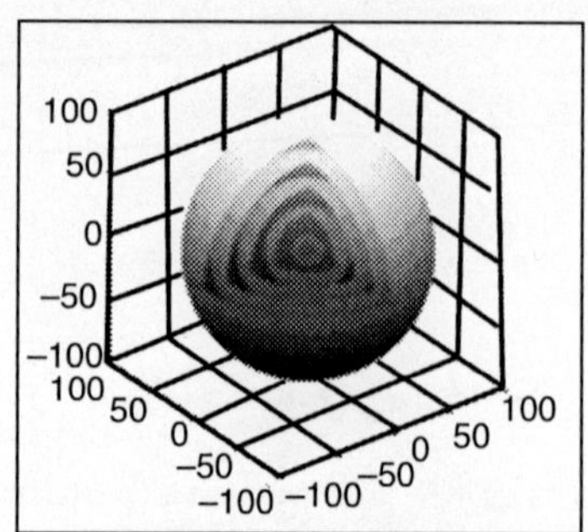

Figure: Cut-away of Spherical Wavefronts, with a Wavelength of 10 units, Propagating from a Point Source.

This equation may be rewritten as

$$(ru)_{tt} - c^2 (ru)_{rr} = 0;$$

the quantity ru satisfies the one-dimensional wave equation. Therefore there are solutions in the form

$$u(t,r) = \frac{1}{r}F(r-ct) + \frac{1}{r}G(r+ct),$$

where F and G are arbitrary functions. Each term may be interpreted as a spherical wave that expands or contracts with velocity c. Such waves are generated by a point source, and they make possible sharp signals whose form is altered only by a decrease in amplitude as r increases. Such waves exist only in cases of space with odd dimensions.

Monochromatic Spherical Wave

A point source is vibrating at a single frequency f with phase = 0 t = 0 at with a peak-to-peak magnitude of $2a$. A spherical wave is propagated from the point. The phase of the propagated wave changes as kr where r is the distance travelled from the source. The magnitude falls off as $1/r$ since the energy falls off as $1/r^2$. The amplitude of the spherical wave at is therefore given by:

$$u(t,r) = Re\left[\frac{a}{r}e^{i(\omega t - kr)}\right]$$

Solution of a General Initial-value Problem

The wave equation is linear in u and it is left unaltered by translations in space and time. Therefore we can generate a great variety of solutions by translating and summing spherical waves. Let $\varphi(\xi,\eta,\zeta)$ be an arbitrary function of three independent variables, and let the spherical wave form F be a delta-function: that is, let F be a weak limit of continuous functions whose integral is unity, but whose support (the region where the function is non-zero) shrinks to the origin. Let a family of spherical waves have centre at (ξ, η, ζ), and let r be the radial distance from that point. Thus

$$r^2 = (x-\xi)^2 + (y-\eta)^2 + (z-\zeta)^2.$$

If u is a superposition of such waves with weighting function φ, then

$$u(t,x,y,z) = \frac{1}{4\pi c}\iiint \varphi(\xi,\eta,\zeta)\frac{\delta(r-ct)}{r}d\xi d\eta d\zeta;$$

the denominator 4πc is a convenience.

From the definition of the delta-function, u may also be written as

$$u(t,x,y,z) = \frac{t}{4\pi}\iint_S \varphi(x+ct\alpha, y+ct\beta, z+ct\gamma)d\omega,$$

where α, β, and γ are coordinates on the unit sphere S, and ω is the area element on S. This result has the interpretation that $u(t,x)$ is t times the mean value of φ on a sphere of radius ct centered at x:

$$u(t,x,y,z) = tM_{ct}[\phi].$$

It follows that

$$u(0,x,y,z) = 0, \quad u_t(0,x,y,z) = \phi(x,y,z).$$

The mean value is an even function of t, and hence if

$$v(t,x,y,z) = \frac{\partial}{\partial t}\left(tM_{ct}[\psi]\right),$$

then

$$v(0,x,y,z) = \psi(x,y,z), \quad v_t(0,x,y,z) = 0.$$

These formulas provide the solution for the initial-value problem for the wave equation. They show that the solution at a given point P, given (t, x, y, z) depends only on the data on the sphere of radius ct that is intersected by the light cone drawn backwards from P. It does *not* depend upon data on the interior of this sphere. Thus the interior of the sphere is a lacuna for the solution. This phenomenon is called Huygens' principle. It is true for odd numbers of space dimension, where for one dimension the integration is performed over the boundary of an interval with respect to the Dirac measure. It is not satisfied in even space dimensions. The phenomenon of lacunas has been extensively investigated in Atiyah, Bott and Gεrding.

MIE SCATTERING

The Mie solution (Mie scattering) to Maxwell's equations (also known as the Lorenz–Mie solution (scattering), the Lorenz–Mie–Debye solution (scattering)) by Gustav Mie describes the scattering of electromagnetic radiation by a sphere. The solution takes the form of an analytical infinite series. The term Mie theory is used on occasion; however, it is misleading because it does not refer to an independent physical theory or law. The phrase "the Mie solution (to Maxwell's equations)" is therefore preferable. Currently, the term "Mie solution" is also used in broader contexts, for example when discussing solutions of Maxwell's equations for scattering by stratified spheres or by infinite cylinders, or generally when dealing with scattering problems solved using the exact Maxwell equations in cases where one can write separate equations for the radial and angular dependence of solutions.

A modern formulation of the Mie solution to the scattering problem on a sphere can be found in many books, *e.g.*, in J.A. Stratton's *Electromagnetic Theory*. In this formulation, the incident plane wave as well as the scattering field is expanded into radiating spherical vector wave functions. The internal field is expanded into regular spherical vector wave functions. By enforcing the boundary condition on the spherical surface, the expansion coefficients of the scattered field can be computed.

For particles much larger or much smaller than the wavelength of the scattered light there are simple and excellent approximations that suffice to describe the behaviour of the system. But for objects whose size is similar to the wavelength (*e.g.*, water droplets in the atmosphere, latex particles in paint, droplets in emulsions including milk, and biological cells and cellular components) this more exact approach is necessary.

The Mie solution is named after its developer, German physicist Gustav Mie. Danish physicist Ludvig Lorenz and others independently developed the theory of electromagnetic plane wave scattering by a dielectric sphere.

The formalism allows the calculation of the electric and magnetic fields inside and outside a spherical object and is generally used to calculate either how much light is scattered, the total optical cross section, or where it goes, the form factor. The notable features of these results are the Mie resonances, sizes that scatter particularly strongly or weakly. This is in contrast to Rayleigh scattering for small particles and Rayleigh–Gans–Debye scattering (after Lord Rayleigh, R. Gans and Peter Debye) for large particles. The existence of resonances and other features of Mie scattering, make it a particularly useful formalism when using scattered light to measure particle size.

MIE SCATTERING CODES

Mie solutions are implemented in a number of codes written in different computer languages such as Fortran, Matlab, Mathematica. These solutions are in terms of infinite series and include calculation of scattering phase function, extinction, scattering, and absorption efficiencies, and other parameters such as asymmetry of parameter or radiation torque. Current usage of "Mie solution" indicate series approximation to solution of Maxwell's equations. There are several known objects which allow such a solution: spheres, concentric spheres, infinite cylinders, cluster of spheres and cluster of cylinders, there are also known series solutions for scattering on ellipsoidal particles. For list of these specialized codes examine these articles

- Codes for electromagnetic scattering by spheres - solutions for single sphere, coated spheres, multilayer sphere, cluster of spheres
- Codes for electromagnetic scattering by cylinders - solutions for single cylinder, multilayer cylinders, cluster of cylinders.

A generalization that allows for a treatment of more general shaped particles is the T-matrix method, which also relies on the series approximation to solutions of Maxwell's equations.

MIE THEORY

Mie theory provides rigorous solutions for light scattering by an isotropic sphere embedded in a homogeneous medium. Extensions of Mie theory include solutions for core/shell spheres and gradient-index spheres. Although these theories are restricted to the case of a perfect sphere, the results have provided insight into the scattering and absorption properties for a wide variety of pigment systems, including non-spherical pigments. In most paper applications, where TiO_2 concentrations are relatively low (< 15% by volume), theoretical calculations predict the relative effects of particle size, particle composition, composition of the surrounding medium and wavelength of light. These trends correlate well with experimental data. In applications with TiO_2 concentrations greater than ~15% by volume, near-field optical interactions between neighbouring particles become significant and can dramatically impact macroscopic optical properties. The optical theories applied in the present study describe the light scattering properties of an isolated spherical particle and therefore cannot be applied to systems in which the particles are crowded together and near-field interactions between particles are significant.

The concepts of geometrical optics (refraction by lenses and reflection by mirrors) that are familiar in the macroscopic world do not adequately describe the interactions of particles with light when the particle size is comparable to the wavelength of the light. Rigorous optical theories such as Mie theory address the full complexity of vector electromagnetic quantities interacting with a particle.. The mathematics of Mie theory is straightforward but tedious, requiring the computation of a potentially large number of series expansions. Digital computers are ideally suited to this task. In the present study, the computer codes BHMIE and BHCOAT provided by Bohren and Huffman have been used to compute the scattering results presented, after slight modification to incorporate computation of the asymmetry parameter.

In general, when the light wavelength is similar to the particle diameter, light interacts with the particle over a cross-sectional area larger than the geometric cross section of the particle. The Mie calculation output provides this scattering cross section, Csca. Often this parameter is divided by the geometric cross-sectional area to give a dimensionless scattering efficiency parameter, Qsca.

$$Qsca = Csca/\pi\ r_2$$

However, in pigment applications, the formulation properties and costs depend on particle volume rather than cross-sectional area. Therefore, a more meaningful efficiency parameter is the scattering coefficient per micron, SCPM, defined as the scattering cross section divided by particle volume.

$$SCPM = Csca/(4\pi\ r_3/3) = 3Qsca/4r$$

Since the intensity of scattered light varies with the scattering angle, the asymmetry parameter must be considered 5 to give the scattering coefficient, S(mm-1), which is the preferred parameter for correlations with experimental data for systems in which multiple scattering is predominant. The relationship is

$$S(mm-1) = 0.75.\ (1 - \cos q).\ SCPM$$

where the asymmetry parameter, cos q, is the average cosine of the scattering angle, weighted by the intensity of the scattered light as a function of angle.

FRAUNHOFER DIFFRACTION THEORY AND MIE SCATTERING THEORY

With the laser diffraction/scattering method, particle size is specified from the light intensity distribution pattern. For this, the correspondence relationship between particle size and light intensity distribution pattern must be known in advance. The Fraunhofer diffraction theory and Mie scattering theory are used to obtain this relationship. That is, these theories are used to calculate what kind of light intensity distribution patterns are produced by particles of various sizes, and this data is stored beforehand on a computer as a parameter table (numerical table) containing a vast amount of information.

A considerable amount of time is required for calculating this parameter table. However, in actual measurement of particle size distribution, the measurement time is not affected at all since there are parameter tables already calculated and stored in computer memory.

Ok, then. Let's take a look at the relationship between the Fraunhofer diffraction theory and Mie scattering theory.

In a word, the Fraunhofer diffraction theory is one of the approximate expressions of the Mie scattering theory. This approximate expression can be used only when the following two conditions are satisfied:

- The particle size is relatively large (at least, 10 times the laser wavelength)
- The scattering angle is small (30° or less).

In a range in which these conditions are satisfied, the Fraunhofer diffraction theory can be used as a fairly valid approximate expression of the Mie scattering theory. Yet, when the particle size is smaller than ten times the laser wavelength, in particular, in measurement in the sub-micron region, the Fraunhofer diffraction theory can no longer be used since a wide-angle light intensity distribution pattern is required. In this case, the Mie scattering theory must always be used.

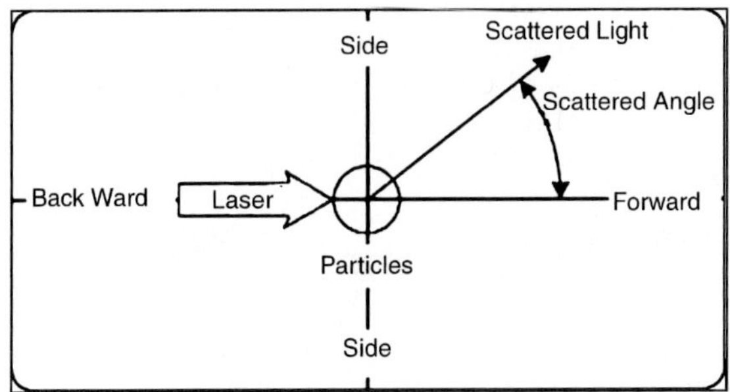

The Mie scattering theory is extremely complex and harder to understand than the Fraunhofer diffraction theory. This requires more complex programming and a fairly fast computer. It was for this reason that the Fraunhofer diffraction theory was only used in the past. On models that use only the Fraunhofer diffraction theory, measurement in the sub-micron region is impossible.

On the other hand, when the particle size is extremely large, the phenomenon of calculation error building up and causing inaccurate calculation results sometimes occurs if the Mie scattering theory is applied as it is to calculate the light intensity distribution pattern. Only in cases such as these, more accurate calculation results can be obtained for the light intensity distribution pattern by using the approximate expression provided by the Fraunhofer diffraction theory.

This is why the Fraunhofer diffraction theory and Mie scattering theory are jointly used at present for calculating the light intensity distribution pattern to ensure that a wide measurement range is covered.

Nevertheless, the Fraunhofer diffraction theory is merely used as a simple approximate expression of the Mie scattering theory only when the particle size is extremely large and the scattering angle is small. In this sense, the expression "jointly used" might be inappropriate since it will invite misunderstanding. It may be more appropriate to simply say that the light intensity distribution pattern is calculated based on the Mie scattering theory for all particle sizes.

So that the light intensity distribution pattern is calculated by using the Mie scattering theory, the refractive index of the particles and medium (liquid medium) must be set. Even when the particle size is the same, if it is smaller than ten times the laser wavelength, the light intensity distribution pattern will be slightly different depending on the refractive index. So, on actual analysers, many parameter tables set with different refractive indices are stored in memory for use. In actual measurement, a different parameter table is selected according to the refractive index to be used as a measurement condition.

On the SALD Series, selection for the refractive index is to be made, and completely separate parameter tables are selected in the case of wet measurement and dry measurement. For wet measurement, the refractive index of water is set to calculate the parameter table. And, for wet measurement, the refractive index of air is set for calculating the parameter table.

COMPUTATIONAL METHODS FOR NANOSCALE OPTICS

Conventional optics is characterized by material boundaries that are smooth on a wavelength scale. When this is the case, light can be understood as being composed of rays which are reected and refracted according to simple laws (Snell's). These simple laws are appropriate for designing conventional optical devices composed of lenses, prisms, mirrors, *etc.* Owing to the continuing progress in fabrication technologies, it has now become possible to build optical devices with highly detailed features on a wavelength scale. Such devices allow far greater exibility in the manipulation of light. Their design and analysis, however, must be based on electromagnetic wave theory and is, in general, more complicated. Although simple cases may be treated, to varying degrees, analytically, the use of computational methods is becoming more and more indispensable.

Computational methods differ in their generality, efficiency, and accuracy. Commercial packages, employing very general methods, can handle a wide range of problems but are usually not very efficient, and thus, turn out to be in- adequate in many cases. Our research focuses on the development of novel computational methods which we tailor to classes of problems and devices in the emerging nanooptics field. One of the results of this research is a public-domain package for the analysis of arbitrarily shaped optical waveguides. In a comparison made with other methods (including both commercial and non-commercial packages) on a benchmark problem, our package has yielded the best ratio of accuracy per computation time and memory.

OPTICAL METROLOGY AND NANOOPTICS DESIGN

The evolution of electric and magnetic fields in any well-defined nanostructure is described by Maxwell's equations. All information about the classical properties of light is contained in the electromagnetic fields, which is used to solve these differential equations. A variety of methods exist to perform rigorous simulations of Maxwell's equations. These include finite-element, finite-difference time-domain, Fourier-modal, and boundary-element approaches. However, especially for large 3D computational domains and highly conductive materials (like silver or gold), the electromagnetic fields of interest are so complicated that reaching accurate approximations becomes challenging for all numerical methods. It has even been suspected that nearly all published numerical results of light scattering off 3D metal nanostructures are of incorrect accuracy. To approach this challenge, we use dedicated finite-element method (FEM)-based simulation tools, which enable significantly improved accuracy and relatively short computation times.

Dedicated FEM Solver for Nanooptics Simulations

In a joint effort between the Zuse Institute Berlin and JCMwave, we are developing the JCMsuite finite-element solver, a dedicated tool for nanooptical simulations. It has been applied successfully to a wide range of electro-magnetic-field computations including metamaterials, photonic-crystal fibres, near-field microscopy, and optical microlithography. The performance of our methods is investigated and compared to competing approaches in a number of benchmark studies. The main ingredients for accurate performance are adaptive methods based on goal-oriented error estimators, higher-order vector elements, and fast direct and iterative numerical methods for solving matrix equations.

Field Enhancement of Plasmonic Nanostructures

Plasmon resonances can be used to tailor local field enhancements near metal nanostructures for specific applications. An example of two gold spheres separated by a small gap. When light of the proper wavelength λ_0 hits the structure, the electric-field intensity in the gap is enhanced by more than five orders of magnitude. These high local fields show promise for applications such as single-biomolecule sensing. Because of the multiscale geometry (gap ~1nm versus λ_0~1μm) and the highly conductive materials, this simple 3D setup presents a challenging problem for most rigorous Maxwell solvers. The simulated field distribution at resonance. FEM allows computation of the field enhancement to a relative accuracy of approximately one percent in a computation time of ~5s, and to an accuracy of about 0.01% in ~2min on a standard PC.

Light Propagation through Apertures in Metal Films

Light interaction with structured metal films (photomasks) plays an essential role in optical lithography. The simulated intensity in a cross section through a Z-shaped nanoaperture (microlithography test case, lateral domain size approximately $10\lambda_0 \times 10\lambda_0$, thickness ~$\lambda_0$). Note the characteristic intensity discontinuities at metal-vacuum interfaces and the smooth distribution after aerial imaging. Results from a convergence check for this test case are display. We compare the accuracies, which are reached at specific numerical settings for the amplitude of a diffraction order, and for the width of a specific feature in the aerial image. The relative error of the observable in the aerial image is at least one order of magnitude larger than the relative near-field error. This is contrary to the widespread misunderstanding that smoothing effects in aerial imaging would 'forgive' inaccurate near-field solutions. In dedicated experimental setups, light propagation through nanoapertures in metal films can lead to unexpected optical phenomena like enhanced transmission and beaming effects. Numerical simulations can help to design the corresponding devices and to understand the underlying physics.

Fast Simulations for Optical Metrology

FEM is the method of choice for accurate and fast simulations of light interaction with nanostructures.

We also use FEM for pattern reconstruction in optical metrology. However, optical inspection in a production environment requires very short computation times for high throughput. When complex patterns are involved, this can be an extremely demanding numerical task. We currently follow two promising approaches to overcome this challenge. First, we split simulations into 'expensive' offline tasks that typically have long computation times and only need to be performed once, and 'cheap' online simulations with very short computation times that are done often, to enable us to generate rigorous solutions to Maxwell's equations in real-time applications. Secondly, parallelized domain-decomposition approaches enable accurate solutions for very large computational domains.

LASERS RAMAN

DIAMOND DEBUTS IN RAMAN LASER

Diamond crystals of sufficient size and quality have been grown and used in compact Raman lasers for the first time.

Commercial availability of diamond crystals of sufficient size and quality has allowed a team in Australia to produce a diamond external cavity Raman laser emitting at 573 nm. A Raman laser's ability to convert the beam quality of the pump laser could see diamond-based Raman lasers used in applications as diverse as terahertz threat detection to ultraviolet lasers and directed energy weapons.

Diamond has long attracted interest as a laser material due to its high thermal conductivity, high Raman gain and broad optical transmission (ranging from 220 nm to 3 μm and 5-100 μm). Such properties make diamond Raman lasers useful as a wavelength converter to almost any spectral region.

The problem facing Mildren and colleagues is the lack of large, high-purity and quality diamond crystals. The solution adopted by the team was to use a low-defect single-crystal diamond produced by industrial diamond manufacturer, Element Six, UK. The diamond was grown by chemical vapour deposition to a size of 5 x 5 x 1.47 mm and exhibited excellent transmission at the pump and output wavelength (573 nm).

Figure: Diamond has Long Attracted Interest as a Laser Material due to its High Thermal Conductivity, High Raman Gain and Broad Optical Transmission (ranging from 220nm - 3 microns and 5-100 microns).

In the laser set-up, the diamond crystal was oriented at Brewster's angle and placed in between two resonator mirrors. The resonator mirrors were positioned as close as possible to the diamond crystal to form a cavity length of 15 mm.

The Raman laser was pumped through the input resonator mirror using a frequency-doubled Nd:YAG laser emitting at 10 ns pulses at 532 nm at a repetition rate of 10 Hz.

"The input mirror is highly transmitting at the pump wavelength and highly reflecting at the Raman shifted wavelengths (555-650 nm)," explained Mildren. *"The pump beam is reflected by the second end-mirror and so makes a double pass of the Raman medium. The second end-mirror is also reflective at 573 nm so that the cavity resonated at this wavelength to give an overall conversion efficiency of 13%."*

SPIN-FLIP LASER RAMAN

PRINCIPLE

Stimulated scattering by electrons in a semiconductor

Incoming light with a frequency ν_p is scattered at a frequency ν_R as :

$$h\,\nu_R = h\,\nu_p \pm \Delta E$$

The spin-flip Raman effect involves a process in which the spin of an electron from the conduction band of the semiconductor crystal is reversed relative to the direction of an external magnetic field B.

$$\Delta E = g.\ m_B.\ B$$

g is the effective gyromagnetic ratio of the electron (which can take high values in a crystalline environment) and m_B is the Bohr magneton.

The high value of g in semiconductors (about 50 for InSb) lead to area sweep of nearly 5×10^{12} Hz for magnetic fields ranging up to 100 kilogauss.

The scattering cross section spin flip Raman increases with a resonant manner when the photon energy pumping approaches the value of the bandgap E_G.

Historical

Stimulated Raman scattering processes have been applied in the development of tunable spin-flip Raman lasers such as disclosed by Chandra K.N. Patel and Earl. D. Shaw, Physical Review B3, 1279 (1971). Typically, previous devices consisted of focusing the optical pumping radiation from either a carbon monoxide laser at 5 μm wavelengths or a carbon dioxide laser at 10 μm wavelengths into a crystal of indium antimonide (InSb) which is held in a cryostat at a temperature of about T = 1K to 30K; a variable magnetic field is applied to the indium antimonide crystal to produce and change the frequency of the spin-flip Raman laser output.

The Raman scattering material which may be indium antimonide crystals shaped as a parallelepiped with dimensions of a few millimetres along each edge. At least two opposite sides are polished planes and parallel to form a Fabry-Perot type optical resonator.

Generally, the polished sides are chosen to be those through which the pumping radiation enters and leaves the crystal. The spin-flip Raman laser operates using a semiconductor in a magnetic field. Each energy level in the conduction band splits into two levels, one with electron spin parallel to the magnetic field and the other with spin antiparallel. There may be transition involving Raman scattering of radiation, in which the electron spin changes its alignment. The wavelength of the laser is varied by changing the magnetic field. Tuning over a range of several micrometers in the infrared is easily possible. Spin-flip Raman lasers have been developed using InSb, $Hg_{0.77}Cd_{0.23}Te$ and $Pb_{0.88}Sn_{0.12}Te$. At least one model of an InSb spin-flip Raman laser is marketed but most work is still in the research stage.

AN ALL-SILICON-BASED RAMAN LASER

The silicon takes the lion's share of industry in micro-electronics but is not used in photonic applications because it does not emit enough light (indirect gap) and is transparent in the infrared.

Researchers at the University of California Los Angeles were discovered, in 2005, a way to exploit the Raman effect and to act a laser on silicon.

The realization of a compact Raman laser all-silicon on a silicon chip so literally opens a new field of research through Haisheng Rong and his colleagues from Intel laboratories of Santa Clara, California, and Jerusalem in Israel.

Experiment

Rong and his colleagues began to realise a waveguide in silicon giving it the form of an "S" using standard techniques of etching and photo-lithography.

The waveguide has an internal cross section of about 1.6 square microns and a length of 4.8 centimeters. They then produces a Raman laser optical cavity by coating one facet of the waveguide with a material highly reflective.

The generation of radiation is from a diode in indium phosphide InP (1536 nm). This wavelength was chosen to be in the area of transparency of silicon. The light thus generated is made coherent in a waveguide in silicon ten μm in width thanks to the Raman scattering.

The manufacturing processes of conventional integrated circuits are used for the waveguide.

The fixation of indium phosphide to silicon is by adherence: a cold oxygen plasma is applied to the 2 parts creating a layer of 25 atoms of oxygen which will keep the silicon and indium phosphide together for a more compact realization.

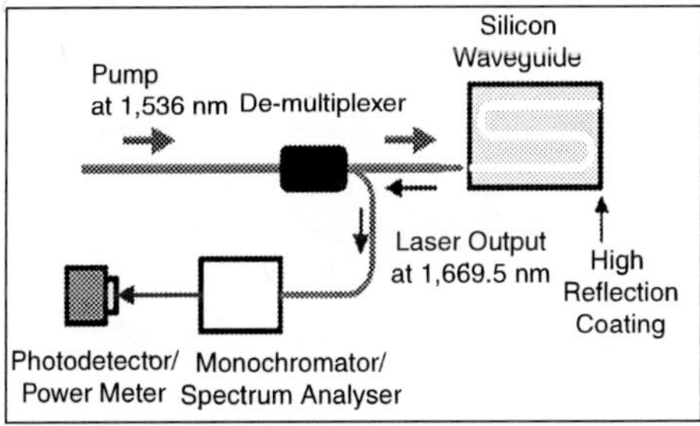

Figure: Set-up for Raman Lasing Experiment.

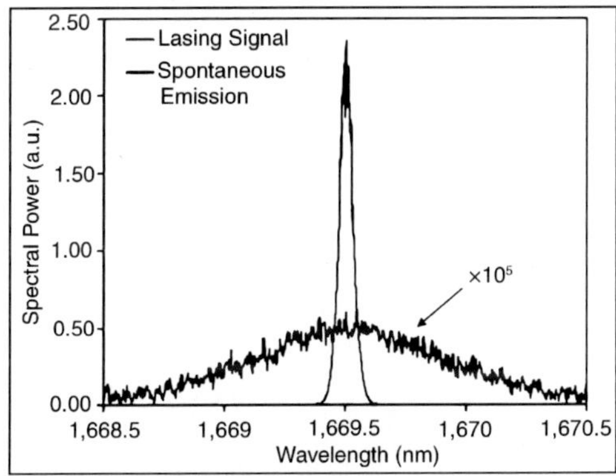

Figure: Silicon Raman Spectra.

Results

By pumping the cavity with pulses from a laser operating at a wavelength of 1536 nanometers (6510 cm^{-1}), the team observed an output signal of a wavelength of 1669.5 nanometers (that is to say à 520 cm^{-1} from pumping laser) when the pump power reached 0.4 microWatts. It is the silicon Raman line at 520 cm^{-1} which is revelated and amplified in the optical cavity shaped as a waveguide

However, this only happened over a period of about 100 nanoseconds delay after which a phenomenon called "two-photon absorption" has stopped the laser activity.

This demonstration of Raman lasing in a silicon waveguide cavity on a single chip represents a significative step towards a more practical, all-silicon-based, c.w., amplifier or laser. We note that the multilayer coatings are not essential for this device, and that single chip resonator architectures, both these alternatives are compatible with CMOS processing. Further reduction in the carrier lifetime or cavity optimization could lead to a c.w. Raman lasing in silicon.

THE FUTURE OF RAMAN LASER WITH PHOTONIC CRYSTALS

Photonic crystals are composed of periodic dielectric or metallo-dielectric nanostructures that affect the propagation of electromagnetic waves (EM) in the same way as the periodic potential in a semiconductor crystal affects the electron motion by defining allowed and forbidden electronic energy bands. Essentially, photonic crystals contain regularly repeating internal regions of high and low dielectric constant. Photons (behaving as waves) propagate through this structure - or not - depending on their wavelength.

Wavelengths of light that are allowed to travel are known as modes, and groups of allowed modes form bands. Disallowed bands of wavelengths are called photonic band gaps. This gives rise to distinct optical phenomena such as inhibition of spontaneous emission, high-reflecting omni-directional mirrors and low-loss-waveguiding, amongst others.

Since the basic physical phenomenon is based on diffraction, the periodicity of the photonic crystal structure has to be of the same length-scale as half the wavelength of the EM waves *i.e.* ~350 nm (blue) to 700 nm (red) for photonic crystals operating in the visible part of the spectrum - the repeating regions of high and low dielectric constants have to be of this dimension. This makes the fabrication of optical photonic crystals cumbersome and complex.

Silicon photonics offers the promise of low-cost and high frequency devices due to its potentially high degree of integration as well as to its large optical bandwidth. Despite the recent advances in the field of silicon photonics, the fabrication of active optical functions and, in particular, the fabrication of lasers remains one of the most difficult task. Until now, except for the approaches that use III-V materials, only the use of a nonlinear effect, the stimulated Raman scattering, has allowed to obtain a laser effect in a rib waveguide etched on a silicon-on- insulator (SOI) substrate.

One of the limitations of these reported Raman scattering studies is that it requires long cavities (> 1 cm) or high Q factor ring cavities with large area (about 1 cm^2) to achieve the gain required by the laser effect.

To overcome this limitation, that is incompatible with the on-chip integration of these components, photonic crystals (PhCs) appear as compact alternatives.

Photonic crystal waveguides in SOI can strongly confine light due to their reduced width and height in a manner similar to photonic wires in which spontaneous and stimulated Raman scattering has already been observed in SOI. Moreover, the existence of slow modes in PhCs, the group velocity of which can be 100 times smaller than the light in vacuum, can theoretically enhance further the spontaneous scattering.

In this work, we demonstrate the enhancement of the spontaneous Raman scattering in PhC waveguides on SOI compared to photonic wires of similar dimension. Contrary to works mentioned before, the present waveguides are short (50 μm long) and made in SOI where the oxide below the PhC has not been removed.

Doing so, we do not rely on the fragile membrane structures that are incompatible with a good thermal dissipation and with the on-chip integration. We show a more than ten-fold enhancement of the Raman scattering efficiency in a W1 photonic crystal waveguide as compared to a single mode ridge waveguide.

RAMAN-ACTIVE OPTICAL AMPLIFICATORS

Telecommunications systems still need more speed from the explosion of the Internet bubble. The optical amplification is a technique always pushing the limits of compromise distance/speed well known in these systems. For the past fifteen years thanks to evolution of manufacturing technologies of powerful lasers, optical amplification by Raman effect is an interesting alternative to other amplification systems.

THE TECHNOLOGY OF TELECOMMUNICATIONS WITH OPTIC FIBRE

An optical fibre is a glass or plastic fibre that carries light along its length. Fibre optics is the overlap of applied science and engineering concerned with the design and application of optical fibres. Optical fibres are widely used in fibre-optic communications, which permits transmission over longer distances and at higher bandwidths (data rates) than other forms of communications. Fibers are used instead of metal wires because signals travel along them with less loss, and they are also immune to electromagnetic interference.

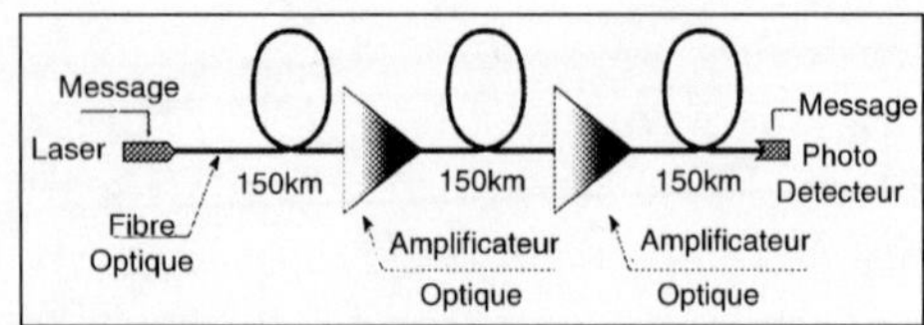

Figure: Standard Optical Telecommunication Link.

The light emitted by the laser signal is amplitude modulated to encode the message to send: a digital message the encoding is simply a series of ignition/extinction of laser light to represent respectively the 0 and 1. This modulated light is injected into the optical fibre and will thus be able to propagate to the output fibre. On output a photodetector translated into electrical signals emitted the message.

The fibres used in long-distance telecommunications applications are always silica *(SiO_2)*, because of the lower optical attenuation. Both multi-mode and single-mode fibres are used in communications, with multi-mode fibre used mostly for short distances, up to 550 m (600 yards), and single-mode fibre used for longer distance links. Because of the tighter tolerances required to couple light into and between single-mode fibres (core diameter about 10 micrometers), single-mode transmitters, receivers, amplifiers and other components are generally more expensive than multi-mode components.

Silica exhibits fairly good optical transmission over a wide range of wavelengths. In the near-infrared (near IR) portion of the spectrum, particularly around 1.5 μm, silica can have extremely low absorption and scattering losses of the order of 0.2 dB/km. A high transparency in the 1.4 μm region is achieved by maintaining a low concentration of hydroxyl groups (OH^-). Alternatively, a high OH^- concentration is better for transmission in the ultraviolet (UV) region.

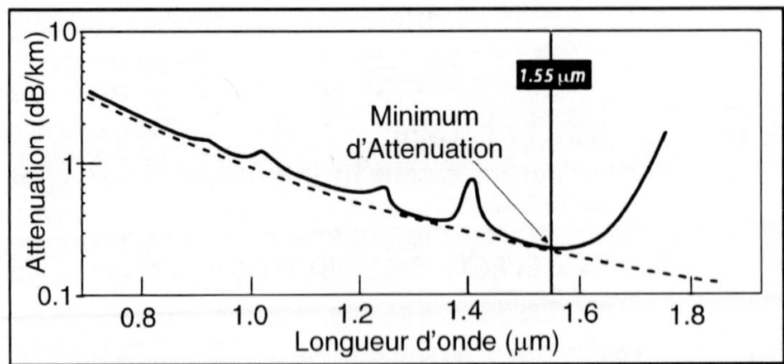

Figure: Optical Attenuation in the Silica Fibre.

Silica can be drawn into fibres at reasonably high temperatures, and has a fairly broad glass transformation range. One other advantage is that fusion splicing and cleaving of silica fibres is relatively effective. Silica fibre also has high mechanical strength against both pulling and even bending, provided that the fibre is not too thick and that the surfaces have been well prepared during processing. Even simple cleaving (breaking) of the ends of the fibre can provide nicely flat surfaces with acceptable optical quality. Silica is also relatively chemically inert. In particular, it is not hygroscopic (does not absorb water).

Optical attenuation of silica presents a minimum at the wavelength λ_s = 1,55 μm as shown down there.

The designers of optical telecommunication systems therefore use for the transmission wavelength of 1.55 μm in order to benefit from the minimum attenuation. It follows that in practice the average distance that you can browse through such fibre without amplification system is of approx 150 km.

To overcome this physical limit most systems initially within a reamplification of signals every 150 km. An optical fibre doped with certain rare earth elements such as erbium can be used as the gain medium of a laser or optical amplifier. Rare-earth doped optical fibres can be used to provide signal amplification by splicing a short section of doped fibre into a regular (undoped) optical fibre line. The doped fibre is optically pumped with a second laser wavelength that is coupled into the line in addition to the signal wave. Both wavelengths of light are transmitted through the doped fibre, which transfers energy from the second pump wavelength to the signal wave.

AMPLIFICATION BY STIMULATED RAMAN EFFECT INTO SILICA FIBRE

Another emerging technique allows it to obtain an optical amplification in the standard silica fibre without introducing dopant. It is the transmission fibre itself will serve as amplifying medium. The energy of the laser pump - this time hold on the wavelength of 1.45 μm - is then transferred to the signal through an optical nonlinear called "stimulated Raman effect". This effect may be caused again under the influence of the pump laser but without the fibre has special characteristics. The average distance was obtained today by Raman amplification without interrupting the fibre for reamplification is about 500 km.

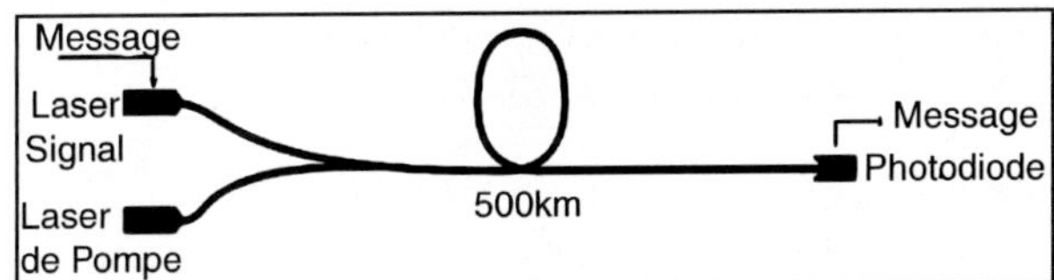

Figure: Optical Link using Raman Effect.

STIMULATED RAMAN

This is indeed the search for optical amplification: under the impact of photons-signals, the matter constituting the fibre with its density of states of phonons, excited by the presence of the pump will promote change in length wavelength photons from the pump the mechanics of the stimulated Raman effect. Thus, if we interact via the Raman effect, a pump photon and a photon signal we obtain two photons at the wavelength signal.

It was thus a phenomenon of "duplication of photons" is the signal wavelength which corresponds to the concept of amplification.

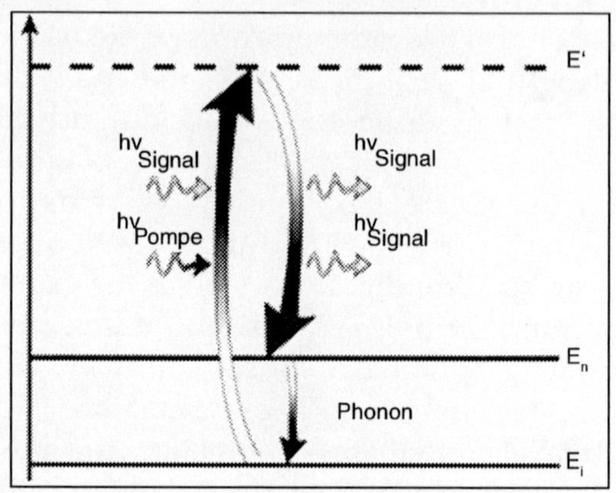

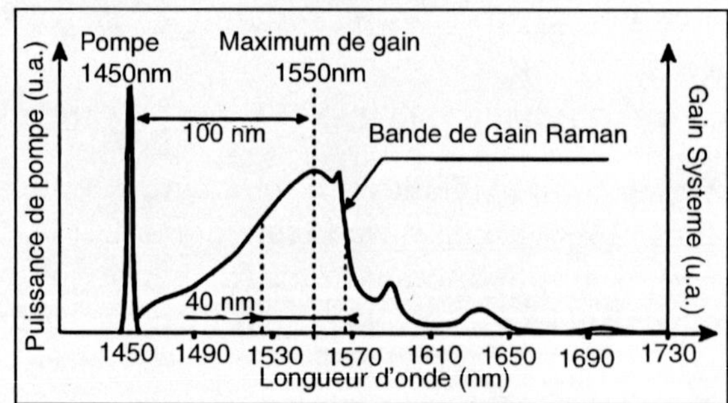

Figure: Principle of the Stimulated Raman Effect Bandwidth for a Raman Amplificator Showing the Vibrational Density of States of the Undoped Silica

APPLICATIONS OF RAMAN

THE ADVANTAGES OF RAMAN SPECTROSCOPY

Whether the goal is qualitative or quantitative data, Raman analysis is one of the few technique which can provide key information, easily and quickly, detailing the chemical composition and the structure of the investigated materials.

Localized or extended (mapping) investigations can be done under microscope by using micro-Raman (Raman micro-probe, Raman laser micro-probe) and, more recently, by using NanoRaman™.

Raman spectroscopy is :

- Non destructive
- Non invasive
- No need of sample preparation
- Information rich
- Need only some milligram (or picogram) of material (gas, liquid or solid).

Raman spectrometers detect also spin waves (magnons) in semi-magnetic crystals.

As for phonons, spin waves with small wave number (near zero, that means at the centrum of the Brillouin zone) are only detected.

Get the full picture of the sample properties:

- Chemical composition & component identification
- Molecular structure characterisation
- Phase transition monitoring
- Components distribution at micron & sub-micron scale
- Stress measurements
- Hyperspectral imaging
- Reaction monitoring
- High throughput screening.

At last, measurements were done usually at room temperature. Nevertheless, they can be done at low or high temperature by using appropriate cells (cryostat or furnace with glass or quartz window(s)).

They also can be done under hydrostatic our uniaxial pressure by using diamond anvil cells.

Such equipments were very convenient to simulate geological condition under the earth's crust.

Application scope of Raman spectroscopy:

- Pharmaceutical industry
- Petrochemistry, industrial chemistry, polymers
- Material science
- Semiconductors, photovoltaic materials
- Life science (human, animal and plants : biology, biochemistry, biophysic, virus, tumour...)
- Geology and mineralogy, gemmology
- Nuclear science, isotopic effects
- Forensics sciences, forensic science laboratories.
- Art, archeology, paints, pigment, stained-glass windows, earthenwares, potteries, inks...
- ... and many more.

Detailed data informations obtained from a Raman spectrum:

- Lines position give informs about compound nature, by comparition with databases.

Variations of Force Constant

Strains induce a frequency shift from the standard value. If the material or the molécule is compressed, the frequency shift is positive. The force constant and the frequency increase, and *vice-versa.*

Isotopic Effects

The isotopic effects concern the nucleus of atoms that make up the compound. Two atoms are called isotopes they have the same number of protons.

The number of protons in the nucleus of an atom is designated by the atomic number.

Two isotopes have the same atomic number. What distinguishes them is that they have a different atomic mass. The atomic mass of an atom is the number of nucleons that contains the nucleus of the atom. The atomic mass difference is due to a different number of neutrons N.

One can cite the example of the widespread natural carbon-12 ^{12}C and carbon-13 ^{13}C.

Raman spectroscopy can reveal the relative proportion of isotopes using the mass of each of them as has been shown a section of the harmonic oscillator. For the same vibration mode, there are as many lines in the Raman spectrum of the compound as there are isotopic combinations (2 isotopes = 3 combinations).

- Ratio between the various peaks intensity informs about the realative composition of the compound sample. With organized structures (like crystals), the ratio informs about directional effects (polarised measurements).
- Wide at half-maximum informs in general about the disorder factor, composition homogeneity or crystalline quality of a structured material.

Mapping Information from Raman Imaging

The new generation of confocal Raman microscopes offers a non-destructive and non-contact method of sample analysis at the sub-micron level.

More particularly, Raman imaging provides the spatial distribution of the various molecular species within the sample, making it possible to produce fast and accurate chemical images.

- Chemical component distribution
- Phase mapping and polymorph identification
- Micro-analysis of stress and molecular orientation
- Identification of functional groups in organic and biological samples.

According to the reported parameter : peak position, integrated intensities, width at half-maximum, Raman imaging informs about composition, stress, relative abundance or crystalline quality.

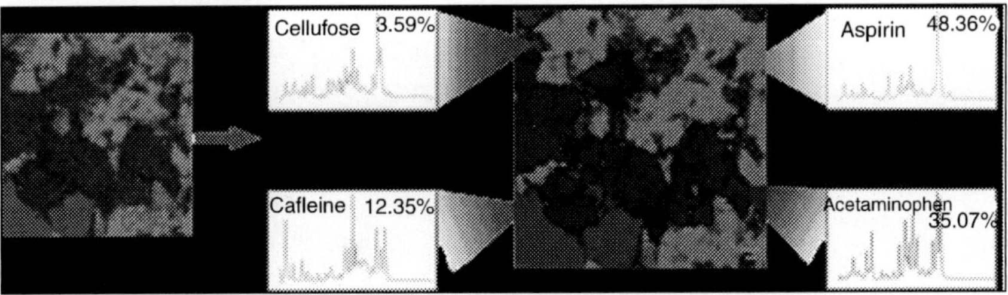

Figure : Raman Image of a Pharmaceutical Tablet showing the Spatial Distribution of the Different Components and their Corresponding Raman Spectra.

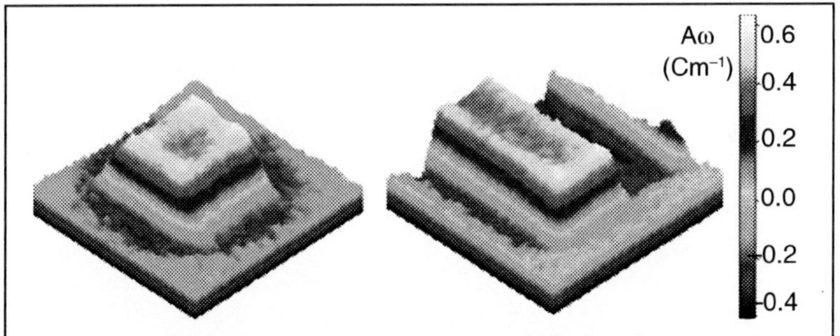

Figure : Shift of the silicon Raman peak, Dw, from the Stress Free Frequency as a Function of the Position on 3 μm Wide Active Square Si Region (left) and line (right) Isolated by PBLOCOS.

COMBINATION WITH MULTIPLE LASER WAVELENGTHS

Depth profiles and surface isolation are best resolved using the true confocal Raman microscope and multiple laser wavelengths.

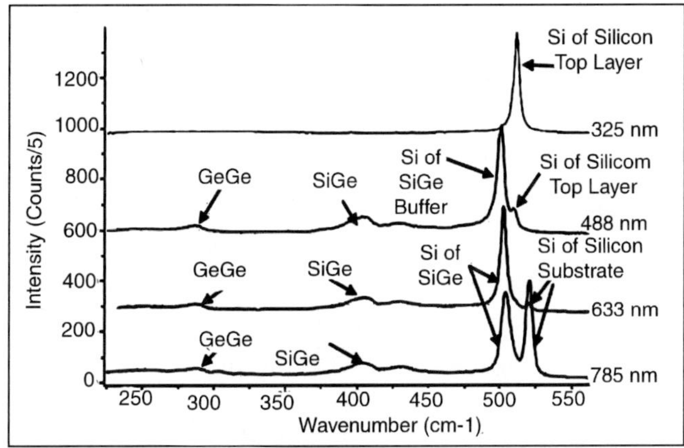

Figure : Spectra Recorded at 325, 488, 633 and 785 nm on Structure of Si/SiGe/on Si Wafer. The Relative Ge Content in the Constant $Si_{1-x}Ge_x$ Layer is Calculated from the Visible Raman Spectrum and the Strain of the Cap Si Layer is Derived from the UV Raman Spectrum.

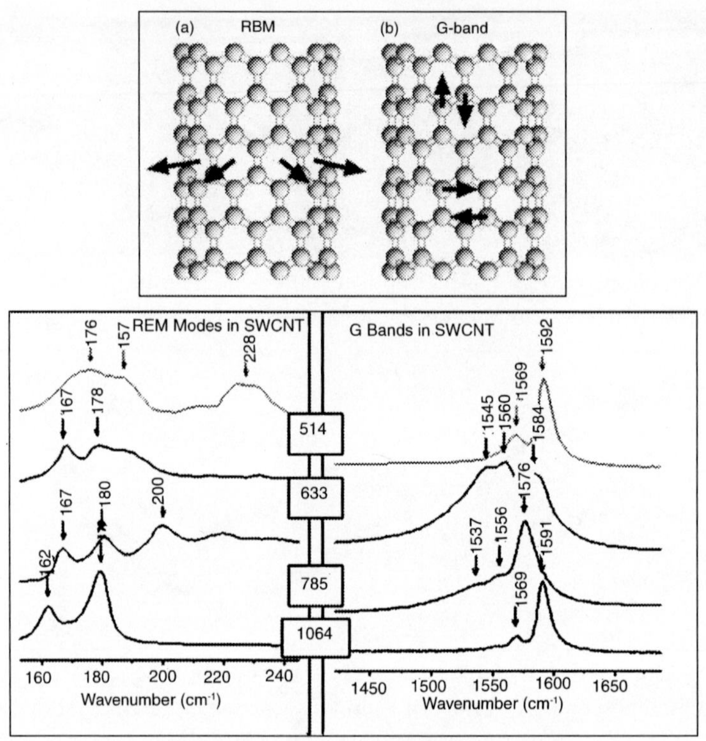

Figure : Carbon Nanotube (SWCNT and MWCNT) Characterisation by Exciting at Different Wavelengths Allows us Sweep Several Resonance Conditions to Probe the Range and Distribution of SWCNT's Diameters.

MicroRaman and Microphotoluminescence Laser Induced

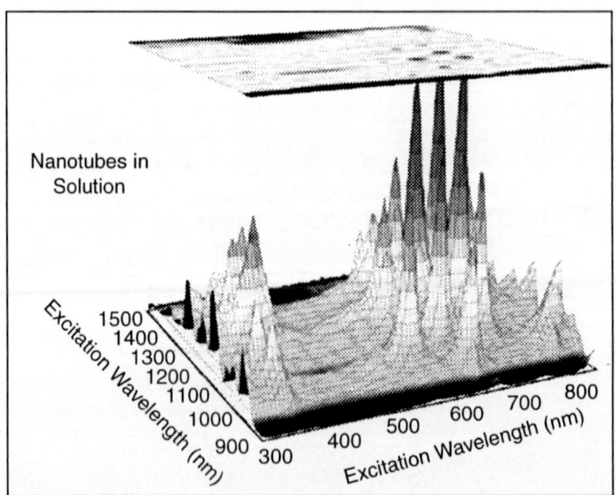

By the nature of the process, laser excitation induces not only Raman effect, but also photoluminescence characterized by lines, bands or continuum.

It is verified experimentally and theoretically that laser excitation by IR or red light induces less photoluminescence than visible or UV light.

Visible photoluminescence measured with a confocal Raman microprobe have similar advantages for spatial and spectral resolution as for microRaman.

The result of all this is that Raman effect and photoluminescence can be recorded simultaneously during the same laser shoot.

RESONANT-CAVITY-ENHANCED PHOTO DETECTOR

Resonant-cavity-enhanced photo detectors (or, RCE photo detectors) enable improved performance over their predecessors by placing the active device structure inside a Fabry–Pérot resonant cavity. Though the active device structure of the RCE detectors remains close to other conventional photodetectors, the effect of the optical cavity, which allows wavelength selectivity and an enhancement of the optical field due to resonance, allows the photo detectors to be made thinner and therefore faster, while simultaneously increasing the quantum efficiency at the resonant wavelengths.

Advantages

The quantum efficiency of conventional detectors is dominated by the optical absorption (electromagnetic radiation) of the semiconductor material. For semiconductors with low absorption coefficients, a thicker absorption regions is required to achieve higher quantum efficiency, but at the cost of the Bandwidth (signal processing) of photodetectors.

A RCE detector improves the bandwidth significantly. The constructive interference of a Fabry–Perot cavity enhances the optical field inside the photodetector at the resonance wavelengths to achieve a quantum efficiency of close to unity. Moreover the optical cavity makes the RCE detectors wavelength selective. This makes RCE photodetectors attractive for low crosstalk wavelength demultiplexing. Improved quantum efficiency gives less power consumption. Higher bandwidth gives faster operation.

The RCE photodetectors have both wavelength selectivity and high speed response making them ideal for wavelength division multiplexing applications. Optical modulators situated in a optical cavity require fewer quantum wells to absorb the same fraction of the incident light, and can therefore operate at lower voltages. In the case of emitters, the cavity modifies the spontaneous emission of light-emitting diodes (LED) improving their spectral purity and directivity.

Thus optical communication systems can perform much faster, with more bandwidth and can become more reliable. Camera sensors could give more resolutions, better contrast ratios and less distortion. For these reasons, RCE devices can be expected to play a growing role in optoelectronics over the coming years.

THEORY OF RCE PHOTO DETECTORS

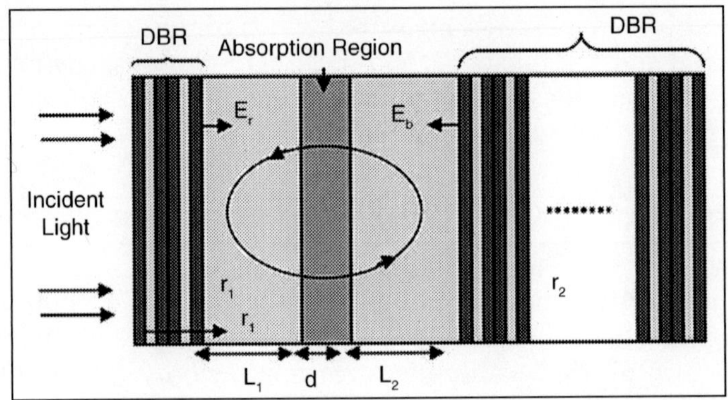

Figure: Schematic of a RCE Photodetector Model.

The RCE photo detectors can provide:

- Higher quantum efficiency,
- Higher detection speed,
- Wavelength selective detection, than compare to a conventional photodiode.

Quantum Efficiency of RCE Photo Detectors

The RCE photodetectors are expected to have higher quantum efficiency η than compare to conventional photodiodes. The formulation of η for RCE devices gives insight to the design criteria.

A generalized RCE photodetector schematic, can give the required theoretical model of photodetection. A thin absorption region of thickness d is sandwiched between two relatively less absorbing region, substrate, of thickness L_1 and L_2.

The optical cavity is formed by a period of $\lambda/4$ distributed Bragg reflector (DBR), made of non-absorbing larger bandgap materials, at the both end of the substrate. The front mirror has a transmittance of t_1 and generally has lower reflectivity than compare to the mirror at back ($R_1 < R_2$). Transmittance t_1 allows light to enter into the cavity, and reflectivity R_1 ($= r_1^2$) and R_2 ($= r_2^2$) provides the optical confinement in the cavity.

The active region and the substrate region have absorption coefficient α and α_{ex} respectively. The field reflection coefficients of the front and the back mirrors are $r_1 e^{-j\phi_1}$ and $r_2 e^{-j\phi_2}$ respectively, where ϕ_1 and ϕ_2 are the phase shifts due to the light penetration into the mirrors.

The optical microcavity allows building up an optical field inside the optical cavity. In compare to conventional detector, where light is absorbed in a single pass through the absorption region, for RCE detectors trapped light is absorbed each time it traverses through the absorption region.

The Quantum efficiency η for a RCE detector is given by:

$$\eta = (1-R_1)(1-e^{-\alpha d})\left[\frac{(e^{-\alpha_{ex}L_1} + r_2^2 e^{-\alpha_{ex}L_2 - \alpha_c L})}{1-2r_1 r_2 e^{-\alpha_c L}\cos(2\beta L + \phi_1 + \phi_2) + (r_1 r_2)^2 e^{-\alpha_c L}}\right]$$

Here $\alpha_c = (\alpha_{ex}L_1 + \alpha_{ex}L_2 + \alpha d)/L$. In practical detector design $\alpha_{ex} << \alpha$, so α_{ex} can be neglected and can be given as:

$$\eta = (1-R_1)(1-e^{-\alpha d})\left[\frac{(1+R_2 e^{-\alpha d})}{1-2\sqrt{R_1 R_2}e^{-\alpha_c d}\cos(2\beta L + \phi_1 + \phi_2) + (R_1 R_2)e^{-\alpha_c d}}\right]$$

The term inside the represents the cavity enhancement effect. This is a periodic function of $2\beta L + \phi_1 + \phi_2$, which has minima at $2\beta L + \phi_1 + \phi_2 = 2m\Pi$. And η enhanced periodically at resonance wavelength that meets this condition. The spacing of the resonant wavelength is given by the Free Spectral Range of the cavity. The peak value of η at resonant wavelength is given as:

$$\eta = (1-R_1)(1-e^{-\alpha d})\left[\frac{(1+R_2 e^{-\alpha d})}{(1-\sqrt{R_1 R_2}e^{-\alpha_c d})^2}\right]$$

for a thin active layer as αd<<1, η becomes:

$$\eta = (1-R_1)\alpha d\left[\frac{(1+R_2 e^{-\alpha d})}{(1-\sqrt{R_1 R_2}e^{-\alpha_c d})^2}\right]$$

This is a significant improve from the quantum efficiency of a conventional photodetector which is given by:

$$\eta = (1-R)\alpha L.$$

This shows that higher quantum efficiency can be achieved for smaller absorption region. The critical design requirements are : a very high back mirror reflectivity and a moderate absorption layer thickness. At optical frequencies metal mirrors have low reflectivity (94%) when used on materials like GaAs. This makes metal mirrors inefficient for RCE detection. Whereas distributed Bragg reflector (DBR) can provide reflectivity near unity, are ideal choices for RCE structures.

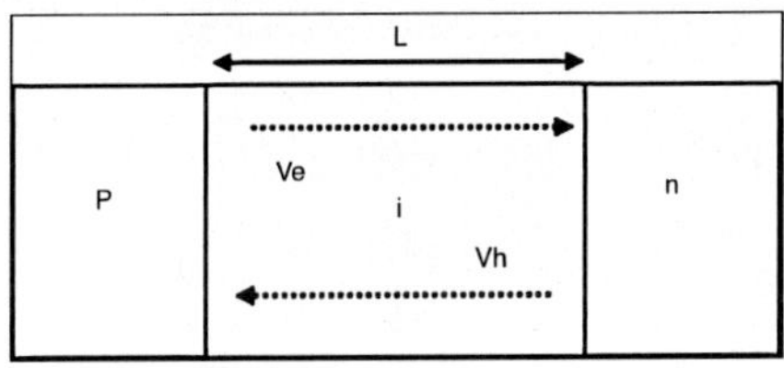

Figure: For a $R_2 = 0.99$ and $\alpha = 10^4$ cm^{-1} with a $R_1 = 0.2$ a η of 0.99 or more can be Achievable for d = 0.7–0.95 μm. Similarly for Different Values of R_1 very High η is possible to Achieve. However, $R_1 = 0$ Limits the Length of Thickness Region, d > 5 μm can Achieve 0.99 η but at the cost of Bandwidth.

Detection Speed of RCE Photodiodes

The detection speed depends upon the drift velocities of the electrons and holes. And between these two holes have slower drift velocity than the electrons. The transit time limited bandwidth of conventional p-i-n photodiode is given by:

$$f_{transit} = 0.45\frac{v_h}{L}$$

However the quantum efficiency is a function of L as:

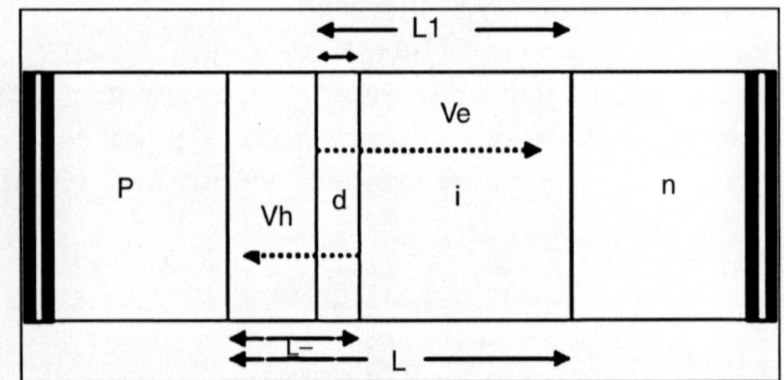

Figure: Schematic Representation of a RCE p-i-n Photodetector

$$\eta = (1 - R)\alpha L.$$

For a high speed detector for a small value of L, as α is very small, η becomes very small (η<<1). This shows for a optimum value of quantum efficiency the bandwidth has to sacrifice.

A p-i-n RCE photodetector can reduce the absorption region to a much smaller scale. In this case the carriers need to traverse a smaller distance as well, L_1 (< L) and L_2 (< L) for electrons and holes respectively.

The length of L1 and L2 can also be optimized to match the delay between the hole and electron drift. And the transition bandwidth becomes:

$$f_{transit} = 0.45\frac{v_h + v_e}{L + d}$$

As in most of semiconductors v_e is more than v_h the bandwidth increases drastically.

It's been reported that for a large device of L = 0.5μm 64 GHz of bandwidth can be achieved and a small device of L = 0.25μm can give 120 GHz bandwidth, where conventional photodetectors have bandwidth of 10° 30 GHz.

Wavelength Selectivity of RCE Photo Detectors

A RCE structure can make the detector wavelength selective to an extent due to the resonance properties of the cavity. The resonance condition of the cavity is given as $2\beta L + \phi_1 + \phi_2 = 2m\Pi$. For any other value the efficiency η reduces from its maximum value, and vanishes when $2\beta L + \phi_1 + \phi_2 = (2m+1)\Pi$.

The wavelength spacing of the maxima of η are separated by the Free Spectral Range of the cavity, given as:

$$FSR = \frac{\lambda^2}{2n_{eff}(L + L_{eff,1} + L_{eff,2})}$$

Where neff is the effective refractive index and Leff,i are the effective optical path lengths of the mirrors.

Finesse, the ratio of the FSR to the FWHM at the resonant wavelength, gives the wavelength selectivity of the cavity.

$$Finesse = \frac{\pi(R_1R_2)^{1/4}e^{-\frac{\alpha d}{2}}}{(1-\sqrt{R_1R_2}e^{-\alpha_c d})^2}$$

This shows that the wavelength selectivity increases with higher reflectivity and smaller values of L.

MATERIAL REQUIREMENTS FOR RCE DEVICES

The estimated superior performance of the RCE devices critically depends on the realization of very low loss active region. This enforces the conditions that: the mirror and the cavity materials must be non-absorbing at the detection wavelength; and the mirror should have very high reflectivity so that it gives highest optical confinement inside the cavity.

The absorption in the cavity can be limited by making the bandgap of the active region smaller than the cavity and the mirror. But a large difference in the bandgap would be a blockage in extraction of photo generated carriers from a heterojunction. Usually a moderate offset is kept within the absorption spectrum.

Different material combinations satisfy all of the above criteria and are therefore usable to the RCE scheme. Some material combinations used for RCE detection are:

- GaAs(M,C)/AlGaAs(M)/InGaAs(A) near 830-920nm.
- InP(C)/$In_{0.53}Ga_{0.47}As$(M)/$In_{0.52}Al_{0.48}As$(M)/$In_{0.53-0.7}GaAs$(A) near 1550nm.
- GaAs(M,C)/AlAs(M)/Ge(A) near 830-920nm.
- Si(M,C)/SiGe(M)/Ge(A) near 1550nm.
- GaP(M)/AlP(M)/Si(A,S) near visible region.

FUTURE OF RCE PHOTODIODES

There are many examples of RCE devices, like p-i-n photodiode, avalanche photodiode, schottky diode are made that verifies the theory successfully. Some of them are in use in practical purposes as well as there is a future prospect in use as modulators, optical logics in wavelength division multiplexing (WDM) systems which could enhance the quantum efficiency, operating bandwidth, wavelength selectivity. RCE detectors are preferable in potential price and performance in commercial WDM systems. RCE detectors have very good potential for implementations in WDM systems and improve the performance significantly. There are various implementations of RCE modulators are made and there is a huge scope for further improvement in performance of those. Other than the photodetectors the RCE structures have lots of other implementations and a very high potential for improved performance. A Light Emitting Diode (LED) can be made to have narrower spectrum and higher directivity to allow more coupling to optical fibre and better utilization of the fibre bandwidth. Optical amplifiers can be made to have more compact, thus lower power required to pump and also at lower cost. Photonic logics will work more efficiently than they do. There will be much less crosstalk, more speed, more gain with simple design.

REFERENCES

- Ashish Bharadwaj: *The Nanoscope : Encyclopedia of Nanoscience and Nanotechnology*, Pentagon Press, Delhi, 2005.
- Kumar, H.D. : *Material Science: Nanotechnology and Applications*, I.K. International, Delhi, 2011
- Levinson, Y. : *Quantum Noise in Mesoscopic Physics*, Kluwer, Dordrecht, 2003.
- Miko Elwenspoek: *Mechanical Microsensors (Microtechnology and MEMS)*, Springer Verlag, Delhi, 2001.
- Mittal, S.K. : *Quantum Optics*, Pearl Books, Delhi, 2010.
- Smith, C.S. : *A Search for Structure*, Cambridge, Mass.: MIT Press 1981.
- Varadan, Vijay K. : *Microsensors, MEMS and Smart Devices*, Wiley, NY, 2001.

8

Photonic Structures and Quantum Fields

Similarities between quantum and classical phenomena are not uncommon in physics albeit quantum and classical physics are grounded on different paradigms. Such analogies have been fruitfully exploited in emerging research areas such as photonic crystals, quantum computing, nano-devices and new forms of light.

Quantum-classical analogies in apparently unrelated fields have been also successfully exploited to mimic at a macroscopic level many quantum phenomena which are currently not of easy access in microscopic quantum systems. In particular, in the past recent years specially-designed waveguide-based photonic structures have proven to be a useful laboratory tool to investigate the optical analogues of a wide variety of coherent quantum effects encountered in different physical contexts, ranging from atomic, molecular, condensed-matter and matter-wave physics. Among others, we mention here the optical analogues of electronic Bloch oscillations and Zener tunneling, dynamic localization, coherent enhancement and destruction of tunneling, adiabatic stabilization of atoms in strong laser fields, Anderson localization, and coherent population transfer. In particular, the use of coupled waveguides with a curved optical axis introduces a kind of "non-inertial forces" for light waves which are capable of mimicking the effects of electric and magnetic fields on the coherent quantum dynamics of a charged particle.

It is the aim of the present contribution to review some of the most relevant quantumoptical analogies, including: coherent destruction of tunneling in driven bistable systems; electronic Bloch oscillations and related effects in crystalline potentials; coherent population transfer and adiabatic passage in laser-driven multilevel atomic or molecular systems; ionization suppression and adiabatic stabilization of atoms in ultra-strong laser fields; quantum decay control and Zeno effect in unstable quantum systems. Besides of general interest, the proposed analogies highlight new and unexpected ways to control light transfer and photon tunneling in integrated photonic structures.

PHOTONIC CRYSTAL SLAB AND STRUCTURES

Photonic crystal (PC) structures have allowed an unprecedented level of control of light in space, frequency, and time. The two major strategies when designing PC slab devices are the introduction of defects, and the exploitation of the critical points of the band dispersion diagram of the perfect crystal. The former approach creates a defect state in the photonic band gap, the defect constituting a cavity for photons. However, these cavities often draw their properties of confinement from an astute management of losses which relies on the careful control of the optical reflections at the boundaries of the cavity. This means that the fabrication of such structures often requires minute adjustments to the structure of the crystal.

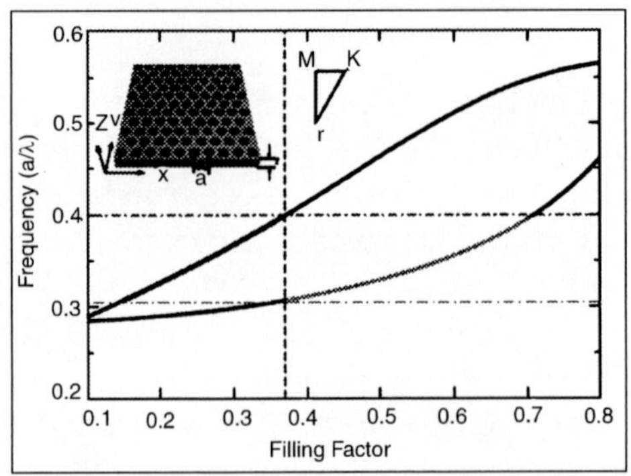

Figure: Frequency of the Band-edges versus the Filling Factor for a PC Slab. a is the Period of the Crystal and λ is the Wavelength of the Mode. The Vertical Dashed Line Corresponds to a Crystal with a Filling Factor $f = 0.37$. The Horizontal Dash-dotted Lines mark the Band Edges of a Crystal with $f = 0.37$.

The second approach considers the perfect crystal (without defects) and builds upon the enhanced local density of electromagnetic states associated with the critical points of the band dispersion diagram. Near these points, the slope of the band is very small and the crystal supports slow light modes (SLMs). Owing to their small group velocity, SLMs are associated with a stronger local density of electromagnetic states than conventional modes with a larger group velocity.

This property makes SLMs an asset in achieving enhanced nonlinear effects or low threshold laser emission. However, the main drawback of the modes of a defect-free crystal is that they are spatially delocalized. This leads to practical problems in the design of compact photonic microstructures. It also implies a more fundamental problem, if the mode size is only limited by its losses, or alternatively, if a large quality factor requires a large photonic crystal structure, how can one design a structure exploiting band-edge SLMs with a high ratio of the quality factor Q to the mode volume V, necessary to foster enhanced radiation dynamics? Srinivasan and coworkers have shown that a band-edge mode could be confined effectively using a graded lattice. In particular they studied the symmetry and Fourier-space properties of such modes. In a related, but different approach, starting from the *delocalized* band-edge mode of a PC slab, we are interested in finding out how, or if, the lateral size of the mode can be reduced without introducing drastic losses. We present a systematic study of the confinement of a band-edge mode in a PC slab with a discussion of spatial versus temporal confinement. We show that band-edge SLMs can be confined spatially without incurring drastic losses. In particular, we show that a high ratio of the quality factor to the field mode volume can be achieved by confining valence band-edge modes, leading to PC designs that provide a suitable platform for cavity-quantum electrodynamics experiments and low-threshold lasing.

Concept of Semiconductor Slab

The PC geometry consists of a semiconductor slab (Indium Phosphide with refractive index n = 3.17), surrounded by air, and patterned with a triangular array of air holes. The period of the lattice is a, the thickness of the slab is t=11a/26, and the air filling factor is $f = 2\pi/\sqrt{3}(r/a)^2$, where r is the radius of the air holes. We consider modes that are even with respect to a mirror plane passing through the centre of the slab (TE-like modes). For these modes, the chosen geometry has a bandgap between the first and the second band. The edge of the valence (first) band is at the K point of the first Brillouin zone (BZ), whereas the edge of the conduction (second) band lies at M. The frequency of the two band edges as a function of f computed by 3D plane wave expansion. One can tune the position of the band edges by changing f. If, for instance, we consider a PC with f = 0.37, we see that its bandgap lies between a/λ = 0.307 and a/λ = 0.400. A second PC with the same lattice parameter, and $0.37<f<0.7$ would have its valence band edge in the bandgap of the first PC. Similarly, a PC with $0.15<f<0.37$ would have its conduction band edge lying in the band gap of the first PC. These ideas relate to the concepts of donor (valence) and acceptor (conduction) defect modes described in detail by Painter and coworkers.

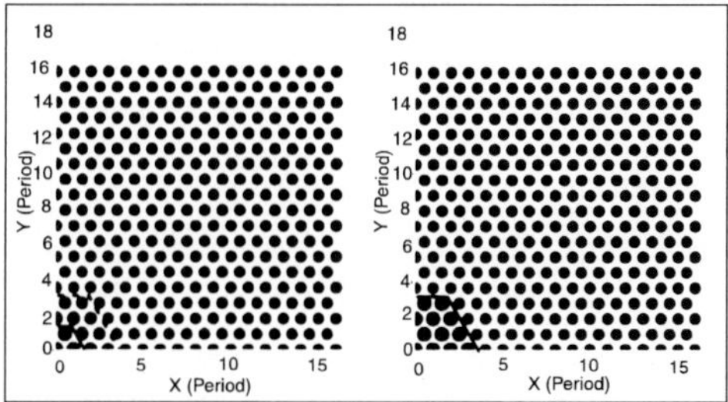

Figure: Examples of Structures: $V^2{}_2$ Cavity (left) and $V^0{}_4$ Cavity (right). Only the Top-right Quadrant (2D Projection of one Octant of the Computational window). The Core has a Hexagonal Shape, with a Higher air Filling Factor than the Mirror (f_{mirror}= 0.37).

In this chapter we seek to explore the influence of the size of the "defect" area on the spatial and energy confinement of the band-edge mode. This will also give us an insight in the evolution of a band-edge mode as it becomes confined.

We shall focus on valence modes, for which light is mostly in the high-index material, thereby ensuring optimal interaction with sources that would be inside the slab. We will show that this approach leads to cavity designs with a large Q/V ratio.

GEOMETRY OF THE PHOTONIC CRYSTAL STRUCTURE

We consider a finite PC slab, with $f = f_{mirror} = 0.37$ comprising 33 and 37 rows of holes along directions x and y respectively. This gives us a crystal that is large enough for us to study cavities with different lateral dimensions, while keeping an overall structure of a practical size. At the centre of this structure we modify the radius of i hexagonal rings of holes to make them bigger (*i.e.*, the filling factor of the central region $f_{core} > f_{mirror}$). Furthermore, we anticipate that the abrupt transition between core and mirror regions will lead to out-of-plane scattering losses.

Therefore, we allow a linear variation of the radius of the holes from the core to the cladding across an additional j hexagonal adaptation rings. We will refer to this cavity as V^j_i, as it supports valence band edge modes. Of course, a cavity with a core consisting of a single hole and a large number of adaptation layers would yield the graded-lattice design of Srinivasan *et al.*

The upper-right quadrant of the in-plane computational domain for two examples of cavities. We use symmetry conditions to reduce the computational domain by a factor of 8.

We use a commercial, parallel implementation of the three-dimensional finitedifference time-domain (FDTD) method to solve rigorously Maxwell's equations for the PC structures. The modal fields, resonant frequency ν, and quality factor Q are found by a sequence of FDTD runs.

From an initial FDTD run with a localized Gaussian excitation, estimates of the resonant frequency v and Q factor are calculated using the filter diagonalization method. The calculation is then repeated and the modal field is extracted by discrete Fourier transform at the approximate frequency v. This process is repeated a number of times with successively longer computation times to refine the modal frequency and spatial profile until the results are converged.

Properties such as in-plane ($Q_{||}$) and out-of-plane ($Q_{\perp}$) quality factors are found by post-processing the fields. An interesting property of these structures is that a good estimate of the quality factor Q of the mode can be obtained with a coarse spatial discretization grid. Indeed, computations with a spatial grid of $a/16$ typically yield an estimate of Q within 10–15%, and sometimes even within a few percent of the results obtained with a higher spatial sampling rate ($\Delta x = a/32$). It is therefore possible to find a good compromise between the speed and the accuracy of the computation. The plots presented in this article were calculated with a cell size of $\Delta x = a/16$.

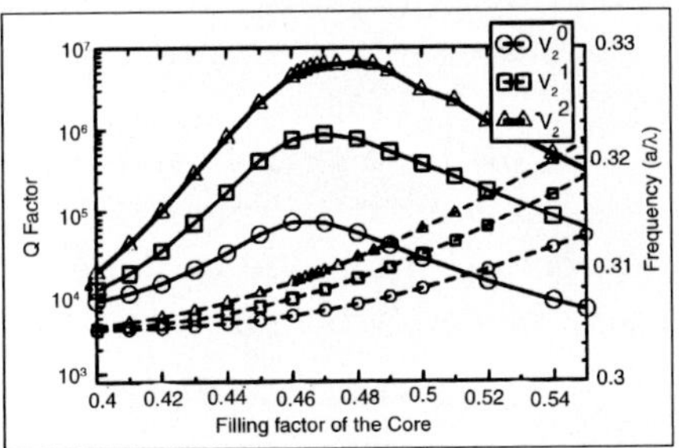

Figure: Quality Factor Q (Solid Lines) and Frequency (Dashed Lines) of the Confined Band-edge mode versus the Filling Factor of the Core for $V^j{}_2$ Cavities with j = 0, 1, 2.

INFLUENCE OF THE ADAPTATION LAYERS

The evolution of Q and normalized frequency $\nu = a/\lambda$ versus the filling factor of the core, for the valence band-edge mode of $V^j{}_2$ cavities. We consider modes with an even and odd symmetry with respect to mirror reflection through the (x, z) and (y, z) planes, respectively (note that there exist also a mode with even-even symmetry).

For the $V^0{}_2$ structure, which corresponds to a core with 2 rings of holes (7 holes in total) surrounded by the mirror region, we see that the quality factor of the core mode has a maximum of 7×10^4 at $f_{core} = 0.46$. The frequency of the mode, on the other hand, increases monotonically with f_{core}. We emphasize that the maximum of Q, which corresponds to the optimal electromagnetic temporal confinement, does not occur when the frequency of the mode is at the centre of the bandgap, but actually quite close to the valence band edge of the mirror PC.

If we soften the transition between core and mirror regions using 1 or 2 adaptation layers, we observe a similar evolution of $Q(f)$ except that the quality factor increases significantly, reaching a maximum around 5.5×10^6 (computed with $\Delta x = a/32$) for $f_{core} = 0.48$ with 2 adaptation layers. Thus, even one or two adaptation layers enables a significant increase in Q.

What is perhaps even more interesting is that the increase in mode volume experienced by the SLM, as more adaptation layers are introduced, is slow enough to ensure that Q/V increases with the number of layers. However, as the number of adaptation layer is increased, the variation of the radius of the holes between two adjacent adaptation layers becomes smaller, thereby making the structure more difficult to fabricate. Therefore, the use of two adaptation layers seems like a convenient trade-off between increasing the quality factor of the cavity, and keeping the fabrication process as simple as possible.

Energy vs. Spatial Confinement

Before looking at the confinement properties of various cavity geometries let us consider a particular design and use it to gain some insight into the electromagnetic processes that govern the confinement of light in this type of cavity.

The optical properties of these cavities are the result of the balance between two effects:mode spatial confinement, which is *maximal* for large filling factors (when the band-edge mode of the core lies deep inside the bandgap of the mirror region), and lattice-mismatch (and by extension mode-mismatch), which is *minimal* when f_{core} is close to f_{mirror}.

In other words, starting from f_{core}=f_{mirror}, as f_{core} increases, Q starts to increase because the mode confinement is improved and *in-plane* losses (IPL) are reduced. The quality factor then reaches a maximum and starts decreasing due to a stronger lattice mismatch which results in stronger *out-of-plane* losses (OPL). The large quality factors obtained for these cavities is also consistent with the analysis presented in which emphasized that a good optical confinement can be achieved through a combination of slow-light (the band-edge mode of the core in our case) and mode-matching at the boundary of the cavity (use of adaptation layers).

To explore the interplay between in-plane and out-of-plane losses, the quality factor and mode volume of the core mode versus the filling factor for the $V^2{}_2$ cavity.

We see that the spatial confinement of the mode is maximal for large values of f_{core} and that the maximum of Q does not coincide with the optimal spatial confinement of the mode. This is a consequence of the three-dimensional nature of the structure and relates to out-of-plane losses. This is confirmed by an analysis of the separate contributions from in-plane and out-of-plane losses to the quality factor. It is clear that $Q\|$ increases as the mode moves deeper into the band gap, while $Q_{\perp}$ decreases as the lateral confinement of the mode improves.

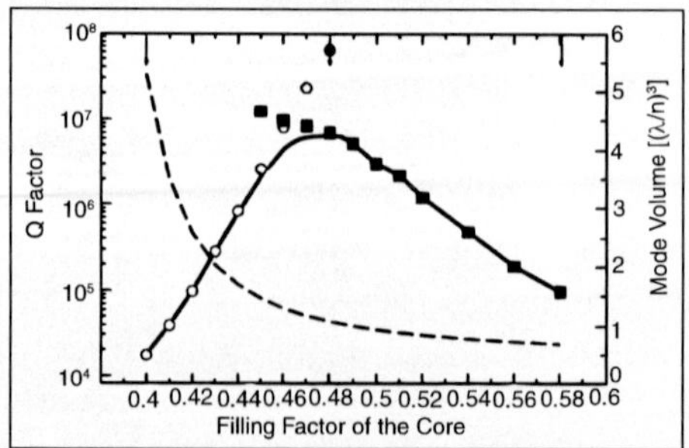

Figure: Quality Factor (solid line) and Mode Volume (dashed line) versus Core Filling Factor for a $V^2{}_2$ Cavity.

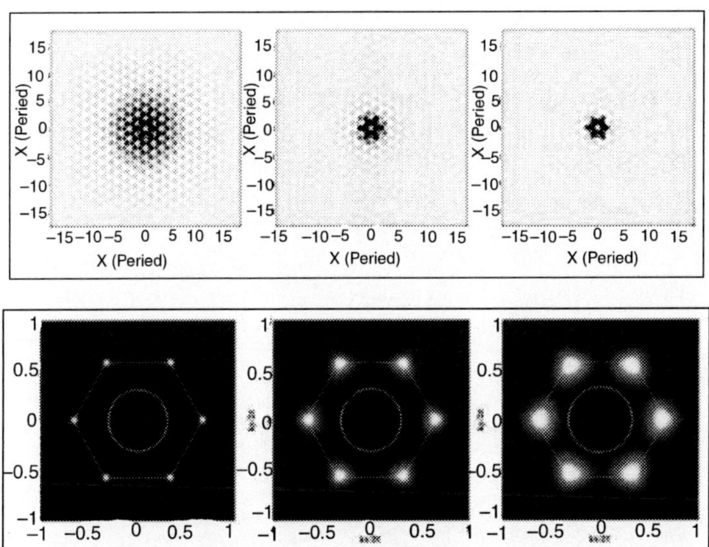

Figure: Near-field Map of H_z (top row) and $|(H_z)|$ (bottom row) for the $V^2{}_2$ Cavity.

Another illustration of the effect of confinement where we plot the near-field maps of the *z*-component of the magnetic field H_z above the slab, and its Fourier transform $|(H_z)|$, for the three values of f_{core} marked by arrows. For f_{core}=0.40 the mode is fairly delocalized because the SLM in the core is close to the band-edge of the mirror region which behaves like a weak mirror. In other words, the SLM is in a weak potential well and its evanescent tail decays very slowly outside of the core. On the other hand, the Fourier space plot shows that the mode spectral signature consists of a well confined peak around the *K* points of the first BZ (with a much weaker contribution from the equivalent symmetry points outside the first BZ which are not shown). As f_{core} increases, the SLM experiences a tighter confining potential, which results in a highly confined mode profile that almost coincides with the physical size of the core. In Fourier space, however, the peaks at the *K* points of the first BZ experience a strong broadening and start to leak into the light cone, thereby causing an increase of out-of-plane losses and the decrease in $Q_\perp$.

Incidentally, the fact that the spectral components of the core mode are quite far away from the light cone suggest that these structures are relatively independent of the substrate index, and could still support a mode with a fairly large quality factor with different implementations, as silicon on insulator, or InP bonded to a dielectric substrate.

Influence of the Size of the Core on the Confinement

We have established that the optical properties of the cavity-confined band-edge mode are the result of a balance between IPL and OPL. Having also described the effect of adaptation layers on the confinement of the mode, we now study the influence of the size of the core on the confinement of the SLM.

We consider cavities with a core consisting of 1 up to 5 rings of holes. To emphasize the role of the size of the core, all cavities have the same number of adaptation layers (2). The evolution of the quality factor versus the filling factor of the core for all these cavities.

For f_{core}*0.43 the quality factor increases with the size of the core. In first approximation, this reflects the mere fact that the bigger the core of the cavity, the longer it takes for photons to undergo successive "bounces" off the edges of the cavity where out-of-plane scattering occurs.

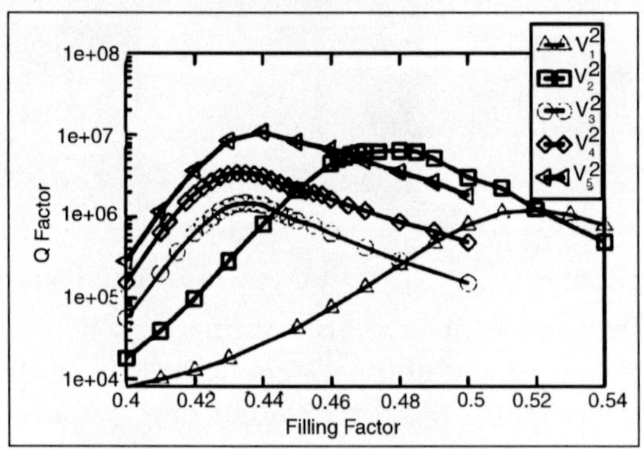

Figure: Quality Factor Q of the Confined Slow-light Mode vs. Filling Factor of the Core for V^2_i Cavities with i = 1, 2, 3, 4, 5.

One can notice the similar evolution of the $Q(f)$ curves for the V^2_3, V^2_4, and V^2_5 cavities. The three curves have their maximum at about the same filling factor, and the slopes are comparable. The only difference comes from the values of Q, which are higher for a larger core, as more energy is stored within the cavity.

Perhaps less trivial is the difference in behaviour of the V^2_1 and V^2_2 cavities compared to cavities with a bigger core. These differences can be explained on the basis and the interpretation present. As the core gets bigger, the frequency of the mode of the core moves closer to the frequency of the valence band-edge of an infinite crystal with $f = f_{core}$. Furthermore, because the frequency of the confined SLM increases more rapidly with f_{core} for larger sizes of the core, the confined SLM moves across the band-gap of the mirror PC over a shorter range of values of f_{core}, leading to a wider "bandwidth" for the cavities with a small core.

As a consequence, the V^2_2, for instance, exhibits a large quality factor (>10^6) over a large range of values of filling factors. This is an interesting property as it ensures that the quality factor will be fairly insensitive to a moderate systematic fabrication error (*e.g.*, lithographic or etching process yielding holes systematically smaller or larger than the optimal, computed filling factor).

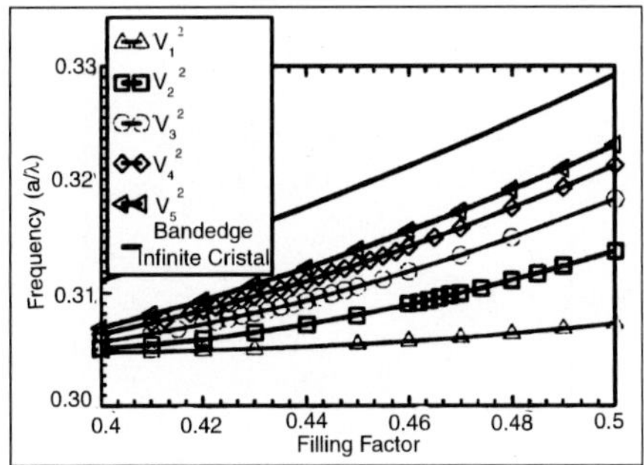

Figure: Frequency of the Confined Slow-light mode vs. Filling Factor of the Core for V^2_i Cavities with i = 1, 2, 3, 4, 5.

Therefore, for a small-core cavity, the filling factor has to reach a larger value before the core mode is deep inside the bandgap of the mirror PC, stage at which OPL becomes the dominant loss mechanism. This explains why the highest quality factor corresponds to a larger filling factor of the core as the core gets smaller. The location of the maximum of *Q(f)* does not vary much between the V^2_3, V^2_4 and V^2_5 cavities because the mode starts to experience the influence of the finite size of the mirror PC.

Indeed, because the size of the PC structure is the same for all cavities, the effective mirror size is smaller for a bigger core size. This explains why the evolution of the quality factor versus the size of the core does not follow the same trend once the core has reached a certain size (3 rings of holes in our example). Additional calculations (not reported here) show that with a wider mirror PC, the part of the *Q(f)* curves to the left of the maximum is shifted upward whereas the part to the right of the maximum remains unchanged. As a consequence the maximum value of *Q* is reached at a smaller filling factor. This is a mere consequence of the reduction of the IPL resulting from the introduction of a wider mirror.

We can also take a closer look at the evolution of the mode volume for cavities of different core size. While the general trend is the same for all cavities, *ie.*, the volume of the mode shrinks as the cavity moves deeper into the bandgap of the surrounding PC, we observe that the size of the core has an impact on how well the mode can ultimately be confined. For small filling factors the mode is highly delocalized and the mode volume is somewhat difficult to quantify as the mode is artificially truncated at the boundaries of the simulation window. On the other hand, as the mode moves deeper into the bandgap and its lateral size is reduced, a comparison of the mode volumes for the different cavities becomes meaningful. Perhaps the most interesting point is that deep in the bandgap, it is not the V^2_1 but the V^2_2 design that yields the smallest mode volume. This illustrates once again the complex interplay between IPL and OPL. In that sense the V^2_2 represents the best compromise between lateral spatial confinement and vertical radiation losses.

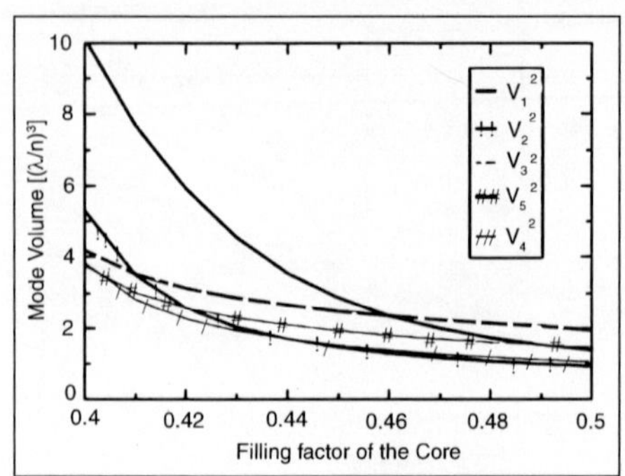

Figure: Normalized Mode Volume of the Confined Slow-light Mode vs. Filling Factor of the Core for $V^2{}_i$ Cavities with i = 1, 2, 3, 4, 5.

ANDERSON LOCALIZATION IN CONDENSED MATTER PHYSICS

In condensed matter physics, Anderson localization, also known as strong localization, is the absence of diffusion of waves in a *disordered* medium. This phenomenon is named after the American physicist P. W. Anderson, who was the first one to suggest the possibility of electron localization inside a semiconductor, provided that the degree of randomness of the impurities or defects is sufficiently large.

Anderson localization is a general wave phenomenon that applies to the transport of electromagnetic waves, acoustic waves, quantum waves, spin waves, *etc.*

This phenomenon is to be distinguished from weak localization, which is the precursor effect of Anderson localization, and from Mott localization, named after Sir Nevill Mott, where the transition from metallic to insulating behaviour is *not* due to disorder, but to a strong mutual Coulomb repulsion of electrons.

In the original Anderson tight-binding model, the evolution of the wave function ψ on the d-dimensional lattice Z^d is given by the Schrödinger equation

$$i\hbar\dot{\psi}_t = H\psi_t \ ,$$

where the Hamiltonian H is given by

$$(H\phi)(j) = E_j\phi(j) + \sum_{k\neq j} V(|k-j|)\phi(k)\ ,$$

with E_j random and independent, and interaction $V(r)$ falling off as r^{-2} at infinity. For example, one may take E_j uniformly distributed in [$-W$, $+W$], and

$$V(|r|) = \begin{cases} 1, & |r| = 1 \\ 0, & \text{otherwise.} \end{cases}$$

Starting with ψ_0 localised at the origin, one is interested in how fast the probability distribution $|\psi_t|^2$ diffuses. Anderson's analysis shows the following:

- if d is 1 or 2 and W is arbitrary, or if $d \geq 3$ and $W/'$ is sufficiently large, then the probability distribution remains localized:

$$\sum_{n \in \mathbb{Z}^d} |\psi_t(n)|^2 |n| \leq C$$

uniformly in t. This phenomenon is called Anderson localization.

- if $d \geq 3$ and $W/''$ is small,

$$\sum_{n \in \mathbb{Z}^d} |\psi_t(n)|^2 |n| \approx D\sqrt{t}\ ,$$

where D is the diffusion constant.

ANALYSIS

The phenomenon of Anderson localization, particularly that of weak localization, finds its origin in the wave interference between multiple-scattering paths. In the strong scattering limit, the severe interferences can completely halt the waves inside the disordered medium.

For non-interacting electrons, a highly successful approach was put forward in 1979 by Abrahams *et al.* This scaling hypothesis of localization suggests that a disorder-induced metal-insulator transition (MIT) exists for non-interacting electrons in three dimensions (3D) at zero magnetic field and in the absence of spin-orbit coupling. Much further work has subsequently supported these scaling arguments both analytically and numerically. In 1D and 2D, the same hypothesis shows that there are no extended states and thus no MIT. However, since 2 is the lower critical dimension of the localization problem, the 2D case is in a sense close to 3D: states are only marginally localized for weak disorder and a small magnetic field or spin-orbit coupling can lead to the existence of extended states and thus an MIT. Consequently, the localization lengths of a 2D system with potential-disorder can be quite large so that in numerical approaches one can always find a localization-delocalization transition when either decreasing system size for fixed disorder or increasing disorder for fixed system size.

Most numerical approaches to the localization problem use the standard tight-binding Anderson Hamiltonian with onsite-potential disorder. Characteristics of the electronic eigenstates are then investigated by studies of participation numbers obtained by exact diagonalization, multifractal properties, level statistics and many others. Especially fruitful is the transfer-matrix method (TMM) which allows a direct computation of the localization lengths and further validates the scaling hypothesis by a numerical proof of the existence of a one-parameter scaling function. Direct numerical solution of Maxwell equations to demonstrate Anderson localization of light has been implemented.

Experimental Evidence

Two reports of Anderson localization of light in 3D random media exist up to date, even though absorption complicates interpretation of experimental results. Anderson localization can also be observed in a perturbed periodic potential where the transverse localization of light is caused by random fluctuations on a photonic lattice. Experimental realizations of transverse localization were reported for a 2D lattice and a 1D lattice. It has also been observed by localization of a Bose-Einstein condensate in a 1D disordered optical potential. Anderson localization of elastic waves in a 3D disordered medium has been reported. The observation of the MIT has been reported in a 3D model with atomic matter waves. Random lasers can operate using this phenomenon.

ANDERSON LOCALIZATION OF LIGHT

Why do some materials conduct, whereas others behave as insulators? In 1958 Philip Anderson had published a paper on this issue, where he made a theoretical prediction which changed the whole area of conduction in solids, and for which Anderson himself had been awarded with the Nobel Prize in 1977-he showed that when sufficient disorder exists in a crystal, the electrons inside the crystal "get stuck", such that the material cannot conduct current anymore. However, this result applies not only to solid state physics-Anderson Localization is a phenomenon which relies on wave interference, and therefore it is general to any wave system in nature, and specifically to light-electromagnetic waves.

In a recent observation how disorder in a periodic structure-photonic lattice-eliminates the diffraction of light inside it, and we find the cross-over to Anderson Localization. In these experiments, we saw the use of optical setups in order to study various aspects of Anderson Localization, and new phenomena related to wave propagation in disordered media, including nonlinear effects.

RANDOM LASER

A random laser is a laser that uses a highly disordered gain medium. A random laser uses no optical cavity but the remaining principles of operation remain the same as for a conventional laser. Random laser action has been observed in many different media, including semiconductor powder, nanostructured and non-nanostructured thin films, laser dyes, ceramics and many more.

Developments in nanoparticles have demonstrated that large amounts of optical scattering can occur when photons are incident. In this way, light can be diffused around a medium in much the same way as it is on white paint and in clouds.

INCOHERENT REGIME

If nano particles are embedded in an optical gain medium, for example, zinc oxide (UV emission - bandgap 3.3 eV), light from a pump source (*e.g.* frequency-tripled Nd:YAG laser) will induce spontaneous emission of light at around 350 nm within the gain medium. These spontaneously emitted photons will then stimulate other radiative transitions in the gain medium to take place, unleashing yet more photons. This is, in many ways analogous to the chain reaction that occurs in the fission of neutrons in a nuclear reactor and has been referred to by R.H. Dicke as an "optical bomb".

ANDERSON LOCALIZATION

Anderson localization is a well-known phenomenon which occurs when electrons become trapped in a disordered metallic structure and this metal goes through a phase transition from conductor to insulator. These electrons are said to be Anderson Localized. The conditions for this localization are that there is a high enough density of scatters in the metal (other electrons, spins, *etc.*) to cause free electrons to follow a single looped path.

In analogy to this, we can imagine photons diffusing through a medium scattering off nanoparticles of diameter 10 - 100 nm. If the Ioffe-Regel criterion, describing the ratio of photon wave-vector k to mean free-path length (of a photon not colliding with anything) l, is met: $kl < 1$, then there is a probability that photons will become trapped in much the same way as electrons are observed to be trapped under Anderson localization. In this way, while the photon is trapped, the scatters may act as an optical cavity. The gain medium in which the scatterers lie will allow for stimulated emission to occur. As in an ordinary laser, if the gain is greater than the losses incurred, the lasing threshold will be broken and lasing can occur.

Photons traveling in this loop will also interfere with each other. The well defined cavity length (1- 10 um) will ensure that the interference is constructive and will allow certain modes to oscillate. The competition for gain permits one mode to oscillate once the lasing threshold has been reached.

RANDOM LASER THEORY

Theory however shows that for multiple scattering in amplifying random media "Anderson" localization of light does not occur at all - even though the calculation of interferences are essential to prove that fact. In contrary so called weak localizations processes can be proven, but it is vividly discussed, whether those mechanisms play the key role in the mode statistics or not.

Recent studies show that these weak localization processes are not the governing phenomena for the onset of Random Lasing. Random Lasing also occurs for kl > 1. This is in agreement with experimental findings. Even though travelling of light on exactly 'closed loops' would explain the occurrence of confined lasing spots intuitively, the question is still open if, *e.g.* the stimulated emission processes are correlated with those processes.

The theory of 'preformed cavities' is however not confirmed.

Typical amounts of gain medium required to exceed the lasing threshold depend heavily on the scatterer density.

Applications

This field is relatively young and as such does not have many realised applications. However, random lasers based on ZnO are promising candidates for electrically pumped UV lasers, biosensors and optical information processing. This is due to the low production cost and that the optimum temperature for substrate production has been observed to be around 500°C for powders. This is in contrast to producing an ordinary laser crystal at temperatures exceeding 700°C.

RANDOM NUMBER GENERATION

A random number generator (RNG) is a computational or physical device designed to generate a sequence of numbers or symbols that lack any pattern, *i.e.* appear random. The many applications of randomness have led to the development of several different methods for generating random data. Many of these have existed since ancient times, including dice, coin flipping, the shuffling of playing cards, the use of yarrow stalks (by divination) in the I Ching, and many other techniques. Because of the mechanical nature of these techniques, generating large amounts of sufficiently random numbers (important in statistics) required a lot of work and/or time. Thus, results would sometimes be collected and distributed as random number tables. Nowadays, after the advent of computational random number generators, a growing number of government-run lotteries, and lottery games, are using RNGs instead of more traditional drawing methods. RNGs are also used today to determine the odds of modern slot machines.

Several computational methods for random number generation exist, but often fall short of the goal of true randomness — though they may meet, with varying success, some of the statistical tests for randomness intended to measure how unpredictable their results are (that is, to what degree their patterns are discernible).

PRACTICAL APPLICATIONS AND USES

Random number generators have applications in gambling, statistical sampling, computer simulation, cryptography, completely randomized design, and other areas where producing an unpredictable result is desirable.

Note that, in general, where unpredictability is paramount — such as in security applications — hardware generators are generally preferred (where feasible) over pseudo-random algorithms.

Random number generators are very useful in developing Monte Carlo-method simulations, as debugging is facilitated by the ability to run the same sequence of random numbers again by starting from the same *random seed*. They are also used in cryptography - so long as the *seed* is secret. Sender and receiver can generate the same set of numbers automatically to use as keys.

The generation of pseudo-random numbers is an important and common task in computer programming. While cryptography and certain numerical algorithms require a very high degree of *apparent* randomness, many other operations only need a modest amount of unpredictability. Some simple examples might be presenting a user with a "Random Quote of the Day", or determining which way a computer-controlled adversary might move in a computer game. Weaker forms of *randomness* are used in hash algorithms and in creating amortized searching and sorting algorithms. Some applications which appear at first sight to be suitable for randomization are in fact not quite so simple. For instance, a system that "randomly" selects music tracks for a background music system must only *appear* random, and may even have ways to control the selection of music: a true random system would have no restriction on the same item appearing two or three times in succession.

"True" Random Numbers vs. Pseudorandom Numbers

There are two principal methods used to generate random numbers. One measures some physical phenomenon that is expected to be random and then compensates for possible biases in the measurement process. The other uses computational algorithms that produce long sequences of apparently random results, which are in fact completely determined by a shorter initial value, known as a seed or key. The latter type are often called pseudorandom number generators.

A "random number generator" based solely on deterministic computation *cannot* be regarded as a "true" random number generator, since its output is inherently predictable. How to distinguish a "true" random number from the output of a pseudo-random number generator is a very difficult problem. However, carefully chosen pseudo-random number generators can be used instead of true random numbers in many applications. Rigorous statistical analysis of the output is often needed to have confidence in the algorithm.

GENERATION METHODS

Physical Methods

The earliest methods for generating random numbers — dice, coin flipping, roulette wheels — are still used today, mainly in games and gambling as they tend to be too slow for most applications in statistics and cryptography.

A physical random number generator can be based on an essentially random atomic or subatomic physical phenomenon whose unpredictability can be traced to the laws of quantum mechanics. Sources of entropy include radioactive decay, thermal noise, shot noise, avalanche noise in Zener diodes, clock drift, the timing of actual movements of a hard disk read/write head, and radio noise. However, physical phenomena and tools used to measure them generally feature asymmetries and systematic biases that make their outcomes not uniformly random. A randomness extractor, such as a cryptographic hash function, can be used to approach a uniform distribution of bits from a non-uniformly random source, though at a lower bit rate.

In 2010, Kanter *et al.* at Bar-Ilan University created a physical random bit generator that operates at a rate of 300 gigabits per second, the fastest thus far created. Various imaginative ways of collecting this entropic information have been devised. One technique is to run a hash function against a frame of a video stream from an unpredictable source. Lavarand used this technique with images of a number of lava lamps. HotBits measures radioactive decay with Geiger–Muller tubes, while Random.org uses variations in the amplitude of atmospheric noise recorded with a normal radio.

Another common entropy source is the behaviour of human users of the system. While people are not considered good randomness generators upon request, they generate random behaviour quite well in the context of playing mixed strategy games. Some security-related computer software requires the user to make a lengthy series of mouse movements or keyboard inputs to create sufficient entropy needed to generate random keys or to initialize pseudorandom number generators.

Computational Methods

Pseudo-random number generators (PRNGs) are algorithms that can automatically create long runs of numbers with good random properties but eventually the sequence repeats (or the memory usage grows without bound). The string of values generated by such algorithms is generally determined by a fixed number called a seed. One of the most common PRNG is the linear congruential generator, which uses the recurrence

$$X_{n+1} = (aX_n + b) \bmod m$$

to generate numbers. The maximum number of numbers the formula can produce is the modulus, m. To avoid certain non-random properties of a single linear congruential generator, several such random number generators with slightly different values of the multiplier coefficient a can be used in parallel, with a "master" random number generator that selects from among the several different generators.

A simple pen-and-paper method for generating random numbers is the so-called middle square method suggested by John Von Neumann. While simple to implement, its output is of poor quality. Most computer programming languages include functions or library routines that purport to be random number generators. They are often designed to provide a random byte or word, or a floating point number uniformly distributed between 0 and 1. Such library functions often have poor statistical properties and some will repeat patterns after only tens of thousands of trials. They are often initialized using a computer's real time clock as the seed, since such a clock generally measures in milliseconds, far beyond the person's precision. These functions may provide enough randomness for certain tasks (for example video games) but are unsuitable where high-quality randomness is required, such as in cryptographic applications, statistics or numerical analysis. Better pseudo-random number generators such as the Mersenne Twister are widely available. Much higher quality random number sources are available on most operating systems; for example/dev/random on various BSD flavours, Linux, Mac OS X, IRIX, and Solaris, or CryptGenRandom for Microsoft Windows.

An example of a simple pseudo-random number generator is the Multiply-with-carry method invented by George Marsaglia. It is computationally fast and has good (albeit not cryptographically strong) randomness properties (note that this example is not thread safe):

```
m_w = <choose-initializer>;/* must not be zero */
m_z = <choose-initializer>;/* must not be zero */

uint get_random()
{
m_z = 36969 * (m_z & 65535) + (m_z >> 16);
m_w = 18000 * (m_w & 65535) + (m_w >> 16);
return (m_z << 16) + m_w;/* 32-bit result */
}
```

Generation from a Probability Distribution

There are a couple of methods to generate a random number based on a probability density function. These methods involve transforming a uniform random number in some way. Because of this, these methods work equally well in generating both pseudo-random and true random numbers. One method, called the inversion method, involves integrating up to an area greater than or equal to the random number (which should be generated between 0 and 1 for proper distributions). A second method, called the acceptance-rejection method, involves choosing an x and y value and testing whether the function of x is greater than the y value. If it is, the x value is accepted. Otherwise, the x value is rejected and the algorithm tries again.

By Humans

Random number generation may also be done by humans directly. However, most studies find that human subjects have some degree of nonrandomness when generating a random sequence of, *e.g.*, digits or letters. They may alternate too much between choices compared to a good random generator.

Post-processing and Statistical Checks

Even given a source of plausible random numbers (perhaps from a quantum mechanically based hardware generator), obtaining numbers which are completely unbiased takes care. In addition, behaviour of these generators often changes with temperature, power supply voltage, the age of the device, or other outside interference. And a software bug in a pseudo-random number routine, or a hardware bug in the hardware it runs on, may be similarly difficult to detect.

Generated random numbers are sometimes subjected to statistical tests before use to ensure that the underlying source is still working, and then post-processed to improve their statistical properties.

Other Considerations

Random numbers uniformly distributed between 0 and 1 can be used to generate random numbers of any desired distribution by passing them through the inverse cumulative distribution function (CDF) of the desired distribution. Inverse CDFs are also called quantile functions. To generate a pair of statistically independent standard normally distributed random numbers (x, y), one may first generate the polar coordinates (r, θ), where $r \sim \div_2^2$ and θ~UNIFORM(0, 2π).

Some 0 to 1 RNGs include 0 but exclude 1, while others include or exclude both.

The outputs of multiple independent RNGs can be combined (for example, using a bit-wise XOR operation) to provide a combined RNG at least as good as the best RNG used. This is referred to as software whitening.

Computational and hardware random number generators are sometimes combined to reflect the benefits of both kinds. Computational random number generators can typically generate pseudo-random numbers much faster than physical generators, while physical generators can generate "true randomness."

Low-discrepancy Sequences as an Alternative

Some computations making use of a random number generator can be summarized as the computation of a total or average value, such as the computation of integrals by the Monte Carlo method. For such problems, it may be possible to find a more accurate solution by the use of so-called low-discrepancy sequences, also called quasirandom numbers. Such sequences have a definite pattern that fills in gaps evenly, qualitatively speaking; a truly random sequence may, and usually does, leave larger gaps.

OPTICAL MICROCAVITY

An optical microcavity is a structure formed by reflecting faces on the two sides of a spacer layer or optical medium. The name *micro*cavity stems from the fact that it is often only a few micrometers thick, the spacer layer sometimes even in the nanometer range. As with common lasers this forms an optical cavity or *optical resonator,* allowing a standing wave to form inside the spacer layer.

The thickness of the spacer layer determines the so-called "cavity-mode", which is the one wavelength that can be transmitted and will be formed as standing wave inside the resonator. Depending on the type and quality of the mirrors, a so-called stop-band will form in the transmission spectrum of the microcavity, a long range of wavelengths, that is reflected and a single one being transmitted. (usually in the centre)

The fundamental difference between a conventional optical cavity and microcavities is the effects that arise from the small dimensions of the system. Quantum effects of the light's electromagnetic field can be observed. For example, the spontaneous emission rate and behaviour of atoms is altered by such a microcavity. One can imagine this as the situation that no photon is emitted, if the environment is a box that is too small to hold it. This leads to an altered emission spectrum, which is significantly narrowed.

There are different means of fabricating microcavities, either by evaporating alternating layers of dielectric media to form the mirrors (DBR) and the medium inside the spacer layer or by modification of semiconductor material or by metal mirrors.

Microcavities have many applications preferably in optoelectronics, where vertical cavity surface emitting lasers VCSEL are probably the best known. Recently, a single photon emitting device was demonstrated by placing a quantum dot in a microcavity. These light sources are interesting for Quantum cryptography and Quantum computers.

An overview is given in Nature.

QUANTIZATION OF THE ELECTROMAGNETIC FIELD

After quantization of the electromagnetic field, the EM (electromagnetic) field consists of discrete energy parcels, photons. Photons are massless particles of definite energy, definite momentum, and definite spin.

In order to explain the photoelectric effect, Einstein assumed heuristically in 1905 that an electromagnetic field consists of parcels of energy $h\nu$, where h is Planck's constant. In 1927 Paul A. M. Dirac was able to weave the photon concept into the fabrics of the new quantum mechanics and to describe the interaction of photons with matter. He applied a technique which is now generally called second quantization, although this term is somewhat of a misnomer for EM fields, because they are, after all, solutions of the classical Maxwell equations. In Dirac's theory the fields are quantized for the first time and it is also the first time that Planck's constant enters the expressions. In its original work, Dirac took the phases of the different EM modes (Fourier components of the field) and the mode energies as dynamic variables to quantize (*i.e.*, he reinterpreted them as operators and postulated commutation relations between them). At present it is more common to quantize the Fourier components of the vector potential.

A quantum mechanical photon state $|\mathbf{k},\mu\rangle$ belonging to mode (k,μ) will be introduced. It will be shown that it has the following properties

$$\begin{aligned} m_{\text{photon}} &= 0 \\ H|\mathbf{k},\mu\rangle &= h\nu|\mathbf{k},\mu\rangle \quad \text{with} \quad \nu = c\,|\,\mathbf{k}\,| \\ P_{\text{EM}}|\mathbf{k},\mu\rangle &= \hbar\mathbf{k}\,|\mathbf{k},\mu\rangle \\ S_z\,|\mathbf{k},\mu\rangle &= \mu\,|\mathbf{k},\mu\rangle, \quad \mu = 1,-1. \end{aligned}$$

These equations say respectively: a photon has zero rest mass; the photon energy is $h\nu = hc|k|$ (k is the wave vector, c is speed of light); its electromagnetic momentum is $\hbar k[\hbar = h/(2\pi)]$; the polarization $\mu = \pm 1$ is the eigenvalue of the z-component of the photon spin.

Second Quantization

Second quantization starts with an expansion of a scalar of vector field (or wave functions) in a basis consisting of a complete set of functions. These expansion functions depend on the coordinates of a single particle. The coefficients multiplying the basis functions are interpreted as operators and (anti)commutation relations between these new operators are imposed, commutation relations for bosons and anticommutation relations for fermions (nothing happens to the basis functions themselves). By doing this, the expanded field is converted into a fermion or boson operator field. The expansion coefficients have been promoted from ordinary numbers to operators, creation and annihilation operators. A creation operator creates a particle in the corresponding basis function and an annihilation operator annihilates a particle in this function.

In the case of EM fields the required expansion of the field is the Fourier expansion.

Electromagnetic Field and Vector Potential

As the term suggests, an EM field consists of two vector fields, an electric field E(r, t) and a magnetic field B(r,t). Both are time-dependent vector fields that in vacuum depend on a third vector field A(r,t) (the vector potential), under the Coulomb gauge condition, through

$$\mathbf{B}(\mathbf{r},t) = \nabla \times \mathbf{A}(\mathbf{r},t)$$
$$\mathbf{E}(\mathbf{r},t) = -\frac{\partial \mathbf{A}(\mathbf{r},t)}{\partial t},$$

here $\nabla \times A$ is the curl of A. The Fourier expansion of the vector potential enclosed in a finite cubic box of volume $V = L^3$ is (bar on top indicates complex conjugate):

$$\mathbf{A}(\mathbf{r},t) = \sum_{\mathbf{k}} \sum_{\mu=-1,1} \left(\mathbf{e}^{(\mu)}(\mathbf{k}) a_{\mathbf{k}}^{(\mu)}(t) e^{i\mathbf{k}\cdot\mathbf{r}} + \overline{\mathbf{e}}^{(\mu)}(\mathbf{k}) \overline{a}_{\mathbf{k}}^{(\mu)}(t) e^{-i\mathbf{k}\cdot\mathbf{r}} \right),$$

where the wave vector k gives the propagation direction of the corresponding Fourier component (a polarized monochromatic wave) of A(r,*t*); the length of the wave vector is $|\mathbf{k}| = 2\pi\nu/c = \omega/c$, with ν the frequency of the mode. The components of the vector k have discrete values (a consequence of the boundary condition that A has the same value on opposite walls of the box),

$$k_x = \frac{2\pi n_x}{L}, \quad k_y = \frac{2\pi n_y}{L}, \quad k_z = \frac{2\pi n_z}{L}, \qquad n_x, n_y, n_z = 0, \pm 1, \pm 2, \ldots$$

The two unit vectors $e^{(\mu)}$ ("polarization vectors") are perpendicular to k. They are related to the orthonormal Cartesian vectors e_x and e_y through a unitary transformation,

$$\mathbf{e}^{(1)} \equiv \frac{-1}{\sqrt{2}}(\mathbf{e}_x + i\mathbf{e}_y) \quad \text{and} \quad \mathbf{e}^{(-1)} \equiv \frac{1}{\sqrt{2}}(\mathbf{e}_x - i\mathbf{e}_y) \quad \text{with} \quad \mathbf{e}_x \cdot \mathbf{k} = \mathbf{e}_y \cdot \mathbf{k} = 0.$$

The k-th Fourier component of A is a vector perpendicular to k and hence is a linear combination of $e^{(1)}$ and $e^{(-1)}$. The superscript μ indicates a component along $e^{(\mu)}$. The Coulomb gauge is imposed,

$$\nabla \cdot A(r,t) = 0$$

which makes A into a transverse field.

Clearly, the (discrete infinite) set of Fourier coefficients $a_{\mathbf{k}}^{(\mu)}(t)$ and $\overline{a}_{\mathbf{k}}^{(\mu)}(t)$ are variables defining the vector potential. In the following they will be promoted to operators.

Quantization of EM Field

The best known example of quantization is the replacement of the time-dependent linear momentum of a particle by the rule

$$p(t) \to -i\hbar\nabla .$$

Note that Planck's constant is introduced here and that the time-dependence of the classical expression is not taken over in the quantum mechanical operator (this is true in the so-called Schrödinger picture).

For the EM field we do something similar. The quantity ε_0 is the electric constant, which appears here because of the use of electromagnetic SI units.

The *quantization rules* are:

$$a_{\mathbf{k}}^{(\mu)}(t) \quad \rightarrow \sqrt{\frac{\hbar}{2\omega V\epsilon_0}}\, a^{(\mu)}(\mathbf{k})$$

$$\overline{a}_{\mathbf{k}}^{(\mu)}(t) \quad \rightarrow \sqrt{\frac{\hbar}{2\omega V\epsilon_0}}\, a^{\dagger(\mu)}(\mathbf{k})$$

subject to the boson commutation relations

$$\begin{aligned}
\left[a^{(\mu)}(\mathbf{k}), a^{(\mu')}(\mathbf{k}')\right] &= 0 \\
\left[a^{\dagger(\mu)}(\mathbf{k}), a^{\dagger(\mu')}(\mathbf{k}')\right] &= 0 \\
\left[a^{(\mu)}(\mathbf{k}), a^{\dagger(\mu')}(\mathbf{k}')\right] &= \delta_{\mathbf{k},\mathbf{k}'}\delta_{\mu,\mu'}.
\end{aligned}$$

The square brackets indicate a commutator, defined by

$$[A,B] \equiv AB - BA$$

for any two quantum mechanical operators A and B. The introduction of Planck's constant is essential in the transition from a classical to a quantum theory. The factor $(2\omega V\varepsilon_0)^{-1/2}$ is introduced to give the Hamiltonian (energy operator) a simple form.

The quantized fields (operator fields) are the following:

$$\begin{aligned}
\mathbf{A}(\mathbf{r}) &= \sum_{\mathbf{k},\mu}\sqrt{\frac{\hbar}{2\omega V\epsilon_0}}\left(\mathbf{e}^{(\mu)}a^{(\mu)}(\mathbf{k})e^{i\mathbf{k}\cdot\mathbf{r}} + \overline{\mathbf{e}}^{(\mu)}a^{\dagger(\mu)}(\mathbf{k})e^{-i\mathbf{k}\cdot\mathbf{r}}\right) \\
\mathbf{E}(\mathbf{r}) &= i\sum_{\mathbf{k},\mu}\sqrt{\frac{\hbar\omega}{2V\epsilon_0}}\left(\mathbf{e}^{(\mu)}a^{(\mu)}(\mathbf{k})e^{i\mathbf{k}\cdot\mathbf{r}} - \overline{\mathbf{e}}^{(\mu)}a^{\dagger(\mu)}(\mathbf{k})e^{-i\mathbf{k}\cdot\mathbf{r}}\right) \\
\mathbf{B}(\mathbf{r}) &= i\sum_{\mathbf{k},\mu}\sqrt{\frac{\hbar}{2\omega V\epsilon_0}}\left((\mathbf{k}\times\mathbf{e}^{(\mu)})a^{(\mu)}(\mathbf{k})e^{i\mathbf{k}\cdot\mathbf{r}} - (\mathbf{k}\times\overline{\mathbf{e}}^{(\mu)})a^{\dagger(\mu)}(\mathbf{k})e^{-i\mathbf{k}\cdot\mathbf{r}}\right),
\end{aligned}$$

where $\omega = c\,|\mathbf{k}| = ck$.

HAMILTONIAN OF THE FIELD

The classical Hamiltonian has the form

$$H = \frac{1}{2}\epsilon_0\iiint_V\left(E(\mathbf{r},t)^2 + c^2B(\mathbf{r},t)^2\right)d^3\mathbf{r} = V\epsilon_0\sum_{\mathbf{k}}\sum_{\mu=1,-1}\omega^2\left(\overline{a}_{\mathbf{k}}^{(\mu)}(t)a_{\mathbf{k}}^{(\mu)}(t) + a_{\mathbf{k}}^{(\mu)}(t)\overline{a}_{\mathbf{k}}^{(\mu)}(t)\right).$$

Substitution of the field operators into the classical Hamiltonian gives the Hamilton operator of the EM field,

$$\begin{aligned}
H &= \frac{1}{2}\sum_{\mathbf{k},\mu=-1,1}\hbar\omega\left(a^{\dagger(\mu)}(\mathbf{k})a^{(\mu)}(\mathbf{k}) + a^{(\mu)}(\mathbf{k})a^{\dagger(\mu)}(\mathbf{k})\right) \\
&= \sum_{\mathbf{k},\mu}\hbar\omega\left(a^{\dagger(\mu)}(\mathbf{k})a^{(\mu)}(\mathbf{k}) + \frac{1}{2}\right)
\end{aligned}$$

By the use of the commutation relations the second line follows from the first. Note again that $\hbar\omega = h\nu = \hbar c|\mathbf{k}|$ and remember that ω depends on k, even though it is not explicit in the notation. The notation ω(k) could have been introduced, but is not common as it clutters the equations.

Digression: Harmonic Oscillator

The second quantized treatment of the one-dimensional quantum harmonic oscillator is a well-known topic in quantum mechanical courses. We digress and say a few words about it. The harmonic oscillator Hamiltonian has the form

$$H = \hbar\omega\left(a^\dagger a + \tfrac{1}{2}\right)$$

where $\omega \equiv 2\pi\nu$ is the fundamental frequency of the oscillator. The ground state of the oscillator is designated by $|\,0\rangle$ and is referred to as *vacuum state.* It can be shown that $a^\dagger$ is an excitation operator, it excites from an n fold excited state to an $n+1$ fold excited state:

$$a^\dagger|\,n\rangle = |\,n+1\rangle\sqrt{n+1} \quad \text{in particular} \quad a^\dagger|\,0\rangle = |\,1\rangle \quad \text{and} \quad (a^\dagger)^n|\,0\rangle \propto |\,n\rangle.$$

Since harmonic oscillator energies are equidistant, the n-fold excited state $|\,n\rangle$ can be looked upon as a single state containing n particles (sometimes called vibrons) all of energy $h\nu$.

These particles are bosons. For obvious reason the excitation operator is called a *creation operator.*

From the commutation relation follows that the Hermitian adjoint de-excites:

$$a|\,n\rangle = |\,n-1\rangle\sqrt{n} \quad \text{in particular} \quad a|\,0\rangle \propto 0,$$

so that

$$a|\,0\rangle = 0.$$

For obvious reason the de-excitation operator is called an *annihilation operator.*

By mathematical induction the following "differentiation rule", that will be needed later, is easily proved,

$$[a,(a^\dagger)^n] = n(a^\dagger)^{n-1} \quad \text{with} \quad (a^\dagger)^0 = 1.$$

Suppose now we have a number of non-interacting (independent) one-dimensional harmonic oscillators, each with its own fundamental frequency ω_i. Because the oscillators are independent, the Hamiltonian is a simple sum:

$$H = \sum_i \hbar\omega_i\left(a^\dagger(i)a(i) + \tfrac{1}{2}\right).$$

Making the Substitution

$$i \to (\mathbf{k},\mu)$$

we see that the Hamiltonian of the EM field can be looked upon as a Hamiltonian of independent oscillators of energy $\omega = |\mathbf{k}|\, c$ and oscillating along direction $e^{(\mu)}$ with $\mu=1, -1$.

Photon States

The quantized EM field has a vacuum (no photons) state $|0\rangle$. The application to it of, say,

$$\left(a^{\dagger(\mu)}(\mathbf{k})\right)^m \left(a^{\dagger(\mu')}(\mathbf{k}')\right)^n |0\rangle \propto |(\mathbf{k},\mu)^m;(\mathbf{k}',\mu')^n\rangle,$$

gives a quantum state of m photons in mode (k,μ) and n photons in mode (k′, μ'). The proportionality symbol is used because the state on the left-hand is not normalized to unity, whereas the state on the right-hand may be normalized.

The Operator:

$$N^{(\mu)}(k) \equiv a^{\dagger(\mu)}(k)a^{(\mu)}(k)$$

is the *number operator*. When acting on a quantum mechanical photon state, it returns the number of photons in mode (k,μ). This also holds when the number of photons in this mode is zero, then the number operator returns zero. To show the action of the number operator on a one-photon ket, we consider

$$\begin{aligned} N^{(\mu)}(k)\,|k',\mu'\rangle &= a^{\dagger(\mu)}(k)a^{(\mu)}(k)\, a^{\dagger(\mu')}(k')|0\rangle = a^{\dagger(\mu)}(k)\left(\delta_{k,k'}\delta_{\mu,\mu'} + a^{\dagger(\mu')}(k')a^{(\mu)}(k)\right)|0\rangle \\ &= \delta_{k,k'}\delta_{\mu,\mu'}|k,\mu\rangle, \end{aligned}$$

i.e., a number operator of mode (k,μ) returns zero if the mode is unoccupied and returns unity if the mode is singly occupied. To consider the action of the number operator of mode (k, μ) on a n-photon ket of the same mode, we drop the indices k and μ and consider

$$N(a^{\dagger})^n\,|0\rangle = a^{\dagger}\left([a,(a^{\dagger})^n] + (a^{\dagger})^n a\right)|0\rangle = a^{\dagger}[a,(a^{\dagger})^n]|0\rangle.$$

Use the "differentiation rule" introduced earlier and it follows that

$$N(a^{\dagger})^n\,|0\rangle = n(a^{\dagger})^n\,|0\rangle.$$

A photon state is an eigenstate of the number operator. This is why the formalism described here, is often referred to as the *occupation number representation*.

PHOTON ENERGY

Earlier the Hamiltonian,

$$H = \sum_{k,\mu} \hbar\omega\left(a^{\dagger(\mu)}(k)a^{(\mu)}(k) + \frac{1}{2}\right)$$

was introduced. The zero of energy can be shifted, which leads to an expression in terms of the number operator,

$$H\,|\,k,\mu\rangle \equiv H\left(a^{\dagger(\mu)}(k)|\,0\rangle\right) = \sum_{k',\mu'} \hbar\omega' N^{(\mu')}(k')a^{\dagger(\mu)}(k)|0\rangle = \hbar\omega\left(a^{\dagger(\mu)}(k)|\,0\rangle\right) = \hbar\omega\,|\,k,\mu\rangle.$$

The effect of H on a single-photon state is

$$H\left|(k,\mu)^m;(k',\mu')^n\right\rangle = \left[m(\hbar\omega)+n(\hbar\omega')\right]\left|(k,\mu)^m;(k',\mu')^n\right\rangle,$$

Apparently, the single-photon state is an eigenstate of H and $\hbar\omega = h\nu$ is the corresponding energy. In the very same way

with

$$\omega = c\,|\,k\,| \quad and \quad \omega' = c\,|\,k'\,|.$$

Example Photon Density

In this chapter the electromagnetic energy density was computed that a 100kW radio station creates in its environment; at 5 km from the station it was estimated to be $2.1\cdot 10^{-10}$ J/m^3. Is quantum mechanics needed to describe the broadcasting of this station?

The classical approximation to EM radiation is good when the number of photons is much larger than unity in the volume

$$\left(\frac{\lambda}{2\pi}\right)^3,$$

where λ is the length of the radio waves. In that case quantum fluctuations are negligible and cannot be heard.

Suppose the radio station broadcasts at ν = 100 MHz, then it is sending out photons with an energy content of $\nu h = 1\cdot 10^8 \times 6.6\cdot 10^{-34} = 6.6\cdot 10^{-26}$ J, where h is Planck's constant. The wavelength of the station is $\lambda = c/\nu = 3$ m, so that $\lambda/(2\pi) = 48$ cm and the volume is 0.111 m^3. The energy content of this volume element is $2.1\cdot 10^{-10} \times 0.111 = 2.3\cdot 10^{-11}$ J, which amounts to

$$3.5\cdot 10^{12} \text{ photons per } \left(\frac{\lambda}{2\pi}\right)^3$$

Obviously, $3.5\cdot 10^{12}$ is much larger than one and hence quantum effects do not play a role; the waves emitted by this station are well into the classical limit.

PHOTON MOMENTUM

Introducing the Fourier expansion of the electromagnetic field into the classical form

$$P_{EM} = \int_0 \iiint_V E(r,t)\times B(r,t)d^3r,$$

yields

$$P_{EM} = V\int_0 \sum_k \sum_{\mu=1,-1} \omega k\left(a_k^{(\mu)}(t)\bar{a}_k^{(\mu)}(t) + \bar{a}_k^{(\mu)}(t)a_k^{(\mu)}(t)\right).$$

Quantization gives

$$P_{EM} = \sum_{k,\mu} \hbar k\left(a^{\dagger(\mu)}(k)a^{(\mu)}(k) + \frac{1}{2}\right) = \sum_{k,\mu} \hbar k N^{(\mu)}(k).$$

The term 1/2 could be dropped, because when one sums over the allowed k, k cancels with –k. The effect of P_{EM} on a single-photon state is

$$P_{EM}|k,\mu\rangle = P_{EM}\left(a^{\dagger(\mu)}(k)|0\rangle\right) = \hbar k\left(a^{\dagger(\mu)}(k)|0\rangle\right) = \hbar k|k,\mu\rangle.$$

Apparently, the single-photon state is an eigenstate of the momentum operator, and $\hbar$ k is the eigenvalue (the momentum of a single photon).

PHOTON MASS

The photon having non-zero linear momentum, one could imagine that it has a non-vanishing rest mass m_0, which is its mass at zero speed. However, we will now show that this is not the case: $m_0 = 0$.

Since the photon propagates with the speed of light, special relativity is called for. The relativistic expressions for energy and momentum squared are,

$$E^2 = \frac{m_0^2 c^4}{1 - v^2/c^2}, \quad p^2 = \frac{m_0^2 v^2}{1 - v^2/c^2}.$$

From p^2/E^2,

$$\frac{v^2}{c^2} = \frac{c^2 p^2}{E^2} \Rightarrow E^2 = \frac{m_0^2 c^4}{1 - c^2 p^2/E^2} \Rightarrow m_0^2 c^4 = E^2 - c^2 p^2.$$

Use

$$E^2 = \hbar^2 \omega^2 \quad \text{and} \quad p^2 = \hbar^2 k^2 = \frac{\hbar^2 \omega^2}{c^2}$$

and it follows that

$$m_0^2 c^4 = E^2 - c^2 p^2 = \hbar^2 \omega^2 - c^2 \frac{\hbar^2 \omega^2}{c^2} = 0,$$

so that $m_0 = 0$.

PHOTON SPIN

The photon can be assigned a triplet spin with spin quantum number $S = 1$. This is similar to, say, the nuclear spin of the ^{14}N isotope, but with the important difference that the state with $M_S = 0$ is zero, only the states with $M_S = \pm 1$ are non-zero.

Define spin operators:

$$S_z \equiv -i\hbar\left(e_x \otimes e_y - e_y \otimes e_x\right) \quad \textit{and cyclically} \quad x \to y \to z \to x.$$

The products between the two orthogonal unit vectors are dyadic products. The unit vectors are perpendicular to the propagation direction k (the direction of the z axis, which is the spin quantization axis). The spin operators satisfy the usual angular momentum commutation relations

$$[S_x, S_y] = i\hbar S_z \quad \text{and cyclically} \quad x \to y \to z \to x.$$

Indeed, use the dyadic product property

$$\left(e_y \otimes e_z\right)\left(e_z \otimes e_x\right) = (e_y \otimes e_x)(e_z \cdot e_z) = e_y \otimes e_x$$

because e_z is of unit length. In this manner,

$$\begin{aligned}\left[S_x, S_y\right] &= -\hbar^2\left(e_y \otimes e_z - e_z \otimes e_y\right)\left(e_z \otimes e_x - e_x \otimes e_z\right) + \hbar^2\left(e_z \otimes e_x - e_x \otimes e_z\right)\left(e_y \otimes e_z - e_z \otimes e_y\right) \\ &= i\hbar\left[-i\hbar\left(e_x \otimes e_y - e_y \otimes e_x\right)\right] = i\hbar S_z.\end{aligned}$$

By inspection it follows that

$$-i\hbar\left(e_x \otimes e_y - e_y \otimes e_x\right) \cdot e^{(\mu)} = \mu e^{(\mu)}, \quad \mu = 1, -1,$$

and therefore μ labels the photon spin,

$$S_z \mid k, \mu\rangle = \mu \mid k, \mu\rangle, \quad \mu = 1, -1.$$

Because the vector potential A is a transverse field, the photon has no forward ($\mu = 0$) spin component.

SHORT RANGE FORCES IN OPTO-MECHANICAL DEVICES

The gradient optical force has been recently explored in optomechanical devices of various geometries. Promising applications including tunable photonics, backaction cooling and amplification and optomechanical signal processing have been proposed. However, these exciting applications rely on transduction with strong gradient force, which is only obtainable with narrow coupling gaps. In this short range regime, besides the forces imposed by real photons, there also exists optical force mediated by virtual photons – the Casimir force. This virtual photon force has been known to affect the fabrication yield and dynamical performance of micro- and nano-mechanical devices. The relative intensity of these two forces therefore has profound impact on the optomechanical performance of optical force driven tunable couplers and tunable photonic filters.

The experimental demonstration of such photon-mediated force has largely benefited from initial theoretical investigations based on numerical simulations and analytical approaches. The theoretical framework for the analysis of gradient optical force has been proven to be very efficient and rather accurate for the investigation of nanophotonic devices that carry guided optical modes. On the other hand,theoretical work on Casimir interactions until recently has been restricted to relatively simple geometries, such as parallel plates or combinations of elementary surfaces. New numerical methods capable of computing the Casimir force in arbitrary geometries have been developed to investigate the strong dependence of the force on geometry and led to a variety of interesting effects. These time-domain simulation based methods are emerging as a convenient way to calculate Casimir forces because they allow for the integration of complex material properties into the simulation engine, which is of importance when dispersive materials are being considered. Such computational methods are also particularly suitable for the investigation of optical forces in nano-optomechanical devices, whose optical properties are nowadays routinely analysed with finite-difference time-domain (FDTD) methods. Free-standing nano-mechanical waveguides constitute fundamental optomechanical elements which can be mechanically actuated by a driving optical force. In coupled waveguides, both the magnitude and sign of the gradient optical force can be controlled. When the separation between the waveguides is reduced to nanoscale dimensions the influence of the Casimir force has to be considered in addition to the gradient force. In this article we determine real and virtual photon forces in coupled free-standing waveguides of typical nanophotonic dimensions, using the finite-difference time-domain and finite-element methods. Because the Casimir force is a broadband force, the chromatic dispersion of silicon is incorporated in the calculation. We find that, compared to metallic reference structures, the Casimir force is reduced by a factor of three in silicon for small waveguide separations and decreases super-exponentially with increasing gap between the two waveguides. Our results show that for gaps greater then 100nm substantial repulsive optical forces can be generated which dominate the Casimir force. When the gap between the waveguides is smaller than 100nm mW levels of optical power are needed to offset the Casimir force.

CASIMIR FORCES

We consider the geometry shown schematically. Two free-standing waveguides are separated by a gap g. The waveguides are of width w and height h and designed to be single-mode. The waveguides are surrounded by air, assuming that the substrate is sufficiently far away so that its effect can be neglected in the following calculations.

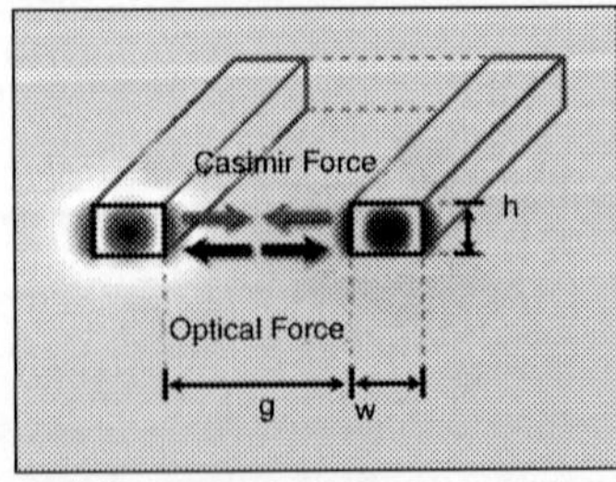

Figure: The Geometry used for the Determination of both the Gradient Optical Force and the Casimir Force. Overlaid is the Profile of the Odd Optical Mode, Responsible for the Generation of Repulsive Gradient Force.

We calculate the Casimir force numerically using the time-domain methodology developed by Rodriguez *et al.*. The numerical procedure provides the Casimir force between arbitrary structures at zero temperature from time-domain electro-magnetic field values. The FDTD method is used to obtain the electrical field distribution resulting from excitation of the coupled waveguides with current sources placed on the surface of the dielectric. This allows us to extract the Maxwell-Stress tensor and subsequently the Casimir force through surface integration of the field correlations in combination with the time-domain function introduced in reference.

The time-domain fields are sampled in close proximity to the waveguide surface. The geometry depicted is surrounded by perfectly matched layers to absorb any electromagnetic fields scattered during the FDTD calculation, leaving sufficient space around the waveguides to avoid boundary effects. We use a fine grid resolution of 5nm to make sure that the results have converged to an average error below 1%. In order to avoid numerical artifacts, we perform three individual Casimir force calculations for each configuration, as suggested in reference. We obtain the Casimir force for the coupled system, the left waveguide only and the right waveguide only, respectively. The forces resulting from the individual waveguides are then subtracted from the force of the coupled beams, in order to correct for discretization errors during the FDTD calculation.

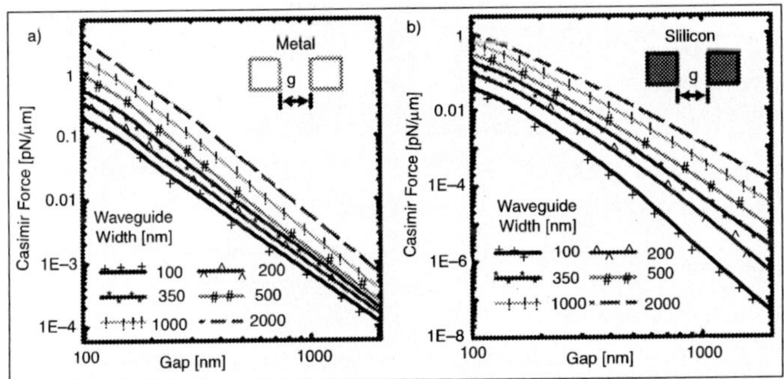

Figure: a): The Casimir Force Calculated for Square Coupled Nano-beams Composed of Perfectly Conducting Metal, in Dependence of Waveguide Separation. b) The Equivalent Calculation for Waveguides Made from Pure Silicon, taking into Account the Frequency Dispersion of Silicon.

To establish a reference, we first consider perfectly conducting metallic waveguides, with a square cross-section of a width varying from 100nm to 2000nm. The separation between the waveguides is considered at a scale relevant to nano-device fabrication. The calculated Casimir force is divided by the length of the waveguide, yielding a force per unit length in that direction. The Casimir force decays exponentially with increasing gap, which is also predicted by the proximity force approximation (PFA). Fitting the results to a power law, we find a decrease of the Casimir force with waveguide separation as $F \sim g^{-4.1}A$, where A is the waveguide area, which agrees well with the PFA. The largest Casimir forces are obtained for small gaps, approaching 3pN/μm at 100nm separation. Due to the exponential decay, Casimir forces are most relevant in short range and thus in the regime corresponding to the preferred operation parameter space of optomechanical devices.

Because the Casimir force correlates with the interaction surface, larger Casimir forces are obtained for wider waveguides. The Casimir force increases linearly with waveguide width for a given gap g in first-order approximation. For gaps approaching 100nm we obtain Casimir forces on the order of pN/μm. Such force values are comparable to gradient optical forces simulated numerically and measured experimentally. Compared to the PFA with perfect metallic plates, the calculated Casimir forces in silicon are reduced by a factor of roughly ten, which is consistent with the results presented by Rodriguez *et.al.*

In the context of optomechanical devices we then consider two coupled silicon beams of square cross-section. The frequency dispersion of silicon is modelled by approximating the dielectric function with a the following Drude-Lorentz function

$$\varepsilon_{Si}(i\omega) = \varepsilon_{\infty} + \omega_0^2 \frac{\varepsilon_0 - \varepsilon_{\infty}}{\omega^2 + \omega_0^2}$$

where $\omega_0 \approx 6.6 \times 10^{15}$rad/s and $\varepsilon 0 = 1.035$.

The Casimir force between two silicon waveguides decays with increasing gap at a rate comparable to the case of waveguides made of perfect conductor. When the waveguides become very small, however, the decay at larger gaps is slightly faster than exponential. Compared to the perfect conductor case, the Casimir force is reduced by a factor of three for small gaps. This reduction factor increases to a factor of 100-200 for 2000nm separation.

Next we consider coupled beam waveguides of 220nm height, which is a typical thickness for conventional silicon waveguides fabricated from silicon-on-insulator substrates.

We keep the waveguide height constant and vary the width of the waveguides for gaps ranging from 100nm to 2000nm. As expected from Casimir force theory, for large gaps, the attractive Casimir force acting between the waveguides does not depend on the waveguide width because the interaction surface between the waveguides remains unchanged. The upper and lower groups of curves represent calculation results for perfectly conducting waveguides and for silicon waveguides, respectively. Clearly visible is the super-exponential decay for the silicon waveguides towards greater separation gaps.

As in the case of square waveguides, for small separations, the Casimir force between silicon beams is about a factor of three smaller than between perfectly conducting beams. This reduction factor increases to a factor of 100-200 for large separations. The reduction in Casimir force is due to the smaller refractive index contrast in silicon structures compared to the index contrast in perfectly conducting structures. The dielectric property of silicon allows the electro-magnetic vacuum fields to penetrate into the waveguides and thus leads to reduced modal confinement.

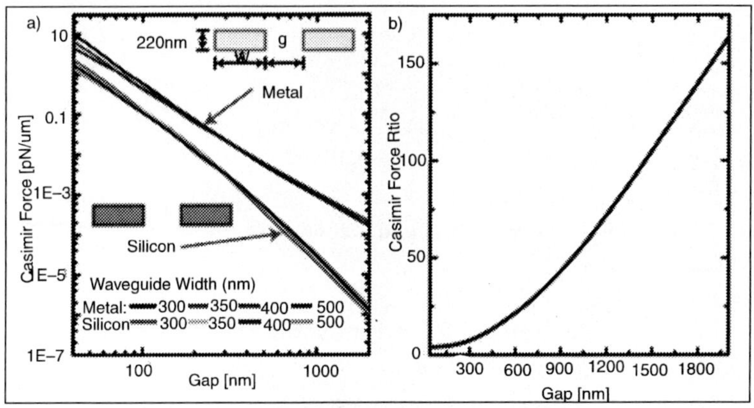

Figure: a) The Casimir Force for Waveguides with a Fixed Height of 220nm in Dependence of Waveguide Separation. b) The Casimir Force Ratio between Metal and Silicon Waveguides.

The ratio between the Casimir force in the metal waveguides with respect to the silicon waveguides. A sharp increase of the ratio for wide gaps is observed, approaching 160 for 200nm gap, which implies that the Casimir force in silicon waveguides is significantly reduced for large separations. Interpreting the numerical results obtained above, we find that the modal guiding properties of the free-standing waveguides can be engineered by tuning the beam width without significantly affecting the magnitude of the Casimir force, provided that the waveguide height is kept constant. This is a convenient situation for top-down fabrication of photonic waveguides since the waveguide height is usually fixed.

Gradient Optical Force

The coupled waveguide system described above supports two optical modes, termed the even and odd modes because of their respective symmetry properties with respect to the direction of propagation. When the distance between the waveguides is varied, the effective refractive mode index changes as well. The resulting change of the eigenenergy of the propagating mode gives rise to an optical gradient force, which was first calculated theoretically by Povinelli *et al.*.

When the even optical mode is excited in the coupled waveguide system, the effective mode index decreases with increasing gap and thus the resulting optical force is attractive. The situation is different for the odd mode, which experiences a minimal effective mode index for a cross-over gap g_c. As a result the optical force is attractive for gaps smaller than g_c and repulsive for gaps greater than g_c.

The optical force between two waveguides can be employed to efficiently actuate mechanical motion of the waveguides. Attractive optical forces pull the waveguides together and repulsive optical forces push the waveguides apart.

While gradient optical forces are mediated by real photons, Casimir forces arise from virtual photons (or vacuum fluctuation). The Casimir force between two silicon waveguides in air is attractive and thus pulls the waveguides together, which can lead to stiction effects when the separation between the waveguides is very small. Nevertheless, the repulsive optical force generated by the odd-mode of coupled waveguides can be exploited to counteract the attractive Casimir force.

Turning to the optical gradient force, we determine the repulsive force for dielectric waveguides that support only a single mode in each beam. The optical force is obtained with the method suggested originally in reference, which involves determining the effective propagation index of the coupled waveguide modes in dependence of waveguide separation. The gap between the waveguides is varied from 50nm to 600nm, which correspond to relevant fabrication distances in coupled free-standing nano-photonic waveguides. The optical force is then proportional to the spatial derivative of the effective index. The effective index is calculated with a commercial finite-element simulation tool, taking into account the frequency dependence of the refractive index of silicon. Because the beam waveguides are intended for operation at telecom wavelengths, we obtain the gradient optical force for a representative wavelength of 1550nm. We calculate the repulsive force as a function of gap for waveguide widths ranging from 300nm to 500nm.

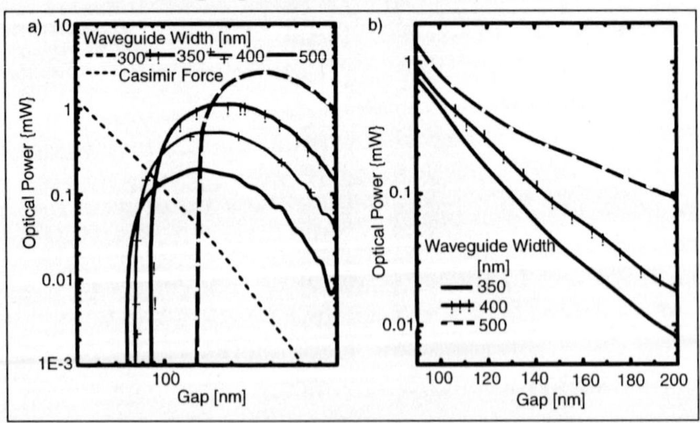

Figure: (a) The Calculated Repulsive Optical Force in Dependence of Waveguide Separation and Waveguide Width, for Waveguides of 220nm Height. (b) The Amount of Optical Power Required for the Cancellation of the Casimir Force.

For reference, the expected Casimir force is also included with the dashed orange line. The repulsive optical force emerges for gaps exceeding a cross-over gap g_c. This cross-over point depends on the waveguide width, where wider waveguides yield smaller g_c. At the same time, wider waveguides yield smaller repulsive forces because the decay of the effective index becomes more pronounced due to reduced evanescent mode tails. The repulsive optical force (normalized to optical input power and waveguide length) is able to overcome the Casimir force beyond a certain compensation gap. For the waveguide widths considered here, the smallest compensation gap is found to be near 85nm for the 400nm wide waveguides.

The compensation gap, the optical force decays rapidly and eventually switches sign to attractive). Both the optical force and the Casimir force decay exponentially for larger gaps, however the optical force remains stronger in this regime.

We determine the optical input power that is required to achieve repulsive gradient force equivalent to the Casimir force for various separation gaps. The compensation of the Casimir force can be achieved with optical input powers below 1mW for gaps of 100nm or larger. The optical power necessary for the compensation increases with increasing waveguide width, because the repulsive gradient optical forces become weaker when the mode is more confined inside the waveguide. For larger gaps the power required for compensation decays more slowly, because the reduction of the optical force is slower than the Casimir effect.

In terms of optomechanical actuation, the Casimir force provides a constant force offset which will reposition the waveguides and thus change the working condition of the optomechanical devices. This effect is significant when the separation between the coupled waveguides is reduced below 100nm and therefore has to be taken into account in the design of functional optomechanical devices.

CASIMIR EFFECT?

The Casimir effect is a small attractive force that acts between two close parallel *uncharged* conducting plates. It is due to quantum vacuum fluctuations of the electromagnetic field.

The effect was predicted by the Dutch physicist Hendrick Casimir in 1948. According to the quantum theory, the vacuum contains virtual particles which are in a continuous state of fluctuation. Casimir realised that between two plates, only those virtual photons whose wavelengths fit a whole number of times into the gap should be counted when calculating the vacuum energy. The energy density decreases as the plates are moved closer, which implies that there is a small force drawing them together.

The attractive Casimir force between two plates of area *A* separated by a distance *L* can be calculated to be,

$$F = \frac{\pi hc}{480L^4} = A$$

where *h* is Planck's constant and *c* is the speed of light.

The tiny force was measured in 1996 by Steven Lamoreaux. His results were in agreement with the theory to within the experimental uncertainty of 5%. Particles other than the photon also contribute a small effect but only the photon force is measurable. All bosons such as photons produce an attractive Casimir force while fermions make a repulsive contribution. If electromagnetism was supersymmetric there would be fermionic photinos whose contribution would exactly cancel that of the photons and there would be no Casimir effect. The fact that the Casimir effect exists shows that if supersymmetry exists in nature it must be a broken symmetry

According to the theory the total zero point energy in the vacuum is infinite when summed over all the possible photon modes. The Casimir effect comes from a difference of energies in which the infinities cancel. The energy of the vacuum is a puzzle in theories of quantum gravity since it should act gravitationally and produce a large cosmological constant which would cause spacetime to curl up. The solution to the inconsistency is expected to be found in a theory of quantum gravity.

CONDUCTING METALS

The Casimir effect can be understood by the idea that the presence of conducting metals and dielectrics alters the vacuum expectation value of the energy of the second quantized electromagnetic field. Since the value of this energy depends on the shapes and positions of the conductors and dielectrics, the Casimir effect manifests itself as a force between such objects.

VACUUM ENERGY

The causes of the Casimir effect are described by quantum field theory, which states that all of the various fundamental fields, such as the electromagnetic field, must be quantized at each and every point in space. In a simplified view, a "field" in physics may be envisioned as if space were filled with interconnected vibrating balls and springs, and the strength of the field can be visualized as the displacement of a ball from its rest position. Vibrations in this field propagate and are governed by the appropriate wave equation for the particular field in question. The second quantization of quantum field theory requires that each such ball-spring combination be quantized, that is, that the strength of the field be quantized at each point in space. At the most basic level, the field at each point in space is a simple harmonic oscillator, and its quantization places a quantum harmonic oscillator at each point. Excitations of the field correspond to the elementary particles of particle physics. However, even the vacuum has a vastly complex structure, so all calculations of quantum field theory must be made in relation to this model of the vacuum.

The vacuum has, implicitly, all of the properties that a particle may have: spin, or polarization in the case of light, energy, and so on. On average, most of these properties cancel out: the vacuum is, after all, "empty" in this sense. One important exception is the vacuum energy or the vacuum expectation value of the energy. The quantization of a simple harmonic oscillator states that the lowest possible energy or zero-point energy that such an oscillator may have is

$$E = \frac{1}{2}\hbar\omega .$$

Summing over all possible oscillators at all points in space gives an infinite quantity. To remove this infinity, one may argue that only differences in energy are physically measurable; this argument is the underpinning of the theory of renormalization. In all practical calculations, this is how the infinity is always handled. In a deeper sense, however, renormalization is unsatisfying, and the removal of this infinity presents a challenge in the search for a Theory of Everything. Currently there is no compelling explanation for how this infinity should be treated as essentially zero; a non-zero value is essentially the cosmological constant and any large value causes trouble in cosmology.

Alternatively, a 2005 paper by Robert Jaffe of MIT states that "Casimir effects can be formulated and Casimir forces can be computed without reference to zero point energies. They are relativistic, quantum forces between charges and currents. The Casimir force (per unit area) between parallel plates vanishes as alpha, the fine structure constant, goes to zero, and the standard result, which appears to be independent of alpha, corresponds to the alpha ! infinity limit," and that "The Casimir force is simply the (relativistic, retarded) van der Waals force between the metal plates."

EFFECTS OBSERVATION

Casimir's observation was that the second-quantized quantum electromagnetic field, in the presence of bulk bodies such as metals or dielectrics, must obey the same boundary conditions that the classical electromagnetic field must obey. In particular, this affects the calculation of the vacuum energy in the presence of a conductor or dielectric.

Consider, for example, the calculation of the vacuum expectation value of the electromagnetic field inside a metal cavity, such as, for example, a radar cavity or a microwave waveguide. In this case, the correct way to find the zero point energy of the field is to sum the energies of the standing waves of the cavity. To each and every possible standing wave corresponds an energy; say the energy of the nth standing wave is E_n. The vacuum expectation value of the energy of the electromagnetic field in the cavity is then

$$\langle E\rangle = \frac{1}{2}\sum_n E_n$$

with the sum running over all possible values of n enumerating the standing waves. The factor of 1/2 corresponds to the fact that the zero-point energies are being summed (it is the same 1/2 as appears in the equation $E = \hbar\omega/2$). Written in this way, this sum is clearly divergent; however, it can be used to create finite expressions.

In particular, one may ask how the zero point energy depends on the shape s of the cavity. Each energy level E_n depends on the shape, and so one should write $E_n(S)$ for the energy level, and $\langle E(s)\rangle$ for the vacuum expectation value. At this point comes an important observation: the force at point p on the wall of the cavity is equal to the change in the vacuum energy if the shape s of the wall is perturbed a little bit, say by δs, at point p. That is, one has

$$F(p) = -\left.\frac{\delta\langle E(s)\rangle}{\delta s}\right|_p .$$

This value is finite in many practical calculations.

Attraction between the plates can be easily understood by focusing on the 1-dimensional situation. Suppose that a moveable conductive plate is positioned at a short distance a from one of two widely separated plates (distance L apart). With $a \ll L$, the states within the slot of width a are highly constrained so that the energy E of any one mode is widely separated from that of the next. This is not the case in open region L, where there is a large number (about L/a) of states with energy evenly spaced between E and the next mode in the narrow slot—in other words, all slightly larger than E. Now on shortening a by da (< 0), the mode in the slot shrinks in wavelength and therefore increases in energy proportional to -da/a, whereas all the outside L/a states lengthen and correspondingly lower energy proportional to da/L (note the denominator). The net change is slightly negative, because all the L/a modes' energies are slightly larger than the single mode in the slot.

DERIVATION OF CASIMIR EFFECT ASSUMING ZETA-REGULARIZATION

In the original calculation done by Casimir, he considered the space between a pair of conducting metal plates at distance apart. In this case, the standing waves are particularly easy to calculate, since the transverse component of the electric field and the normal component of the magnetic field must vanish on the surface of a conductor. Assuming the parallel plates lie in the xy-plane, the standing waves are

$$\psi_n(x,y,z;t) = e^{-i\omega_n t} e^{ik_x x + ik_y y} \sin(k_n z)$$

where ψ_s stands for the electric component of the electromagnetic field, and, for brevity, the polarization and the magnetic components are ignored here. Here, k_x and k_x are the wave vectors in directions parallel to the plates, and

$$k_n = \frac{n\pi}{a}$$

is the wave-vector perpendicular to the plates. Here, n is an integer, resulting from the requirement that ψ vanish on the metal plates. The energy of this wave is

$$\omega_n = c\sqrt{k_x{}^2 + k_y{}^2 + \frac{n^2\pi^2}{a^2}}$$

where c is the speed of light. The vacuum energy is then the sum over all possible excitation modes

$$\langle E\rangle = \frac{\hbar}{2}\cdot 2\int \frac{A dk_x dk_y}{(2\pi)^2}\sum_{n=1}^{\infty}\omega_n$$

where A is the area of the metal plates, and a factor of 2 is introduced for the two possible polarizations of the wave. This expression is clearly infinite, and to proceed with the calculation, it is convenient to introduce a regulator. The regulator will serve to make the expression finite, and in the end will be removed. The zeta-regulated version of the energy per unit-area of the plate is

$$\frac{\langle E(s)\rangle}{A} = \hbar\int \frac{dk_x dk_y}{(2\pi)^2}\sum_{n=1}^{\infty}\omega_n \mid \omega_n \mid^{-s}.$$

In the end, the limit $s \to 0$ is to be taken. Here s is just a complex number, not to be confused with the shape discussed previously. This integral/sum is finite for s real and larger than 3.

The sum has a pole at $s = 3$, but may be analytically continued to $s = 0$, where the expression is finite.

The above expression is easily simplified to:

$$\frac{\langle E(s)\rangle}{A} = \frac{\hbar c^{1-s}}{4\pi^2}\sum_n \int_0^{\infty} 2\pi q dq \left| q^2 + \frac{\pi^2 n^2}{a^2}\right|^{(1-s)/2}$$

where polar coordinates $q^2 = k_x^2 + k_y^2$ were introduced to turn the double integral into a single integral. The in front is the Jacobian, and the 2π comes from the angular integration. The integral is easily performed and converges if Re > 3, resulting in

$$\frac{\langle E(s)\rangle}{A} = -\frac{\hbar c^{1-s}\pi^{2-s}}{2a^{3-s}}\frac{1}{3-s}\sum_n \mid n \mid^{3-s}.$$

The sum clearly diverges at s in the neighbourhood of zero, but if the damping of large-frequency excitations corresponding to analytic continuation of the Riemann zeta function to $s = 0$ is assumed to make sense physically in some way, then one has

$$\frac{\langle E\rangle}{A} = \lim_{s\to 0}\frac{\langle E(s)\rangle}{A} = -\frac{\hbar c\pi^2}{6a^3}\zeta(-3).$$

But $\zeta(-3) = 1/120$ and so one obtains

$$\frac{\langle E \rangle}{A} = \frac{-\hbar c \pi^2}{3 \cdot 240 a^3}.$$

The analytic continuation has evidently lost an additive positive infinity, somehow exactly accounting for the zero-point energy outside the slot between the plates, but which changes upon plate movement within a closed system. The Casimir force per unit area F_c / A for idealized, perfectly conducting plates with vacuum between them is

$$\frac{F_c}{A} = -\frac{d}{da}\frac{\langle E \rangle}{A} = -\frac{\hbar c \pi^2}{240 a^4}$$

where

$\hbar$ (hbar, $\hbar$) is the reduced Planck constant,

c is the speed of light,

a is the distance between the two plates.

The force is negative, indicating that the force is attractive: by moving the two plates closer together, the energy is lowered. The presence of $\hbar$ shows that the Casimir force per unit area F_c/A is very small, and that furthermore, the force is inherently of quantum-mechanical origin.

NOTE: In Casimir's original derivation, a moveable conductive plate is positioned at a short distance a from one of two widely separated plates (distance L apart). The 0-point energy on *both* sides of the plate is considered. Instead of the above *ad hoc* analytic continuation assumption, non-convergent sums and integrals are computed using Euler-Maclaurin summation with a regularizing function (*e.g.*, exponential regularization) not so anomalous as $|\omega_n|^{-s}$ in the above.

More Recent Theory

Casimir's analysis of idealized metal plates was generalized to arbitrary dielectric and realistic metal plates by Lifshitz and his students. Using this approach, complications of the bounding surfaces, such as the modifications to the Casimir force due to finite conductivity, can be calculated numerically using the tabulated complex dielectric functions of the bounding materials. Lifshitz' theory for two metal plates reduces to Casimir's idealized $1/a^4$ force law for large separations a much greater than the skin depth of the metal, and conversely reduces to the $1/a^3$ force law of the London dispersion force (with a coefficient called a Hamaker constant) for small a, with a more complicated dependence on a for intermediate separations determined by the dispersion of the materials.

Lifshitz' result was subsequently generalized to arbitrary multilayer planar geometries as well as to anisotropic and magnetic materials, but for

several decades the calculation of Casimir forces for non-planar geometries remained limited to a few idealized cases admitting analytical solutions.

For example, the force in the experimental sphere–plate geometry was computed with an approximation (due to Derjaguin) that the sphere radius R is much larger than the separation a, in which case the nearby surfaces are nearly parallel and the parallel-plate result can be adapted to obtain an approximate R/a^3 force (neglecting both skin-depth and higher-order curvature effects).

However, in the 2000s a number of authors developed and demonstrated a variety of numerical techniques, in many cases adapted from classical computational electromagnetics, that are capable of accurately calculating Casimir forces for arbitrary geometries and materials, from simple finite-size effects of finite plates to more complicated phenomena arising for patterned surfaces or objects of various shapes.

Measurement

One of the first experimental tests was conducted by Marcus Sparnaay at Philips in Eindhoven, in 1958, in a delicate and difficult experiment with parallel plates, obtaining results not in contradiction with the Casimir theory, but with large experimental errors. Some of the experimental details as well as some background information on how Casimir, Polder and Sparnaay arrived at this point are highlighted in a 2007 interview with Marcus Sparnaay.

The Casimir effect was measured more accurately in 1997 by Steve K. Lamoreaux of Los Alamos National Laboratory, and by Umar Mohideen and Anushree Roy of the University of California at Riverside. In practice, rather than using two parallel plates, which would require phenomenally accurate alignment to ensure they were parallel, the experiments use one plate that is flat and another plate that is a part of a sphere with a large radius.

In 2001, a group (Giacomo Bressi, Gianni Carugno, Roberto Onofrio and Giuseppe Ruoso) at the University of Padua (Italy) finally succeeded in measuring the Casimir force between parallel plates using microresonators.

Regularisation

In order to be able to perform calculations in the general case, it is convenient to introduce a regulator in the summations. This is an artificial device, used to make the sums finite so that they can be more easily manipulated, followed by the taking of a limit so as to remove the regulator.

The heat kernel or exponentially regulated sum is

$$\langle E(t)\rangle = \frac{1}{2}\sum_n \hbar\,|\omega_n|\exp(-t\,|\omega_n|)$$

where the limit $t \to 0^+$ is taken in the end. The divergence of the sum is typically manifested as

$$\langle E(t)\rangle = \frac{C}{t^3} + \text{finite}$$

for three-dimensional cavities. The infinite part of the sum is associated with the bulk constant C which *does not* depend on the shape of the cavity. The interesting part of the sum is the finite part, which is shape-dependent. The Gaussian regulator

$$\langle E(t)\rangle = \frac{1}{2}\sum_n \hbar|\omega_n|\exp(-t^2|\omega_n|^2)$$

is better suited to numerical calculations because of its superior convergence properties, but is more difficult to use in theoretical calculations. Other, suitably smooth, regulators may be used as well. The zeta function regulator

$$\langle E(s)\rangle = \frac{1}{2}\sum_n \hbar|\omega_n||\omega_n|^{-s}$$

is completely unsuited for numerical calculations, but is quite useful in theoretical calculations. In particular, divergences show up as poles in the complex s plane, with the bulk divergence at $s = 4$. This sum may be analytically continued past this pole, to obtain a finite part at $s = 0$.

Not every cavity configuration necessarily leads to a finite part (the lack of a pole at $s = 0$) or shape-independent infinite parts. In this case, it should be understood that additional physics has to be taken into account. In particular, at extremely large frequencies, metals become transparent to photons (such as X-rays), and dielectrics show a frequency-dependent cutoff as well. This frequency dependence acts as a natural regulator. There are a variety of bulk effects in solid state physics, mathematically very similar to the Casimir effect, where the cutoff frequency comes into explicit play to keep expressions finite.

Generalities in Casimir Effect

The Casimir effect can also be computed using the mathematical mechanisms of functional integrals of quantum field theory, although such calculations are considerably more abstract, and thus difficult to comprehend. In addition, they can be carried out only for the simplest of geometries. However, the formalism of quantum field theory makes it clear that the vacuum expectation value summations are in a certain sense summations over so-called "virtual particles".

More interesting is the understanding that the sums over the energies of standing waves should be formally understood as sums over the eigenvalues of a Hamiltonian. This allows atomic and molecular effects, such as the van der Waals force, to be understood as a variation on the theme of the Casimir effect. Thus one considers the Hamiltonian of a system as a function of the arrangement of objects, such as atoms, in configuration space. The change in the zero-point energy as a function of changes of the configuration can be understood to result in forces acting between the objects.

In the chiral bag model of the nucleon, the Casimir energy plays an important role in showing the mass of the nucleon is independent of the bag radius. In addition, the spectral asymmetry is interpreted as a non-zero vacuum expectation value of the baryon number, cancelling the topological winding number of the pion field surrounding the nucleon.

Wormholes of Exotic Matter

Exotic matter with negative energy density may be required to stabilize a wormhole. Morris, Thorne and Yurtsever pointed out that the quantum mechanics of the Casimir effect can be used to produce a locally mass-negative region of space-time, and suggested that negative effect could be used to stabilize a wormhole to allow faster than light travel.

In physics, a wormhole is a hypothetical topological feature of spacetime that would be, fundamentally, a "shortcut" through spacetime. For a simple visual explanation of a wormhole, consider spacetime visualized as a two-dimensional (2D) surface. If this surface is folded along a third dimension, it allows one to picture a wormhole "bridge". A wormhole is, in theory, much like a tunnel with two ends each in separate points in spacetime. This concept has been used extensively in science fiction.

Dynamic Casimir Effect

A similar analysis can be used to explain Hawking radiation that causes the slow "evaporation" of black holes (although this is generally visualised as the escape of one particle from a virtual particle-antiparticle pair, the other particle having been captured by the black hole).

The dynamical Casimir effect is the production of particles and energy from an accelerated boundary, often referred to as a *moving mirror* or *motion-induced radiation.*

Constructed within the framework of quantum field theory in curved spacetime, the dynamical Casimir effect has been used to better understand acceleration radiation; *i.e.* the Unruh effect.

Moving mirrors create entropy, particles, energy and gravitational-like effects. In analogy to the event horizon of a black hole, an accelerated mirror amplifies quantum field vacuum fluctuations.

An experimental verification of the dynamical Casimir effect was first achieved in May 2011 at Chalmers University of Technology, in Gothenburg, Sweden.

Repulsive Forces in Casimir Effect

There are few instances wherein the Casimir effect can give rise to repulsive forces between uncharged objects. In a seminal paper, Evgeny

Lifshitz showed (theoretically) that in certain circumstances (most commonly involving liquids), repulsive forces can arise. This has sparked interest in applications of the Casimir effect toward the development of levitating devices.

An experimental demonstration of the Casimir-based repulsion predicted by Lifshitz was recently carried out by Munday *et al.* Other scientists have also suggested the use of gain media to achieve a similar levitation effect, though this is controversial because these materials seem to violate fundamental causality constraints and the requirement of thermodynamic equilibrium (Kramers-Kronig relations). Casimir and Casimir-Polder repulsion can in fact occur for sufficiently anisotropic electrical bodies; for a review of the issues involved with repulsion see.

Local Density of States

Local density of states (LDOS) is a physical quantity that describes the density of states, but space-resolved. In materials science, this term is useful when interpreting the data from an STM, since this method is capable of imaging electron densities of states with atomic resolution. According to crystal's structure, this quantity can be predicted by computational methods, as for example with density functional theory.

SPONTANEOUS EMISSION ENHANCEMENT NEAR NANOPARTICLES

Spontaneous photon emission from an atom in the presence of different environments has received considerable attention over the last few years, with particular emphasis in geometries that give rise to either strong enhancement or inhibition of the emission.

In particular, strong enhancement has been reported for decay from atoms in photonic crystals, sometimes accompanied by a pronounced angular focusing effect.

Inhibition of spontaneous emission takes place when the emission lies in the gap of a photonic crystal. Other geometries including dielectric slabs and optical cavities have also been discussed. Strong features occur when the emission frequency is near the plasmon frequency of the nearby material, giving rise to very intense enhancement.

In this work, we will focus on the enhancement of spontaneous emission by placing the atom close to metallic nanoparticles when the emission resonates with plasmon modes of the particles.

The effect is similar to the properties of the spectra obtained in surface-enhanced Raman spectroscopy (SERS). Also, we are reporting strong focusing of the near field accompanying the emission for atoms inside ellipsoidal dielectric nanoparticles.

CAVITY QUANTUM ELECTRODYNAMICS

Cavity quantum electrodynamics (cavity QED) is the study of the interaction between light confined in a reflective cavity and atoms or other particles, under conditions where the quantum nature of light photons is significant. It could in principle be used to construct a quantum computer.

The case of a single 2-level atom in the cavity is mathematically described by the Jaynes-Cummings model, and undergoes vacuum Rabi oscillations $|e\rangle|n-1\rangle \leftrightarrow |g\rangle|n\rangle$, that is between an excited atom and n-1 photons, and a ground state atom and n photons.

If the cavity is on resonance with the atomic transition, a half-cycle of oscillation starting with no photons coherently swaps the atom qubit's state onto the cavity field's, $(\alpha|g\rangle+\beta|e\rangle)|0\rangle \leftrightarrow |g\rangle(\alpha|0\rangle+\beta|1\rangle)$, and can be repeated to swap it back again; this could be used as a single photon source (starting with an excited atom), or as an interface between an atom or trapped ion quantum computer and optical quantum communication.

Other interaction durations create entanglement between the atom and cavity field; for example, a quarter-cycle on resonance starting from $|e\rangle|0\rangle$ gives the maximally entangled state (a Bell state) $(|e\rangle|0\rangle+|g\rangle|1\rangle)/\sqrt{2}$. This can in principle be used as a quantum computer, mathematically equivalent to a trapped ion quantum computer with cavity photons replacing phonons.

The history of CQED started, more than 50 years ago, with a seminal remark by Purcell. The radiation properties of an atom can be changed by controlling the boundary conditions of the electromagnetic field with mirrors or cavities. CQED experiments initially measured modifications of spontaneous emission rates or spatial patterns in low-quality cavities. They evolved to higher and higher atom-cavity couplings and photon storage times. Most of them are now in the so-called regime of 'strong coupling', in which the coherent interaction of a single atom with one photon stored in a very-high-quality cavity, a modern equivalent of Einstein's photon box, overwhelms the incoherent dissipative processes.

CQED experiments implement a situation so simple that their results can be cast in terms of the fundamental postulates of quantum theory. They are thus appropriate for tests of basic quantum properties: quantum superposition, complementarity or entanglement. In the context of quantum information processing, the atom and the cavity are long-lived qubits, and their mutual interaction provides a controllable entanglement mechanism – an essential requirement for quantum computing or teleportation applications. CQED is therefore a fertile ground for quantum information processing. In addition, the ability to manipulate mesoscopic fields, containing a few to a few tens of photons, made it possible to explore the fuzzy boundary between the quantum and the classical worlds, unveiling the decoherence mechanisms that confine the quantum weirdness at a microscopic scale. CQED comes in two flavours: microwave and optical. Both situations achieve the strong-coupling regime, with different and complementary features. In the microwave domain, very excited 'Rydberg' states interact with superconducting millimetre-wave cavities. Dissipation is very low, and the pace of the atom-field entanglement process is slow. An exquisite degree of control is reached, making it possible to tailor complex multi-qubit entangled states. In Europe, this line of research is mainly pursued by H. Walther in Munich, and S. Haroche with one of us (JMR) in Paris. In the optical domain, low-lying atomic levels interact with room-temperature optical cavities. The interaction is much faster, as is the dissipation. This, however, turns out to be an asset: optical photons can be efficiently coupled in or out of the cavity. Optical CQED thus provides a natural and essential interface between flying photonic qubits for the transmission of quantum information and stationary atomic qubits for the storage of quantum information. In Europe, this regime of CQED is studied by one of us (GR), while other groups are setting up new experiments.

REFERENCES

- Barker, J.R., S. Roy, S. Babiker, and A. Asenov: *Proceedings of the International Conference on Quantum Devices and Circuits,* Singapore, World Scientific, 1997.
- Bharadwaj, Ashish: *The Nanoscope : Encyclopedia of Nanoscience and Nanotechnology,* Pentagon Press, Delhi, 2005.
- Geoffrey N.: *Introduction to Nonlinear Optics,* Cambridge University Press, Delhi, 2011.
- Isaac L. Chuang: *Quantum Computation and Quantum Information,* Cambridge University Press, 2000.
- Khare, R.P. : *Fiber Optics and Optoelectronics,* Oxford University Press, Delhi, 2004.
- Levinson, Y. : *Quantum Noise in Mesoscopic Physics,* Kluwer, Dordrecht, 2003.

Bibliography

Aggarwal, S.K. : *A Textbook of Optics*, Wisdom Press, Delhi, 2012.

Ashish Bharadwaj: *The Nanoscope : Encyclopedia of Nanoscience and Nanotechnology*, Pentagon Press, Delhi, 2005.

Avadhanalu, M. N. : *A Text Book Of Optics*, S. Chand Publisher, Delhi, 2006.

Awasthi, K.P : *Advanced Magnetic Nanostructures*, Cyber Tech Publications, Delhi, 2008.

Barker, J.R., S. Roy, S. Babiker, and A. Asenov: *Proceedings of the International Conference on Quantum Devices and Circuits*, Singapore, World Scientific, 1997.

Bharadwaj, Ashish: *The Nanoscope : Encyclopedia of Nanoscience and Nanotechnology*, Pentagon Press, Delhi, 2005.

Bhatia, A.S. : *Nanowires and Nanobelts*, Deep and Deep, Delhi, 2011.

Bose, G.K. : *A-Z Optics*, Centrum Press, Delhi, 2009.

Brooker, Geoffrey: *Modern Classical Optics*, Oxford University Press, Delhi, 2008.

Carlsson, S.-B., T. Junno, H. Xu, L. Montelius: *Extended abstracts of the 3rd International Workshop on Quantum Functional Devices*, Tokyo, R&D Association for FED, 1997.

Cerofolini, G.F.: *Nanoscale Devices*, Springer-Verlag, Berlin, 2009.

Drexler, K. Eric: *Nanosystems: Molecular Machinery, Manufacturing, and Computation*, New York: Wiley, 1992.

Evans, C.: *Surface Characterization Techniques*, New Delhi: Long Life Publication, 2003.

Geckeler, K.E. : *Biomedical Applications of Nanostructured Materials*, Macmillan Publishers India, Delhi, 2011.

Geoffrey N.: *Introduction to Nonlinear Optics*, Cambridge University Press, Delhi, 2011.

Ghatak, Ajoy and K. Thyagarajan: *Fiber Optics and Lasers : The Two Revolutions*, Macmillan Publishers India, Delhi, 2006.

Ghosh, Krishnapada and Anandamoy Manna: *Textbook Of Physical Optics*, Macmillan, Delhi, 2007.

Govind P. Agrawal: *Nonlinear Fiber Optics*, Academic Press, Boston, 2007

Gupta, Ashish: *Understanding of Nano Science and Technology*, Global Vision, Delhi, 2006.

Hall, J Storrs : *Nanofuture : What's Next for Nanotechnology*, Manas, Delhi, 2006.

Herzberg, G. : *Molecular Spectra and Molecular Structure*, Van Nostrand Reinhold Company, New York, 1979.

Imry, Y. : *Introduction to Mesoscopic Physics*. Oxford University Press, New York, 1997.

Ipsita Roy: *Bio-Nano-Geo Sciences : The Future Challenge*, Ane Books, Delhi, 2009.

Isaac L. Chuang: *Quantum Computation and Quantum Information*, Cambridge University Press, 2000.

Jenkins, Francis A. and Harvey E. White: *Fundamentals of Optics*, Tata McGraw Hill, Delhi, 2011.

Jha, D.K. : *Contemporary Optics*, Discovery Publishing House, Delhi, 2009.

Joseph, Subita: *The Illustrated Encyclopaedia of Nano-Photonics*, Dominant Pub, Delhi, 2008.

Kaushik, N.L. : *Nanotechnology and Micromachines*, Pearl Books, Delhi, 2007.

Khare, R.P. : *Fiber Optics and Optoelectronics*, Oxford University Press, Delhi, 2004.

Kuhn T.: *"The Structure of Scientific Revolutions."* University of Chicago Press Chicago, 1962.

Kumar, H.D. : *Material Science: Nanotechnology and Applications*, I.K. International, Delhi, 2011

Levinson, Y. : *Quantum Noise in Mesoscopic Physics*, Kluwer, Dordrecht, 2003.

Ljubisa Ristic: *Sensor Technology and Devices*, Artech House, Delhi, 1994.

Load, B.B. : *Lasers and Non-Linear Optics*, New Age International, Delhi, 2011.

Mahan, G.D. : *Many-Particle Physics*, Plenum Press, New York, 1990.

Mark Fox: *Quantum Optics : An Introduction*, Oxford University Press, Delhi, 2009.

Merkle R.: *"On Physics, Fundamentals, and Nanorobots."* Institute for Molecular Manufacturing, 2001.

Miko Elwenspoek: *Mechanical Microsensors (Microtechnology and MEMS)*, Springer Verlag, Delhi, 2001.

Mittal, S.K. : *Quantum Optics*, Pearl Books, Delhi, 2010.

Nakazato, K.: *Extended Abstracts of 15th Symposium on Future Electron Devices*, Tokyo, R&D Association for FED, 1996.

Pandit, Rajnish: *Basics of Optics*, Anmol, Delhi, 2008.

Plauger P.J.; *Software Tools in Pascal*; Addison-Wesley, 1981.

Ranganathan, S.: *Handbook of Nanoparticles*, Chennai: T.T.Ranganathan Clinical research Foundation, 2008.

Schubert, A. & Zsindely, S.: *Nanoscience and Nanotechnology on the Balance*, USA: Oxford,1997.

Sharma, P.K. : *NanoMEMS Physics : Applications and Principles*, Vista International Pub, Delhi, 2011.

Shew, A.: *Discovering the Nanoscale*, Amsterdam: IOS Press, 2004.

Singh, Seema: *Optics and Atomic Physics*, Campus Books, Delhi, 2009.

Sinha, D.N.: *Nanotubes and Nanowires*, Patna: Indian Academy of Sciences, 2005.

Smith, C.S. : *A Search for Structure*, Cambridge, Mass.: MIT Press 1981.

Srivastava, N.K. : *Electromagnetic Theory and Optics*, ABD Pub., Delhi, 2007.

Sujata Agrawa: *Small Wonder : The Making of the Nano*, Westland, Delhi, 2010.

Suri, Shalini : *Nano Technology : Basic Science to Emerging Technology*, APH, Delhi, 2006.

Tuchin, Valery: *Tissue Optics: Light Scattering Methods And Instruments For Medical Diagnosis*, New Age International, Delhi, 2010.

Varadan, Vijay K. : *Microsensors, MEMS and Smart Devices*, Wiley, NY, 2001.

Vasudev Bhatnagar: *Nanostructure Science and Technology*, Campus Books, Delhi, 2009.

Index

A

Adaptation Layers, 238, 239, 240, 241.

B

Black Body Radiation, 37, 38.
Black Hole Evaporation, 37, 46.

C

Casimir Forces, 2, 262, 264, 266, 269, 273.
Charge Transfer Interaction, 200.
Complex Metal Optics, 200.
Constitutive Relations, 29, 30.
Conventional Lens, 107, 111, 114.
Conventional Optical Limitations, 21.
Coordinate Transformations, 19, 23, 25, 26, 114.
Crepuscular Rays, 191, 196.

D

Dielectric Probes, 96.

E

Electromagnetic Field, 3, 24, 36, 84, 97, 138, 139, 148, 156, 157, 176, 199, 253, 254, 259, 268, 269, 270, 271.
Electromagnetic Radiation, 3, 4, 19, 38, 51, 122, 220, 242.
Electromagnetic Spectrum, 4, 110.
Evanescent Tails, 65, 67.

F

Field Optical Microscopy, 85, 89, 92, 94, 95, 96, 101, 134, 135, 139, 148.
Field Spectroscopy, 88, 152.

H

Hawking Radiation, 33, 35, 36, 37, 38, 40, 42, 43, 44, 46, 47, 48, 49, 52, 53, 275.
High Resolution Scanning, 2.

L

Lasers Raman, 215.
Light Microscopy, 135, 140.
Light Scattering, 133, 176, 191, 192, 193, 196, 210, 214.

M

Metal Optics, 200, 201, 202, 203.
Mie Scattering Theory, 176, 211, 212, 213.
Mimicking Celestial Mechanics, 19.
Moving Light, 53.
Multidimensional Universe, 21.
Multiple Laser Wavelengths, 227.

N

Nano Plasmonics, 197.
Nanooptics Design, 213.
Nanoparticle Systems, 183, 187, 188, 189.
Nanoscale Optical Microscopy, 134.
Nanoscale Optics, 213.
Nanoscale Resolution, 135.
Nanotechnology, 1, 3, 13, 120, 122, 133, 234.
Nonlinear Optical Absorption, 82.
Numerical Methods, 31, 156, 164, 177, 214, 262.

O

Optical Black Holes, 17, 18, 51, 52.
Optical Design Geometry, 25.

Optical Devices, 13, 23, 56, 74, 82, 83, 84, 85, 107, 156, 184, 222.
Optical Elements, 5, 6, 21, 22, 23, 25, 102, 175.
Optical Fibre, 12, 49, 50, 51, 60, 62, 72, 87, 88, 102, 103, 104, 136, 208, 254, 270, 271.
Optical Metrology, 213, 215.
Optical Microscopy, 85, 86, 87, 89, 92, 94, 95, 96, 101, 110, 112, 134, 135, 139, 147, 148, 197.
Optical Sensors, 131, 132.
Optical Superresolution, 102.

P

Perfect Lens, 15, 24, 107, 108, 109, 114.
Photo Detectors, 229, 230, 232.
Photometric Measurements, 153, 154.
Photon Energy, 217, 254.
Photon Momentum, 259.
Photon Spin, 254, 261.
Photonic Crystal Essentials, 128.
Photonic Crystal Structure, 220, 236.
Photonic Crystals, 12, 113, 120, 127, 128, 129, 130, 233, 234, 245.
Photonic Structures, 130, 235.
Plasmon Resonances, 152, 170, 177, 180, 181, 199, 214.
Producing Black Holes, 20.

Q

Quantum Dots Lasers, 124.

R

Raman spectroscopy, 88, 117, 120, 224, 225.
Random Laser Theory, 247.
Random Number Generation, 248, 252.
Rectangular Waveguides, 60, 61, 69, 71.

S

Scalar Wave Equation, 204, 206.
Scanning Optical Microscope, 96.
Seamless Manufacturing System, 201.
Spectroscopy, 88, 117, 120, 123, 149, 150, 151, 152, 153, 224, 225.
Spherical Waves, 206.
Strip Waveguides, 61, 71.
Surface Plasmon Polaritons, 116, 197, 203.

T

Tensor Densities, 26.
Time Multiplexing, 104, 105, 106.
Transformation Optics, 14, 15, 16, 17, 18, 19, 21, 22, 50.

U

Unruh Radiation, 33, 35, 40.

V

Vacuum Energy, 268, 269, 270, 271.
Vacuum Interpretation, 32.
Volumetric Light Scattering, 194.

W

Waveguide Couplers Information, 64.
Waveguide Directional Couplers, 62.
Waveguide Gratings, 71, 72.
Waveguide Optics, 56.
Waveguide Power Dividers, 64.